Pharmacotherapeutic Botanicals for Cancer Chemoprevention

Manish Kumar • Ashita Sharma •
Praveen Kumar
Editors

Pharmacotherapeutic Botanicals for Cancer Chemoprevention

Springer

Editors
Manish Kumar
SD College
Barnala, Punjab, India

Ashita Sharma
Chandigarh University
Mohali, Punjab, India

Praveen Kumar
Technion – Israel Institute of Technology
Technion City, Haifa, Israel

ISBN 978-981-15-5998-3   ISBN 978-981-15-5999-0 (eBook)
https://doi.org/10.1007/978-981-15-5999-0

This Springer imprint is published by the registered company Springer Nature Singapore Pte Ltd.
The registered company address is: 152 Beach Road, #21-01/04 Gateway East, Singapore 189721, Singapore

Preface

Cancer is a major cause of mortality all over the globe leading to range of disabilities and premature deaths among human population. In 2015, around 8.8 million died due to cancer, and this figure is predicted to upsurge by about 70% over the next two decades. Experimental and epidemiological studies have revealed that exposure to environmental and lifestyle factors are major contributors towards cancer incidences along with inborn genetic defects (10% only). Currently, chemotherapy is considered as standard medical treatment for cancer, but it has been observed that resistance of cancer cells to various chemotherapeutic agents has increased widely. Also, many therapeutic agents have serious effects on normal cells, thus increasing the side effects of chemotherapy. 80–90% of fatalities associated with cancer are because of the abovementioned reasons. This stimulated scientific community to search more specific and less toxic anticancer agents that can overcome cancer cell's resistance and enhance efficacy of existing medicinal agents. Fortunately, phytochemicals from plants with varied chemical structures and moieties have been proved to be effective against several types of cancers as these can avert or restrict the cancer growth by means of diverse cellular and molecular approaches. Various phytochemicals or natural compounds have been identified for potent anti-cancerous abilities and are approved by FDA for cancer management. This book is the compilation of recent studies providing deep understanding of diverse signaling pathways being targeted by phytochemicals along with recent advancements and future directions.

The first chapter, compiled by Kumar and Thakur, broadly outlines the relevance of plant products as therapeutic agents against cancer. In the second chapter, Arora and coworkers throw light on the concept of cancer chemoprevention and role of phytochemicals in modulating signaling cascades in cancer. The toxicity associated with currently available drugs used against cancer is precisely discussed in the third chapter of the book, suggesting an integrative approach for understanding the complex interaction of drugs with various signaling networks. The chapter also emphasized importance of personalized therapy in effective cancer management. Importance of natural products as chemosensitizers for adjunct therapy in cancer management has been discussed in detail by Sinha et al. In the next chapter, Bhattacharjee et al. have reviewed various studies regarding epigenetic potential of phytochemicals and highlighted significance of in silico studies in epitherapeutic drug designing. In the sixth chapter, Mandal et al. have discussed various ways in which phytochemicals are known to modulate cancer cell metabolism and their microenvironment. The seventh chapter by Chattopadhyay is focussed on the concept of nutrigenomics and nutrigenetics w.r.t. cancer chemoprevention. In the eighth chapter, Kalia and coworkers emphasized on the nano-enabled delivery vehicles to improve bioavailability of phytochemical based anticancer drugs. In the next chapter, Mandal et al. have broadly discussed the biphasic effects of various phytochemicals and its importance in cancer therapy. In the next chapter, mode of action and effects of phytochemicals from Zingiberaceae are largely discussed. In the eleventh chapter, authors discussed the significance and applications of 3D cell culture techniques and their usefulness in cancer research. The twelfth chapter of the book throws light on various animal models that can be used in drug discovery before preclinical trials. The last chapter of the book summarized significance of natural plant products and their derivatives in clinical trials for treatment of cancer. The book is compilation of recent studies in the field of cancer research and thus can aid scientific community working in the field to have deeper and broader insight to various aspects of cancer therapy.

It is to be noted that the authors of various chapters are responsible and answerable for any scientific queries and questions. We are highly grateful to the scientists, who have contributed their research and made this compilation a unique collection of recent studies in the field. The book can serve as a handbook for researchers working in the field of cancer chemoprevention.

Barnala, India — Manish Kumar
Mohali, India — Ashita Sharma
Haifa, Israel — Praveen Kumar

Contents

1 Phytochemicals in Cancer Chemoprevention: A Brief Perspective . . . 1
Praveen Kumar and Anita Thakur

2 Role of Phytochemicals in Modulating Signaling Cascades in Cancer Cells . . . 11
Davinder Singh, Avinash Kumar, Astha Bhatia, Hasandeep Singh, Sahiba Kukreja, Balbir Singh, Saroj Arora, and Rohit Arora

3 Diverse Cancer Therapeutic Interactions: Complexities in Cancer Management . . . 47
Prashant Tarale, Pravin Naoghare, Jyoti Tagde, Amit Bafana, and Saravanadevi Sivanesan

4 Natural Products as Chemosensitizers for Adjunct Therapy in Cancer Management . . . 67
Suchisnigdha Datta, Priyanka Saha, Subhabrata Dey, and Dona Sinha

5 Dietary Phytochemicals as Epigenetic Modulators in Cancer Prevention: Emerging Research Trends, Gaps, and Future Perspectives . . . 121
Ritwija Bhattacharya, Shrinjana Dhar, Aniruddha Mukhopadhyay, and Pritha Bhattacharjee

6 Modulation of Cancer Cell Metabolism and Microenvironment by Phytochemicals . . . 143
Amarjeet Shrama, Priyanshi, and Chandi C. Mandal

7 Role of Nutrigenetics and Nutrigenomics in Cancer Chemoprevention . . . 167
Indranil Chattopadhyay

8 Nano-Delivery Carriers for Enhanced Bioavailability of Antitumor Phytochemicals . . . 189
Anu Kalia and Gagandeep Kaur

9 Biphasic Effects of Phytochemicals and their Relevance to Cancer Therapeutics 197
Urvashi Vijayvergia, Shreetama Bandyopadhayaya, and Chandi C. Mandal

10 Potential Pharmacotherapeutic Phytochemicals from Zingiberaceae for Cancer Prevention 221
Indrani Manna, Debalina Das, Sejuty Mondal, and Maumita Bandyopadhyay

11 3 Dimensional Cell Culture Techniques in Cancer Research 283
K. C. Dhanya and Aditya Menon

12 Animal Models Systems of Cancer for Preclinical Trials 299
Sneha Soni, Shreetama Bandyopadhayaya, and Chandi C. Mandal

13 Cancer Chemoprevention by Natural Plant Products and Their Derivatives: Clinical Trials 325
Jatinder Kaur Katnoria, Savita, Amritpreet Kaur, Akanksha Bakshi, and Avinash Kaur Nagpal

Editors and Contributors

About the Editors

Manish Kumar is an Assistant Professor of Botany at S.D. College, Barnala, Punjab, India. He holds M.Sc. (Hons.), M.Phil. & Ph.D. in Botany from Guru Nanak Dev University, Punjab, India. He has over 3 years of teaching and 9 years of research experience in cancer chemoprevention, cancer cell therapeutics, environmental mutagenesis, and medicinal plants. He has published more than 35 research articles in journals and books of international repute with a cumulative impact factor of more than 35. He is an editorial board member of various national and international journals.

Ashita Sharma is currently working as an Associate Professor of Environmental Sciences at Chandigarh University, Gharuan, Mohali, India. She has completed her M.Sc. (Hons.) and Ph.D. in Environmental Sciences from Guru Nanak Dev University, Amritsar, India. Dr. Sharma has over 6 years of teaching experience and 7 years of research experience in soil remediation, environmental genotoxicity/carcinogenicity, and biomonitoring. During her research, she has published numerous articles in reputed international journals and books. She is a member of various professional societies.

Praveen Kumar is currently working as a Postdoctoral Researcher at the Technion, Israel Institute of Technology Haifa, Israel. He has completed his Ph.D. from the National Institute of Immunology, New Delhi, India. He has over 9 years of research experience in molecular cell biology and cancer metabolism. His specializations include cancer cell metabolomics, cell cycle metabolomics, and metabolomics based drug screening.

List of Contributors

Rohit Arora Department of Biochemistry, Sri Guru Ram Das University of Health Sciences, Amritsar, Punjab, India

Saroj Arora Department of Botanical and Environmental Sciences, Guru Nanak Dev University, Amritsar, Punjab, India

Amit Bafana Directors Research Cell, CSIR-National Environmental Engineering Research Institute (CSIR-NEERI), Nagpur, India

Akanksha Bakshi Department of Botanical and Environmental Sciences, Guru Nanak Dev University, Amritsar, Punjab, India

Shreetama Bandyopadhayaya Department of Biochemistry, Central University of Rajasthan, Ajmer, Rajasthan, India

Maumita Bandyopadhyay Plant Molecular Cytogenetics Laboratory, Centre of Advanced Study, Department of Botany, Ballygunge Science College, University of Calcutta, Kolkata, India

Astha Bhatia Department of Botanical and Environmental Sciences, Guru Nanak Dev University, Amritsar, Punjab, India

Pritha Bhattacharjee Department of Environmental Science, Ballygunge Science College, University of Calcutta, Kolkata, India

Ritwija Bhattacharya Department of Environmental Science, Ballygunge Science College, University of Calcutta, Kolkata, India

Indranil Chattopadhyay Department of Life Sciences, Central University of Tamil Nadu, Thiruvarur, Tamil Nadu, India

Debalina Das Plant Molecular Cytogenetics Laboratory, Centre of Advanced Study, Department of Botany, Ballygunge Science College, University of Calcutta, Kolkata, India

Suchisnigdha Datta Department of Receptor Biology and Tumor Metastasis, Chittaranjan National Cancer Institute, Kolkata, India

Subhabrata Dey Department of Clinical and Translational Research, Chittaranjan National Cancer Institute, Kolkata, India

K. C. Dhanya Department of Microbiology, St. Mary's College, Thrissur, Kerala, India

Shrinjana Dhar Department of Environmental Science, Ballygunge Science College, University of Calcutta, Kolkata, India

Anu Kalia Electron Microscopy and Nanoscience Laboratory, Department of Soil Science, College of Agriculture, Punjab Agricultural University, Ludhiana, Punjab, India

Jatinder Kaur Katnoria Department of Botanical and Environmental Sciences, Guru Nanak Dev University, Amritsar, Punjab, India

Amritpreet Kaur Department of Botanical and Environmental Sciences, Guru Nanak Dev University, Amritsar, Punjab, India

Gagandeep Kaur Department of Microbiology, College of Basic Sciences and Humanities, Punjab Agricultural University, Ludhiana, Punjab, India

Sahiba Kukreja Department of Biochemistry, Sri Guru Ram Das University of Health Sciences, Amritsar, Punjab, India

Avinash Kumar Department of Botanical and Environmental Sciences, Guru Nanak Dev University, Amritsar, Punjab, India

Praveen Kumar Technion – Israel Institute of Technology, Technion City, Haifa, Israel

Chandi C. Mandal Department of Biochemistry, Central University of Rajasthan, Ajmer, Rajasthan, India

Indrani Manna Plant Molecular Cytogenetics Laboratory, Centre of Advanced Study, Department of Botany, Ballygunge Science College, University of Calcutta, Kolkata, India

Aditya Menon Origin Diagnostics and Research, Karunagappally, Kerala, India

Sejuty Mondal Plant Molecular Cytogenetics Laboratory, Centre of Advanced Study, Department of Botany, Ballygunge Science College, University of Calcutta, Kolkata, India

Aniruddha Mukhopadhyay Department of Environmental Science, Ballygunge Science College, University of Calcutta, Kolkata, India

Avinash Kaur Nagpal Department of Botanical and Environmental Sciences, Guru Nanak Dev University, Amritsar, Punjab, India

Pravin Naoghare Environmental Impact and Sustainability Division, CSIR-National Environmental Engineering Research Institute (CSIR-NEERI), Nagpur, India

Priyanshi Department of Biochemistry, Central University of Rajasthan, Ajmer, Rajasthan, India

Priyanka Saha Department of Receptor Biology and Tumor Metastasis, Chittaranjan National Cancer Institute, Kolkata, India

Savita Department of Botany, Hindu College, University of Delhi, New Delhi, Delhi, India

Amarjeet Shrama Department of Biochemistry, Central University of Rajasthan, Ajmer, Rajasthan, India

Balbir Singh Department of Pharmaceutical Sciences, Guru Nanak Dev University, Amritsar, Punjab, India

Davinder Singh Department of Botanical and Environmental Sciences, Guru Nanak Dev University, Amritsar, Punjab, India

Hasandeep Singh Department of Pharmaceutical Sciences, Guru Nanak Dev University, Amritsar, Punjab, India

Dona Sinha Department of Receptor Biology and Tumor Metastasis, Chittaranjan National Cancer Institute, Kolkata, India

Sarvanadevi Sivanesan Health and Toxicity Cell, CSIR-National Environmental Engineering Research Institute (CSIR-NEERI), Nagpur, India

Sneha Soni Department of Biochemistry, Central University of Rajasthan, Ajmer, Rajasthan, India

Jyoti Tagde Environmental Impact and Sustainability Division, CSIR-National Environmental Engineering Research Institute (CSIR-NEERI), Nagpur, India

Prashant Tarale Health and Toxicity Cell, CSIR-National Environmental Engineering Research Institute (CSIR-NEERI), Nagpur, India

Anita Thakur Technion – Israel Institute of Technology, Technion City, Haifa, Israel

Urvashi Vijayvergia Department of Biochemistry, Central University of Rajasthan, Kishangarh, Rajasthan, India

Phytochemicals in Cancer Chemoprevention: A Brief Perspective

1

Praveen Kumar and Anita Thakur

Abstract

Cancer is one of the leading causes of deaths globally. There are various treatment options available to cure cancer such as radiotherapy, chemotherapy, and immunotherapy. Despite being the primary choices of usage, current cancer therapies suffer with tremendous side effects with yet poor patient survival. Further, with the development of drug resistance in cancer cells, there is requirement to develop new therapeutics against cancer. A number of studies either on plant extracts or purified phytochemicals have shown promise towards cancer therapy directly or in combination with existing drugs. Several plant-derived compounds have been reported with anti-proliferative activities against various types of cancers by modulating complex cellular pathways. Since, phytochemicals are generally regarded as safe and easily available to consume; they are perceived as therapeutic agents with much less side effects. In this chapter we are presenting a very brief summary of relevance of plant products as therapeutic agents against cancer.

Keywords

Cancer · Cancer chemoprevention · Phytochemicals · Bioavailability · Biphasic effects

1.1 Introduction

Cancer is a highly morbid disease causing 9.6 million deaths worldwide in 2018. Lung cancer, breast cancer, and colorectal cancer are the major cancer types and among top five cancers resulting in the mortality [1, 2]. Various carcinogens present in the environment cause cancer initiation, e.g. tobacco consumption, ionizing

P. Kumar (✉) · A. Thakur
Technion – Israel Institute of Technology, Technion City, Haifa, Israel

M. Kumar et al. (eds.), *Pharmacotherapeutic Botanicals for Cancer Chemoprevention*, https://doi.org/10.1007/978-981-15-5999-0_1

radiations, and industrial chemicals. Cancer can also be caused by certain infections and it can also be genetic. There are several progressive changes that occur in cells undergoing tumorigenesis in spatio-temporal manner referred to as the hallmarks of cancer. Cancer hallmarks include modulation of signaling pathways for sustained proliferation (e.g., PI3K signaling), evading tumor suppressors (p53 and pRB regulation), activation of invasion and metastasis, enabling replicative immortality, angiogenesis induction, and resisting death [3]. All of these features help cancer cells to proliferate rapidly resulting in unchecked growth.

Cancer chemoprevention is the process of chronic administration of a natural, synthetic, or a biological agent in order to suppress, delay, or reduce the occurrence of malignancy [4, 5]. The interest in cancer chemoprevention is increased by the studies elucidating a better mechanistic understanding of the cancer biology, which led to the development of new drugs. Cancer chemo-preventive agents have been broadly classified into hormonal, dietary, medications, and vaccines [5]. Anti-estrogens such as selective estrogen receptor modulators (SERMs) are used against breast cancer [6]. Medications like aspirin, metformin, and statins have been implicated with anticancer effects. Several dietary agents such as vitamins and phytochemicals have shown anti-proliferative activity against various cancers. There are vaccines against infections that cause cancers, e.g. hepatitis B virus [5]. Cancer exhibits various progressive stages, viz. initiation, promotion, conversion, progression, invasion, and metastasis. [7]. Most of the drugs target the last stages of the cancer progression. There are severe limitations to target the advanced stages, e.g. severe side effects, high cost, and single target. For some tumors (sarcomas and leukemias), single target is enough for chemoprevention; however, for others multiple targets are required. Therefore most of advanced therapeutic strategies suggest targeting more than one stage including multiple targets for cancer treatment with the aid of computational models [8].

Remarkably high proliferation of cancer cells is contributed by diverse mechanisms altered in these cells. These alterations also make cancer cells vulnerable to therapeutic interventions. Most of the cancer drugs available in the market target the abnormally functional pathways to halt the proliferation of cancer cells. The various classes of anticancer drugs such as kinase inhibitors, metabolic inhibitors, immune checkpoint inhibitors, radiation therapy (summarized in great details in the review [9]) are important to mention here but are beyond the scope of this chapter. Since these drugs also target the normal pathways, patients suffer from severe side effects from prolonged therapy [10]. Further, currently prescribed therapy promises only a very limited life span for cancer patients. Therefore, despite achievement of significant advances in medical research aiming at cancer drugs, there is still shortage of a promising therapy, which can be prescribed to cure or prevent cancer without causing significant side effects. Since, cancer is a tremendously heterogeneous disease, one uniform treatment regimen might not be suitable for all the cancer types. Drug resistance to the existing therapies curbs the patient life span drastically necessitating the need for the discovery of alternative drug choices. Plant and plant-derived products are very promising as an alternative and complementary to the currently available drug options.

1.2 Phytochemicals in Cancer Therapy

About 25–28% of the currently available drug compounds are derived from higher plants [11] demonstrating the potential of plant products in cancer therapeutics and further encourages the scope of research to develop additional drugs from plants. Phytochemicals are plant derived non-nutritive chemicals synthesized as defense mechanism against harsh environment including various pathogens [2]. The phytochemicals exhibiting various medicinal properties are classified as polyphenols, terpenoids, and alkaloids, which are also a part of human diet.

There is a large number of publications that describe the potential of plant metabolites against cancers [2, 7, 12]. Many of plant products exhibit a great deal of activity against various types of cancers (resveratrol). There are several known phytochemicals exhibiting a specific activity against a tumor, e.g. epigallocatechin-3-gallate exhibits an increase in caspase-3, p27, and calpain I activities in human Jurkat T, prostate cancer (PC-3, LNCaP) cells, and breast cancer (MCF-7) [13]. Capsaicin showed inhibition of growth and reversal of transformed phenotype in H-ras MCF-10A as reviewed in [2].

1.3 Potential of Phytochemicals for Cancer Chemoprevention

Phytochemicals exhibit a number of modulatory effects on the cancer cells that make them suitable as drug candidates. Plant products act as blocking agents to suppress the interaction of molecules leading to carcinogenic phenotype, e.g. DNA damage or production of ROS.

1.3.1 Phytochemicals as ROS Scavengers and Redox Modulators

Rapidly proliferating cancer cells result in the production of several free radical (H_2O_2, OH^- O_2^-) known as reactive oxygen species (ROS). While ROS promote the cancer cell proliferation, they are deleterious for normal cells inducing DNA damage and cell death. There are several studies demonstrating the protection provided by phytochemicals against ROS. For example, Tea polyphenols normalized the levels of superoxide dismutase (SOD) and catalase; *Curcuma longa* normalized SOD and CAT in mice [14] and rats [15], respectively. Some phytochemicals exploit the fact that cancer cells have high levels of ROS. These phytochemicals help elevate ROS even further and cause cell death, e.g. gallic acid mediated cell death due to elevated ROS in DU145 human prostate cancer cells [16].

1.3.2 Phytochemicals in DNA Damage and Repair

Cancer cells exhibit continued ROS production, which leads to cancer driving mutations in cancer cells and DNA damage in normal cells. Phytochemicals exhibit dual function of DNA damage and DNA repair depending upon the genomic stability of the cells and can be selective against cancer. For example, curcumin induced both mitochondrial and nuclear DNA damage after 72 h incubation in G2 hepatoma cells [17]. Also, there are several reports exhibiting DNA damaging effect like *Tinospora* extracts and turmeric [2], sulforaphane exhibited DNA protection on certain embryonic derived cells [18].

1.3.3 Phytochemicals Control Gene Expression

Oncogenes and tumor suppressor genes dictate whether the outcome of a cellular phenotype would be cancerous or not. p53 tumor suppressor gene is a master regulator of cell division and is mutated in a large number of cancer types. Further, tumor cells hyperactivate various oncogenes, which are responsible for uncontrolled tumor growth. Plant metabolites exhibit activities to restore or upregulate tumor suppressor functions and downregulating the oncogene function to keep tumor growth in check. For example, in a study using 26 medicinal plants on MDR leukemia cells, many plant extracts such as *Leonotis leonurus*, *Hypoestes aristata*, *Salvia apiana* showed increased p53 expression and cells death and lower levels of RAS and EGFR [19]. Similarly growth inhibition of breast cancer cells such as myoblasts and MCF7 was observed using plant extracts which either upregulate p53 expression or downregulate oncogene function or both [2].

1.3.4 Phytochemicals Modulate Phase 1 and Phase 2 Enzymes

Upon ingestion, a xenobiotic compound undergoes detoxification process in liver. Three sets of enzymes are involved in detoxification process: phase I enzymes mainly cytochrome p450 (oxidoreduction step); phase II includes glutathione S-transferase (conjugation step); and phase III, as example Multidrug Resistance Protein (MRP) (excretion step) [20]. Typically, phase I enzymes make the non-polar xenobiotic compound more hydrophilic thereby often increasing the toxicity of the xenobiotic. Phase II enzymes conjugate with the products of phase I reactions rendering them less toxic. Phase III enzymes help excretion of the waste products. Dietary plants contain a number of anti-oxidants that through nuclear (factor erythroid 2) related factor 2 (nrf2) pathway induce the production of cytoprotective antioxidant enzymes such as glutathione S transferase and superoxide dismutase [21]. Many plant products have been reported to decrease expression of phase I and increase the expression of phase II enzymes promoting more cytoprotection. For example, *Brassica oleracea* increases GST enzymes in nrf2 background mice [22].

1.3.5 Phytochemicals Inhibit Cell Proliferation

Cancer cells typically exhibit very high proliferation rates, which is responsible for various manifestations of the cancers. Therefore, major therapeutic aim is to reduce the proliferation of cancer cells. A number of phytochemicals have been reported that reduce the proliferation rate of the cancerous cells. For example, MCL (Mantle cell lymphoma) growth was inhibited in a dose dependent treatment with curcumin by suppressing cyclin D1, NFkB, and Survivin protein expression, which caused G1 phase arrest [23]. Breast cancer cell line MCF-7 was growth inhibited by treatment with phenolic compound rich cranberry extract effected by reduction in CDK4 and cyclin D1 levels [24].

1.3.6 Phytochemicals Promote Autophagy

Autophagy, also known as macroautophagy, is a highly conserved and regulated process that targets proteins and damaged organelles for lysosomal degradation to maintain cellular homeostasis at a basal state, as well as during cellular stress [25]. The role of autophagy in cancer is complex and is primarily dictated by tumour type and stage. Numerous studies have associated its role constrains to tumor initiation in normal tissue and to tumor promotion and maintenance in certain tumor types. Furthermore, several synthetic autophagy modulators have been identified as potential candidates for cancer treatment. Emerging evidence has allied phytochemicals targeting the autophagic pathway as promising agent against various malignancies with minimal side effects. Paclitaxel, a taxane class diterpenoid, triggers early autophagy in both normoxic and hypoxic conditions in breast cancer cells and is associated with apoptosis. [26, 27]. In addition, diverse phytochemicals derived from natural sources, such as curcumin, ursolic acid, apigenin (4′,5,7-trihydroxyflavone), resveratrol, quercetin, thymoquinone, celastrol, and γ-tocotrienol, also have attracted attention as potential autophagy modulators and therefore can help to overcome chemoresistance and radioresistance [28].

1.3.7 Phytochemicals Exhibit Anti-inflammatory Effects

Onset of cancer is associated with a systemic inflammatory response. The inflammatory response possibly interferes with drug metabolism as inflammation hinders cytochrome P450 activity. Therefore anti-inflammatory drugs are given to the cancer patients. However, they also present patient with side effects over prolonged usage [29]. There are several phytochemicals that are known to inhibit known pathways (NFkB, Cox II and iNOS) thereby lowering the inflammation (reviewed in [2]). Quercetin inhibited NFkB in mouse derived inflamed intestinal epithelial cells and reduced inflammation [30].

1.3.8 Phytochemicals Modulate Tumor Metabolism

In the recent years it is being increasing clear that tumor cells exhibit remarkably different metabolism compared to the normal cells due to requirement of unique pathways operating in tumor cells in order to proliferate at enormous rates. Therefore recently, targeting the altered cancer cell metabolism is a very attractive target for the development of cancer therapy. Cancer cells display addiction to glycolytic pathway even in the presence of oxygen called as the Warburg effect [31, 32]. Various phytochemicals were shown to modulate cancer cell metabolism, e.g. resveratrol was shown to partially reverse Warburg effect and resulted in more oxygen consumption and caused cell death [33]. Curcumin and docetaxel treatment lead to an altered glucose, lipid, and glutathione metabolism in cancer cells [34]. A recent study demonstrated phloretin specifically inhibited GLUT2 and resulted in cell cycle arrest in breast cancer cell line, MDA-MB-231, but not in a normal cell line, MCF-10A [35].

1.3.9 Phytochemicals Modulate Gut Microbiota to Prevent Cancer

The role of gut microbiota in cancer is being gradually discovered. Healthy gut microbiota leads to healthy metabolism and offers protection from several diseases including cancer. Dietary phytochemicals such as polysaccharides and phenolic compounds were found to regulate the gut microbiota under stress conditions which led in the reduction of stress related diseases such as inflammatory bowel disease, cancer obesity, and risk of cancer by self-regulation of microbiota [7, 36, 37].

1.4 Limitations of Phytochemicals as Therapeutics for Chemoprevention

As we covered the potential of phytochemicals briefly as anticancer compounds and their potential in chemotherapy, phytochemicals do suffer with certain limitations.

1.4.1 Complex Mixture of Metabolites

Phytochemicals are typically isolated using various extraction procedures employing a variety of solvent systems. When ingested as crude water extract, the efficacy of the mixture might be limited due to less relative abundance of the active compound. Therefore, the separation of active ingredients is needed to achieve the desirable effects. However, it is not always easy to separate and identify the mixture and is a very complicated process requiring a specific expertise in compound separation and identification.

1.4.2 Bioavailability of Phytochemicals

Despite the demonstrated therapeutic potential of phytochemicals, they resulted in limited efficiency in preclinical or clinical trials. One of the biggest reasons that can affect the therapeutic potential is the bio-availability of the active compounds to the target tissues. There are several barriers that affect the bioavailability of the active compounds: indigestibility, rate of metabolism and kinetic stability, interaction with other molecules and phytochemicals [38]. Upon administration, the phytochemicals are subjected to liberation, absorption, distribution, metabolism, and elimination processes. Using computational models depending upon the necessary parameters related to the compounds, better bioavailability could be achieved. To enhance the bioavailability of the compounds, active compounds should be separated, purified, tested and suitable target delivery strategies should be employed. For example, delivery of pH sensitive nanoparticles coated with paclitaxel were more effective compared to neutral nanoparticles [39].

1.4.3 Biphasic Effects

Many plant derived products exhibit biologically opposite effects at different concentrations (hormesis), i.e. they are biphasic as the response changes according to the concentration of the phytochemical. As an example, Genistein exhibits biphasic effects on a number of cell lines. Genistein at lower concentrations (1 μM) promotes cell proliferation, while higher concentration (10 μM) is cytostatic for estrogen dependent MCF-7 cells [40]. Therefore, a titration of dose for phytochemicals is required for desired effects.

1.5 Phytochemicals in Clinical Trials and the Therapies

Although there are enormous reports on anticancer properties of phytochemicals against a variety of cancers, only a few get to the clinical trials and are prescribed for the treatment [41]. There are several plant derived compounds that show promise for cancer treatment in preclinical studies in various models tested, e.g. curcumin, genistein, allicin, etc. Phytochemicals serve three purposes in clinical trials: to improve the effects of chemotherapy and radiotherapy, to reduce the side effects of the drugs, and to check the unwanted drug interactions. Despite the limited preliminary usage, the phytochemicals have been used in clinical trials, e.g. curcumin is used in phase II clinical trial for advanced pancreatic cancers [42]. There are many plant products, which have made it to the different preclinical and clinical trials based on their antitumor and anti-proliferative effects by virtue of the pleiotropic effects on numerous pathways in the cells. This has been nicely summarized in many detailed reviews [41, 43].

1.6 Future Directions: Filling in the Gaps

Despite the availability of reports on phytochemicals proven to be effective against various cancers, the studies were not very well organized and typically used crude plant extracts in a certain solvent systems are documented in popular traditional medicine systems such as Ayurveda medicine system and Chinese medicine system.

There are several reports indicating the ability of phytochemicals to act as signaling molecules against cancers, reviewed in [2]. However, there are only limited studies that properly document the isolation of bioactive phytochemicals in preclinical studies to be used in clinical trials [41]. The complex mixture of the crude plant extracts needs to be separated using suitable separation techniques, such as HPLC or other separation methods. Individual bioactive components should be tested for their biological activities against cancer in preclinical studies and their dosage should be established. High-throughput testing of compounds will enable faster results and should be considered wherever possible. Further, combinations of phytochemicals in addition to the individual screening should be done to check the effects. To enhance the bioavailability of the compounds for optimal effects targeted delivery strategies such as nanoparticles; liposomes or similar strategies should be tested and developed. Further, combination of different phytochemicals or existing therapies should be examined for better efficacy.

1.7 Conclusions

Plants not just constitute our daily food but are sources of valuable phytochemicals with demonstrated potential for cancer chemoprevention in different in vitro, in vivo, preclinical and clinical studies. There are several factors making phytochemicals an attractive choice for the development of an alternative or synergistic approach with conventional cancer therapies. However, to develop a suitable drug from plants organized studies are required. Identification of the bioactive constituents, separation, in vitro and animal testing are required before the clinical studies. The selected plant derived drug candidates should be tested for several pharmacotherapeutic parameters such as pharmacokinetics, drug metabolism, stability, drug–drug interactions dosage, and other required screening.

Conflict of interest Authors declare no conflict of interest

References

1. WHO (2018) Latest global cancer data. WHO, Geneva
2. Kaur V, Kumar M, Kumar A et al (2018) Pharmacotherapeutic potential of phytochemicals: implications in cancer chemoprevention and future perspectives. Biomed Pharmacother 97:564–586. https://doi.org/10.1016/j.biopha.2017.10.124
3. Hanahan D, Weinberg RA (2011) Hallmarks of cancer: the next generation. Cell 144:646–674. https://doi.org/10.1016/j.cell.2011.02.013

4. Steward WP, Brown K (2013) Cancer chemoprevention: a rapidly evolving field. Br J Cancer 109:1–7. https://doi.org/10.1038/bjc.2013.280
5. Benetou V, Lagiou A, Lagiou P (2015) Chemoprevention of cancer: current evidence and future prospects. F1000Research. https://doi.org/10.12688/f1000research.6684.1
6. Vogel VG (2014) Role of hormones in cancer prevention. Am Soc Clin Oncol Educ B. https://doi.org/10.14694/edbook_am.2014.34.34
7. Koh Y-C, Ho C-T, Pan M-H (2020) Recent advances in cancer chemoprevention with phytochemicals. J Food Drug Anal 28:14–37. https://doi.org/10.1016/j.jfda.2019.11.001
8. Brouwer AF, Meza R, Eisenberg MC (2017) Parameter estimation for multistage clonal expansion models from cancer incidence data: a practical identifiability analysis. PLoS Comput Biol. https://doi.org/10.1371/journal.pcbi.1005431
9. Falzone L, Salomone S, Libra M (2018) Evolution of cancer pharmacological treatments at the turn of the third millennium. Front Pharmacol 9:1300
10. Nurgali K, Jagoe RT, Abalo R (2018) Adverse effects of cancer chemotherapy: anything new to improve tolerance and reduce sequelae? Front Pharmacol 9:245
11. Fridlender M, Kapulnik Y, Koltai H (2015) Plant derived substances with anti-cancer activity: from folklore to practice. Front Plant Sci 6:799
12. Wang H, Oo Khor T, Shu L et al (2012) Plants vs. cancer: a review on natural phytochemicals in preventing and treating cancers and their druggability. Anti Cancer Agents Med Chem 12:1281–1305. https://doi.org/10.2174/187152012803833026
13. Nam S, Smith DM, Dou QP (2001) Ester bond-containing tea polyphenols potently inhibit proteasome activity in vitro and in vivo. J Biol Chem. https://doi.org/10.1074/jbc.M004209200
14. Hu Y, Guo D-H, Liu P et al (2011) Bioactive components from the tea polyphenols influence on endogenous antioxidant defense system and modulate inflammatory cytokines after total-body irradiation in mice. Phytomedicine 18:970–975. https://doi.org/10.1016/j.phymed.2011.02.012
15. Singh H (2011) Hepatoprotective activity of turmeric and garlic against 7-12, dimethylbenzanthracene induced liver damage in wistar albino rats. European J Med Plants. https://doi.org/10.9734/ejmp/2011/587
16. Yordi EG, Pérez EM, Matos MJ, Villares EU (2012) Antioxidant and pro-oxidant effects of polyphenolic compounds and structure-activity relationship evidence. Nutr Well-Being Heal. https://doi.org/10.5772/1864
17. Cao J, Liu Y, Jia L et al (2007) Curcumin induces apoptosis through mitochondrial hyperpolarization and mtDNA damage in human hepatoma G2 cells. Free Radic Biol Med 43:968–975. https://doi.org/10.1016/j.freeradbiomed.2007.06.006
18. Benedict AL, Knatko EV, Dinkova-Kostova AT (2012) The indirect antioxidant sulforaphane protects against thiopurine-mediated photooxidative stress. Carcinogenesis 33:2457–2466. https://doi.org/10.1093/carcin/bgs293
19. Saeed MEM, Meyer M, Hussein A, Efferth T (2016) Cytotoxicity of South-African medicinal plants towards sensitive and multidrug-resistant cancer cells. J Ethnopharmacol. https://doi.org/10.1016/j.jep.2016.04.005
20. Jancova P, Anzenbacher P, Anzenbacherova E (2010) Phase II drug metabolizing enzymes. Biomed Pap. https://doi.org/10.5507/bp.2010.017
21. Thomas H. Rushmore, A.-N. Tony Kong (2005) Pharmacogenomics, regulation and signaling pathways of phase I and II drug metabolizing enzymes. Curr Drug Metab. https://doi.org/10.2174/1389200023337171
22. McWalter GK, Higgins LG, McLellan LI, et al (2004) Transcription factor Nrf2 is essential for induction of NAD(P)H:quinone oxidoreductase 1, glutathione S-transferases, and glutamate cysteine ligase by broccoli seeds and isothiocyanates. J Nutr. https://doi.org/10.1093/jn/134.12.3499s
23. Shishodia S, Amin HM, Lai R, Aggarwal BB (2005) Curcumin (diferuloylmethane) inhibits constitutive NF-κB activation, induces G1/S arrest, suppresses proliferation, and induces apoptosis in mantle cell lymphoma. Biochem Pharmacol. https://doi.org/10.1016/j.bcp.2005.04.043

24. Sun J, Hai Liu R (2006) Cranberry phytochemical extracts induce cell cycle arrest and apoptosis in human MCF-7 breast cancer cells. Cancer Lett. https://doi.org/10.1016/j.canlet.2005.10.027
25. Kimmelman AC (2011) The dynamic nature of autophagy in cancer. Genes Dev 25 (19):1999–2010
26. Horwitz SB (1994) Taxol (paclitaxel): mechanisms of action. Ann Oncol 5(Suppl 6):S3–S6
27. Notte A, Ninane N, Arnould T, Michiels C (2013) Hypoxia counteracts taxol-induced apoptosis in MDA-MB-231 breast cancer cells: role of autophagy and JNK activation. Cell Death Dis. https://doi.org/10.1038/cddis.2013.167
28. Sohn EJ, Park HT (2017) Natural agents mediated autophagic signal networks in cancer. Cancer Cell Int 17:110
29. Roxburgh CSD, McMillan DC (2014) Cancer and systemic inflammation: treat the tumour and treat the host. Br J Cancer 110(6):1409–1412
30. Ruiz PA, Braune A, Hölzlwimmer G, et al (2007) Quercetin inhibits TNF-induced NF-κB transcription factor recruitment to proinflammatory gene promoters in murine intestinal epithelial cells. J Nutr. https://doi.org/10.1093/jn/137.5.1208
31. Liberti MV, Locasale JW (2016) The Warburg effect: how does it benefit cancer cells? Trends Biochem Sci 41(3):211–218
32. Warburg O (1956) On the origin of cancer cells. Science. https://doi.org/10.1126/science.123.3191.309
33. del Blanquer-Rosselló MM, Hernández-López R, Roca P et al (2017) Resveratrol induces mitochondrial respiration and apoptosis in SW620 colon cancer cells. Biochim Biophys Acta. https://doi.org/10.1016/j.bbagen.2016.10.009
34. Bayet-Robert M, Morvan D (2013) Metabolomics reveals metabolic targets and biphasic responses in breast cancer cells treated by curcumin alone and in association with docetaxel. PLoS One. https://doi.org/10.1371/journal.pone.0057971
35. Wu KH, Ho CT, Chen ZF, et al (2018) The apple polyphenol phloretin inhibits breast cancer cell migration and proliferation via inhibition of signals by type 2 glucose transporter. J Food Drug Anal. https://doi.org/10.1016/j.jfda.2017.03.009
36. Li Y, Hao Y, Gao B, et al (2019) Chemical profile and in vitro gut microbiota modulatory, anti-inflammatory and free radical scavenging properties of chrysanthemum morifolium cv. Fubaiju. J Funct Foods. https://doi.org/10.1016/j.jff.2019.04.053
37. Malik A, Sharma D, Malireddi RKS, et al (2018) SYK-CARD9 signaling axis promotes gut fungi-mediated inflammasome activation to restrict colitis and colon cancer. Immunity. https://doi.org/10.1016/j.immuni.2018.08.024
38. Estrela JM, Mena S, Obrador E, et al (2017) Polyphenolic phytochemicals in cancer prevention and therapy: bioavailability versus bioefficacy. J Med Chem. https://doi.org/10.1021/acs.jmedchem.6b01026
39. Shenoy D, Little S, Langer R, Amiji M (2005) Poly(ethylene oxide)-modified poly(β-amino ester) nanoparticles as a pH-sensitive system for tumor-targeted delivery of hydrophobic drugs. 1. In vitro evaluations. Mol Pharm. https://doi.org/10.1021/mp0500420
40. Jodynis-Liebert J, Kujawska M (2020) Biphasic dose-response induced by phytochemicals: experimental evidence. J Clin Med 9:718. https://doi.org/10.3390/jcm9030718
41. Choudhari AS, Mandave PC, Deshpande M et al (2020) Phytochemicals in cancer treatment: From preclinical studies to clinical practice. Front Pharmacol 10:1614
42. Dhillon N, Aggarwal BB, Newman RA, et al (2008) Phase II trial of curcumin in patients with advanced pancreatic cancer. Clin Cancer Res. https://doi.org/10.1158/1078-0432.CCR-08-0024
43. Russo M, Spagnuolo C, Tedesco I, Russo GL (2010) Phytochemicals in cancer prevention and therapy: truth or dare? Toxins 2(4):517–551

Role of Phytochemicals in Modulating Signaling Cascades in Cancer Cells

2

Davinder Singh, Avinash Kumar, Astha Bhatia, Hasandeep Singh, Sahiba Kukreja, Balbir Singh, Saroj Arora, and Rohit Arora

Abstract

Dietary and environmental factors have been known to affect human health, for good or for bad. Diet plays a crucial role in modulating the disease state in an individual. Among the different chronic diseases, cancer has been the most researched topic due to its wide associations and high mortality rate. Efforts have been made to elucidate molecular mechanisms capable of altering pathways of carcinogenesis. Though chemical drugs have proved their worth as anticancer drugs, the side effects far exceed the benefit they confer. In contrast, plants are considered as the highest source of phytochemicals that offer great potential of acting as an anticarcinogenic agent with minimal side effects. They help in upregulating cytoprotective genes, encoding carcinogenic detoxifying enzymes, and antioxidant enzymes. Higher consumption of berries, vegetables, and whole grains has been associated with reduced cancer risk and other chronic diseases. In general, it has been shown that phytochemicals modulate oncogenic processes by their antioxidant and anti-inflammatory activities and their capacity to replicate the chemical structure and hormone production. Phytochemicals act as anticancer agents that target signaling pathways at different levels, such as transcriptional and post-transcription control, protein activation, and intercellular communication. These compounds have been known to modulate coding as well as non-coding RNAs such as microRNAs and short non-coding RNAs. Other

D. Singh · A. Kumar · A. Bhatia · S. Arora
Department of Botanical and Environmental Sciences, Guru Nanak Dev University, Amritsar, Punjab, India

H. Singh · B. Singh
Department of Pharmaceutical Sciences, Guru Nanak Dev University, Amritsar, Punjab, India

S. Kukreja · R. Arora (✉)
Department of Biochemistry, Sri Guru Ram Das University of Health Sciences, Amritsar, Punjab, India

M. Kumar et al. (eds.), *Pharmacotherapeutic Botanicals for Cancer Chemoprevention*, https://doi.org/10.1007/978-981-15-5999-0_2

mechanistic pathways for chemoprevention involve the initiation of cell cycle arrest and apoptosis and disruption of signal transduction pathways mainly of mitogen-activated protein kinases, phosphoinositide 3-kinase, protein kinases C, glycogen synthase kinase contributing to pathological cyclooxygenase-2, activator protein-1, nuclear factor kB, and c-myc expression. The Warburg effect is another interesting target for chemotherapeutics. A logical approach for chemoprevention is to address malfunctioning molecules along the compromised signal transduction pathway in cancer. Another crucial chemotherapeutic strategy is to counter the pathways involved in controlling the invasiveness and angiogenesis of tumors. Therefore, cell signaling cascades and their influencing factors have become important targets of chemoprevention, and in this direction the plant extracts are showing promising leads.

Keywords

Cancer · Phytochemicals · Signaling cascade · Proteins · Plants

2.1 Introduction to Cancer Pathophysiology

Transition of a normal phenotype into a cancerous phenotype requires multiple genetic and epigenetic modifications, with each phase resulting in some kind of growth and/or cell survival benefit. Important changes in this multistage cycle include: mutations in tumor suppressor genes (TSGs), (Proto)oncogenic activation, apoptosis and telomerase control deregulation, enhanced angiogenesis, and tissue invasion. Mutations in the germline cells results in the failure of the activities of tumor suppressor genes and the subsequent introduction of oncogenes in somatic cells. Somatic mutations involves mutations, rearrangements between intra-chromosomes, and changes in copy numbers. In addition, genomic instability at a chromosomal level, such as chromosomal translocations and microsatellite instability (MI), characterizes malignant transformation.

Somatic mutations are often related to "passenger mutations" and "driver mutation" but passenger mutations do not specifically participate in oncogenesis and do not offer a growth gain [1]. A driver mutation, in contrast, gives cancer cells significant survival and increase benefit and can lead to clonal expansion. The number of driver mutations is expected to be five or more per common adult epithelial cancer, but fewer incidents in hematologic cancer are required.

Oncogenes are mutated in order to enable the constitutive expression of genes in conditions were wild type is not. For example, a valine changes to glutamate at codon 599, which is the most common activatory mutation of BRAF in human cancer, a residue in the kinase domain. In adjacent residues, phosphorylation (Thr598 and Ser601) is usually the control of the activation loop. Glutamate substitution in codon 599 mimics and constitutes an active enzyme even in the absence of stimuli that normally cause phosphorylation in threonine or serine residues to its analogous form. On the other hand, genetic changes affect tumor suppressor genes in the opposite way where the gene product/function decreases due to mutations. The possible reasons for these inactivations include insertions or deletions of different sizes, formation of truncated protein due to mutations, missense mutation of necessary residue, or gene expression silencing due to epigenetic modifications (methylation, acetylation, etc.). At the normal physiological levels, mutations in both the oncogene and tumor suppressor genes occur equally. By inducing cell growth or by halting cell death, they contribute to the neoplastic cycle by increasing the number of tumors. This can be achieved by activating genes that promote cell cycle by inhibiting normal apoptotic processes or by enhanced angiogenesis. Another class of genes defined as resilience genes facilitates tumorigenesis in a different fashion. This includes the base-excision repair (BER), mismatch repair (MMR), and nuclear excision repair (NER), which are accountable for the repair of fine errors throughout regular DNA replication and mutagen exposure. There are some other genes that regulate translocations, mitotic recombination, and chromosomal segregations such as BRCA1, ATM, and BLM. The sporadic tumors may result in mutations in these three genetic classes that lead to hereditary cancer predictions or to single somatic cells, respectively. The first somatic mutation in an oncogene or tumor suppressor gene transforms a normal cell into neoplastic cell. The subsequent somatic mutations contribute to further cycles of clonal expansion (and thus growth of the tumor). Mutations of these genes are predisposed to disease, not to cancer per se: the neoplastic cycle is a leading example, because a mutation that can support cancer has already occurred in all their cells.

2.2 Cancer and Phytochemicals (Dietary and Non-dietary)

Although there is no single drug that can cure cancer completely but certain types of cancer may be avoidable. Preventing carcinogenic chemicals—or at least minimizing their intake—can reduce the chances of cancer occurrence, but such key avoidance can be hard to implement without full knowledge of the associated risk factors. In addition, reducing certain risk factors can require significant lifestyle changes that are not easy to implement.

It is estimated that 10–70% (average 35%) of mortality in humans is due to diet [2]. The findings are mainly based on the epidemiological and statistical concerning risk related dietary factors. There are numerous epidemiological, medical, and laboratory research evidences that relate the risk of cancer to dietary factors, however, the exact percentage is not known.

Thus, many dietary constituents increase cancer risk, however, there is an accumulation of scientific proofs to support a negative correlation between daily intake of fruits and vegetables and the risk of particular cancers. World Health Organization (WHO), the American Cancer Society, the American Institute of Cancer Research (AICR), and the National Cancer Institute (NCI) have provided dietary regulations to reduce the cancer prevalence. Some nutritional supplements and modified diets to prevent cancer are under clinical trials. In the near future, it is possible that the people may only need to take formulated medicines containing chemotherapeutic products extracted from edible plants to prevent or postpone its onset of cancer [3]. A detailed mechanistic evaluation of the fruit and vegetable products that prevent cancer is needed before their use in dietary supplements or in human intervention trials.

Phytochemicals are non-nutritional plant based components of diet with effective anticarcinogenic and antimutagenic activities. It is not practicable to classify structure-activity interactions, or to determine their fundamental molecular mechanisms, due to the enormous structural complexity of phytochemicals. A simpler solution is to examine their impact on the cancer-associated signal transduction networks.

Numerous population-based studies have demonstrated that the nutrients (carbohydrates, proteins, fat, fiber, antioxidant, vitamins, and trace minerals) of vegetable and fruits have the potential to act as anticancer agent. Vitamins and precursors present in green leafy vegetables and yellow/orange fruits have shown promising anticancer activity. The spotlight recently shifted to non-nutritious phytochemicals. The NCI (National Cancer Institute) reported more than 1000 specific phytochemicals having cancer-preventive efficacy in laboratory studies. It is estimated that almost 100 different phytochemicals can be present in single serving of vegetables. A multistage progression known as carcinogenesis is developed through different molecular and cellular changes. From the study of studies including experimentally induced carcinogenesis in rodents revealed that the tumor growth consist of different but closely linked stages, i.e. initiation, promotion, and progression. While these phases are an over-simplification of carcinogenesis, when contemplating potential opportunities for chemoprevention, it is useful to think at these levels.

According to the standard classification of chemopreventive agents, there are two main categories—blocking agents and suppressing agents. Chemicals which prevent entry of carcinogens to target sites, i.e., from metabolic activation or from eventually interacting with necessary cellular macromolecules (e.g., DNA, RNA, and proteins) while the suppressing agents are those which hinders the malignant transformation of already initiated cells, either in the stage of promotion or progression. The premalignant phase of multistep carcinogenesis can be prevented or reversed by certain chemopreventive phytochemicals. These can also interrupt the production and growth of precancer cells into malignant ones or at least delay them. Recent advances in the mechanistic studies of carcinogens at the cellular and molecular level have shown that blocking and suppressing classification is an over-simplification. For more precise classification the numerous cellular molecules and activities that could be potential targets of chemopreventive agents should be described [4–6]. Consequently, the ability of any phytochemical to inhibit tumor growth should be regarded as the result of combining multiple distinct sets of intracellular effects instead of a specific biological response.

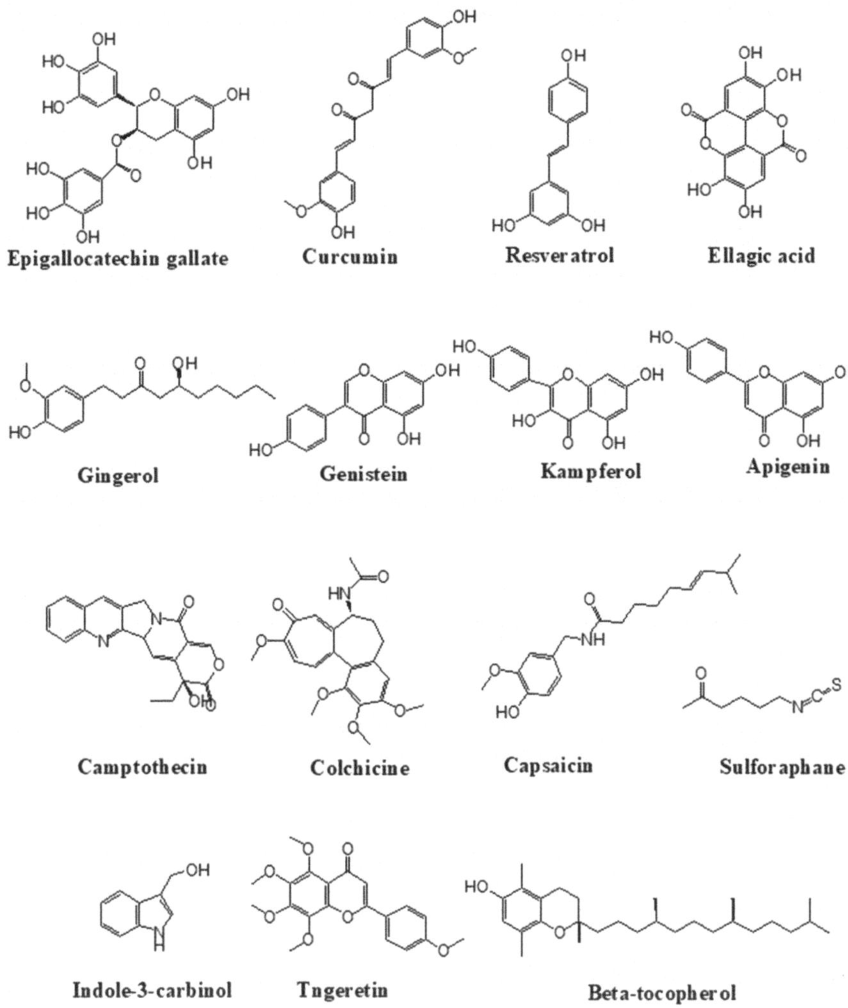

Fig. 2.1 Some chemopreventive phytochemicals

Figure 2.1 shows the chemical structures and nutritional origins of common dietary phytochemicals considered to have chemopreventive potential. Such chemopreventive phytochemicals inhibit cellular and molecular activities including carcinogenic activation/detoxification through xenobiotic-metabolizing enzymes, apoptosis, DNA damage repair, cell cycle progression; angiogenesis, metastasis, oncogene or tumor suppressor gene expression, and hormonal regulation.

2.3 Molecular Targets Involved in Cancer Progression

In past two or three decades, a marked improvement has been made in the field of corealting of biochemical processes with multistage carcinogenesis cycle and now we realized the role of phytochemicals in changing the mechanisms. The significant development in the field of carcinogenesis includes molecular and cellular genetics such as discovery of oncogenes, tumor suppressor genes, unique carcinogen metabolizing genes, DNA repair proteins and enzymes, and regulators of apoptosis and cell cycle. All these advancement in the field of studying carcinogenesis has provided the new and better outlook of the mechanism involved in neoplastic transformation. Advances have been made in diagnosing and identifying the factors that mediate tumor formation, invasion, metastasis, and angiogenesis. Advances in identifying factors that mediate tumor invasion, metastasis, and angiogenesis have also been developed.

The major factors of intracellular network of signaling which preserves homeostasis is serine threonine kinase group of proline like mitogen activated kinase (MAPKs). Abnormal activation of the MAPK pathway may lead to unregulated cell growth, contributing to malignancy. MAPKs pathway has major role in proper regulated growth of a cell but abnormality in the activation or silencing of pathway or its improper downstream signaling, transcription factor can lead to unregulated cell growth, contributing towards malignancy. Many phytochemicals turn on or off the basic signaling molecule(s), depending upon the nature of the signaling cascade which they bind, preventing cell from excessive proliferation and growth [4–11] (Fig. 2.2). MAPKs is a cell signaling kinase just like other kinases such as protein kinase c (PKC) and phosphatidylinositol 3-kinase (PI3K) which acts as an active targets of certain chemopreventive phytochemicals. Such upstream kinases stimulate a distinct array of transcription factors, including the nuclear factor-kB (NF-kB) and the activator protein 1 (AP1) activator enzyme.

2.4 Signaling Cascades

2.4.1 NF-κB and AP1

The NF-kB signaling system concentrates on the signalosome multiprotein complex with IkB kinase (IKK), contributing to IkB phosphorylation, ubiquitylation, and eventual 26S proteasome degradation. The NF-kB is activated and translocated into the nucleus, then it attaches different genes to similar promoter regions. IKK complex is triggered by the NF-kB-inducing kinase (NIK). NF-KB-inducing kinase (NIK) triggers an IKK complex. NIK is regulated by upstream signaling of extracellular signal-regulated kinase (ERK), MAPK/ERK kinase (MEK1/2), and p38 MAPK by a group of mitogen-activated protein kinases (MAPKs), such as MAPK kinase-1 (MEKK1). Previous literature shows that the AKT signaling pathway also controls the activation of NF-kB [12, 13]. PI3K stimulates AKT/protein kinase B by 3-phosphoinositide-dependent kinase-1 (PDK1) protein phosphorylation. Genistein

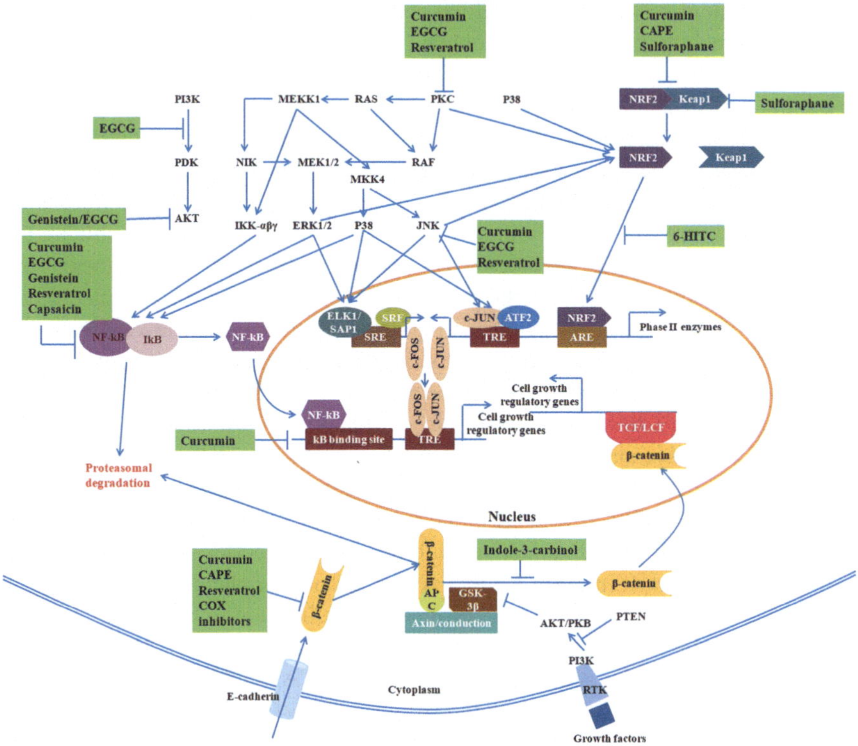

Fig. 2.2 Effect of phytochemicals on cell signaling cascades in cancer

directly prevents the activation of AKT and AKT-mediated NF-kB activation [12, 13]. PI3K and AKT can be inhibited by epigallocatechin gallate (EGCG) [14]. Crosstalk is present between the signaling pathways of AKT and NF-kB—AKT phosphorylation leads to the activation of NF-kB by initiating the activity of IKK [15]. Curcumin, EGCG, and resveratrol also target the IKK for chemoprevention [16–20]. The proteins of MAPK family also control the expression of API which are heterogenous group of dimeric proteins composed of c-FOS, c-JUN, and ATF family members. This results in activation of target genes transcriptionally. The specific form of isoform of PKC is activated by the external stimulus like ultraviolet radiation and phorbol ester, they can stimulate the signaling pathway p21 RAS-ERK with the help of MEK1/2. The upstream kinase MEKK mediated the activation of c-JUN N-terminal kinase (JNK), MAPK kinase-4 (MKK4) and p38.

2.4.1.1 Plant Products in the Modulation of NF-κB and AP1 Signaling

Curcumin—a yellow pigment present in the turmeric rhizome (*Curcuma longa* L.) and related species—is one of the most widely studied phytochemicals with regard to the potential for chemoprevention *Curcuma longa* L. A turmeric rhizome contains a yellow pigment called curcumin which is thoroughly studied phytochemicals for its chemopreventive activity. Topical application of curcumin blocked PMA-induced activation of both NF-kB and Ap1 [21]. The activity of NF-kB can be inhibited by decreased translocation of p65 into nucleus and decreased phosphorylation of IKBα which blocks the degradation of NF-kB [22]. Curcumin blocks the TNF-α-induced nuclear translocation and the hydrogen-peroxide-induced activation of NF-kB in a human myeloid leukemia cell line by suppressing phosphorylation and eventual deterioration by curcumin therapy [23]. The PMA is responsible for the activation of NFKB and attenuation of hydrogen peroxide by treatment with curcumin. [6]-Gingerol—a phenolic compound responsible for ginger spice flavor (*Zingiber officinale* Roscoe)—has been report to suppress tumor development and PMA-induced decarboxylase (ODC) activity and Tnf-α output in the skin tissue of the mouse [24]. Capsaicin is thought to be a co-carcinogen in animal's study due to its unpleasant properties, but somehow recent literature shows that the compound has chemopreventive and chemoprotective effects [25–27]. The topical application of capsaicin blocked the development of PMA-induced mouse-skin tumors and the activation of NF-kB [28]. Due to the blockage of IkBα degradation and translocation to the nucleus of NF-kB skin tumors are inhibited. Epigallocatechin gallate (EGCG) is a polyphenol present in green tea that is antioxidant and chemopreventive. PMA stimulated mouse epidermal cell line (JB6) shows inhibition in malignant transformation of cells by inducing blockage of Ap1 or NFKB activation [29, 30]. EGCG inhibited the constitutive activation of NF-kB and STAT-3 this completely blocks the development of vascular endothelial growth factor (VEGF) but not Akt pathway [31].

Genistein, an isoflavone derived from soy is suspected to possess chemopreventive property against breast and prostate cancer. Genistein blocks the PMA-induced AP1 expression through c-FOS and ERK activation in certain human mammary cell lines. Genistein therapy revoked NF-kB DNA interaction in human hepatocarcinoma cells activated by hepatocyte growth factor [32]. In androgen-sensitive (LNCaP) and androgen-insensitive (PC3) cells, genistein at apoptogenic concentrations blocked the H_2O_2, or TNF-α activated NF-kB by reducing IkBα phosphorylation and NF-kB nuclear translocation [33]. Resveratrol treatment inhibited PMA-induced cyclooxygenase-2 (COX-2) expression and catalytic behavior, via the cyclic-AMP response component (CRE), in human mammary epithelial cells. The levels of prostate-specific antigen (PSA) and P65 are downregulated with p53, p300/CBP and activation of APAF1 are related with treatment of androgen sensible prostate cancer (LNCaP) cell line with resveratrol [34]. Resveratrol-induced apoptosis was associated with p53 phosphorylation in mouse JB6 epidermal cells, which appeared to be regulated by Erk and p38 activation [35]. Resveratrol targets the suppression of NF-kB dependent transcription, nuclear translocation in p65, and TNF-α-induced phosphorylation in myleiod cells [36]. TNF-dependent activation of MAPK kinases (MEK) and JNK and TNF-dependent activation of caspases were also inhibited by resveratrol [36]. Caffeic acid phenethyl ester (CAPE), sulforaphane, silymarin, apigenin, emodin, quercetin, and anethole are compounds can lead to the blockage of chemical prevention and cytostaic effect by the activation of AP1 [37].

2.4.2 NRF–KEAP1 Complex

NRF2 is an antibody controlling transcription factor for many detoxification enzymes or antioxidants. Kelch-like ECH-associated protein 1 (KEAP1) is a cytoplasmic NRF2 repressor which prevent it from translocating into the nucleus. The domains which are rich in both glycine and hydrophilic regions are present in NEH2 domain of NRF2 and they communicate and interact with each other. The numerous cysteine residues are contained in KEAP1.These cysteine residues are vulnerable to oxidation or covalent modifications by phase II inducers [38]. Oxidation and covalent modification leads to the separation of NRF2 from KEAP1. In addition, NRF2 phosphorylation of serine and threonine residues is thought to allow dissociations of NRF2 from KEAP1 and subsequent translocations into the kinase by kinases such as PI3K, PKC, JNK, and ERK [39]. The NRF2 translocation can both be activated and prevented by p38. The center is composed of a heterodimer that attaches to the antioxidant-responsive component (ARE) for the induction of gene expression. The term is derived from musculoaponeurotic fibrosarcoma. NRF2/MAF main genes encode either Phase II or antioxidants enzymes, including glutathione S-transferase-α2 (GSTA2) NAD(P)H: quinone oxidoreductase (NQO1) and heme-oxygenase-1 (HO-1) cysteine ligase Ś-GCLC and glutamate-cysteine ligase modified (GCLM). In combination with an NRF2 and ARE-binding element, PI3K also phosphorylates the CCAAT/enhancer-binding protein-β (C/EBPβ), inducing its nucleus translocation and the CCAAT gene C/EBP-β answer component (XRE) [40]. ARE is activated by human PI3K neuroblastoma cell transfection, attenuated by a PI3K or NRF2 pharmacy inhibitors [41]. The phenethyl ester (CAPE) of curcumin and caffeic acid disrupts the complex NRF2-KEAP1 which leads to increased NRF2 binding to ARE [42, 43]. By covalent binding to its thiol groups, Sulforaphane interacts directly with KEAP191. The NRF2 nuclear translocation, which ultimately stimulates ARE, is activated by 6-(methylsulfinyl)hexyl isothiocyanate (6-HITC), a Japanese version of horseradish wasabi [44].

2.4.2.1 Plant Products in the Modulation of NRF-KEAP1 Signaling

HepG2 exposure to green tea extract induces a phase II expression of enzymes through ARE [45]. This upregulation was supported by ERK2 and JNK1 activation and immediate early c-JUN and c-FOS gene activation. Follow-up experiments have shown EGCG to activate the enzyme production of Phase II in HepG2 cells transcriptionally, calculated from the ARE Reporter-Gen Assay [46]. An evaluation of the oligonucléotide microarray gene-expression profiles indicated that NqO1, Gst, and Gcs were increased in the small intestines of wild-type mice, whereas Nrf2-null mice displayed significantly lower rates of those enzymes [47]. Morimitsu and colleagues identified the analog sulforaphane (6-methyl sulphinylhexyl isothiocyanate, 6-HITC) as a key GST inducer in Japanese horseradish, wasabí (wasabia japonica or eutrema wasabi Max) during the extensive screening of vegetable extracts in cultured rats' liver epithelial cells RL-34 [44]. Oral administration of 6-HITC succeeded in inducing more than sulforaphane detoxification enzymes in phase II hepatic induction, whereas NRF2-null mice reversed this induction [44]. The Nrf2 complex, which was correlated with a significant increase in Ho-1 development and transcription in pork renal epithelial cells, both the curcumin and CAPE induced the Nrf2 expression [42]. Curcumin-induced Ho-1 gene induction seems to involve MAPK which is upstream of the NRF2. It is notable that the occurrence of α, β-unsaturated ketone moieties in both curcumin and CAPE, which can serve as accepters to Michael reaction that can alter cysteine thiols found in Keap1. Sulforaphane also interacts explicitly with thiol groups of Keap-1.

2.4.3 β-Catenin

β-catenin (β-cat) mediates pathways of signaling through both growth factor and WNT. The association of a WNT ligand with its transmembrane receptor—a "frizzled receptor"—recruit disheveled protein which is inactivated by phosphorylation at serine-9 by the glycogen synthase kinase-3β (GSK-3β). On the other side, the association of growth factor with receptor tyrosine kinase (RTK), which in turn causes phosphorylation of AKT/protein kinase B (PKB), is activated by phosphatidylinositol 3-kinase (PI3K). GSK-3β is also inactivated by serine-9 phosphorylation through phosphorylated AKT. The AKT-mediated GSK-3β inactivation is blocked by a tumor suppressor protein phosphatase and a tensin homolog on chromosome 10 (PTEN). GSK-3β—a component of multiprotein complex consisting of adenomatous polyposis coli (APC), axin and conductin—regulates the intracellular fate of β-catenin, which functions as a component of cell–cell adhesion machinery, in its membrane-bound form and as a signaling molecule in its free cytosolic form. GSK-3β. In the absence of a growth factor, the cytosolic β-Catenin phosphorylates GSK-3β are targeted for ubiquitylation followed by proteasomal degradation, with the amino end serine or the threonine residue being amino-terminal residues. The inactivation of GSK-3β contributes to the cytosolic stabilization of the β-catenin in reaction to the above stimulus. In contrast to GSK-3β inactivation, the mutation of APC or axin and β-catenin also induces cytoplasm stabilization. This stabilized cytosolic β-catenin further binds to the T-cell factor (TCF) in the nucleus. This complex (β-catenin-TCF/LEF) activates a gene transcript that regulates the process of cellular growth.

2.4.3.1 Plant Products in the Modulation of β-Catenin Signaling

Throughout their molecular chemical prevention process, many dietary phytochemicals demonstrated to decrease the β-centered signaling pathway. In the multiple intestinal neoplasms (min/+) mouse model, curcumin and CAPE inhibited tumorigenesis and decreased β-catenin expression [48]. Curcumin also reduced β-catenin's cellular levels by the cleavage caused by caspases [49]. Resveratrol also regulates the production of β-catenin in human colon cancer cell line

[50]. The pattern of β-catenin mutation has been altered by indole-3-carbinol in chemically induced colonic rat tumors, inhibited adhesion, transition and invasion of human cultured breast carcinoma cells along with upregulation of E-cadherin and β-catenin [51, 52]. Tangeretin from citrus has shown a similar effect [53]. COX inhibitors were found also to suppress signaling of β-catenin and transcriptional activity of β-catenin—TCF/LEF [54–56]. As COX-2 upregulation induces tumorigenesis, and β-catenin controls COX-2 expression, β-catenin signaling control may be a potential molecular goal for dietary phytochemicals chemoprevention.

2.4.4 MAPK

The MAPK cascade is an intracellular signaling pathway that controls numerous cellular functions such as cell proliferation, cell cycle, etc. The cascade is described as a linear signaling pathway mediated by receptor tyrosine kinases at the cell surface. Activation of Ras is the first step in the MAPK cascade. Upon Ras activation, Raf (A-Raf, B-Raf, or Raf-1) is recruited by Ras binding into the cell membrane and triggered in a complex process involving phosphorylation and several cofactors that are not fully understood. Raf proteins activate MEK1 and MEK2 directly by phosphorylation of multiple serine residues. MEK1 and MEK2 are themselves tyrosine and threonine/serine dual-specific kinases which ultimately result in activation of phosphorylating threonine and tyrosine residues in Erk1 and Erk2. While MEK1/2 has no identified targets other than Erk proteins, Erk has multiple targets like Elk-1, c-Ets1, c-Ets2, and Erk2 [57].

2.4.4.1 Plant Products in the Modulation of MAPK Signaling

Bromelain, a vital ingredient of *Ananas comosus*, induced apoptosis in MCF-7 cells apparent by cell aggregation in the G1 process, formation of apoptotic body and fragmentation of nucleus [58]. *Camellia sinensis*'s main polyphenols are theaflavins and thearubigins. In malignant melanoma A375 cells, mostly JNK and p38MAPK but not ERK is triggered as a consequence of treatment with theaflavins and thearubigins [59]. Combinatorial strategy involves application of E-piplartine along with curcumin enhances cytotoxicity in tumor cell. Piplartine alone arrested the G1 phase of cell cycle by inactivating cdk2 and destabilizing cyclin D1 which outpace the arrest of G2/M induced by curcumin. Piplartine and the mixture of curcumin disrupted the activation of ERK1/2 and Raf-1, cell cycle progression and enhanced expression of HSP-90 [60]. Plumbago zeylanica a traditional Indian plant produces a molecules called 3-hydroxy-20(29)-ene27,28-dioic acid (PZP) which shows a potential to induce apoptosis in breast cancer cells regulated by mitochondria [61]. It blocked cell adhesion, prevented wound healing migration and invasion of MDA-MB-231 cells by downregulating metalloproteinase-2 and metalloproteinase-9 expression, and stimulated MAPK protein levels (JNK, ERK1/2, PI3 K, Akt, MMP-2, MMP-9, VEGF, and HIF-1). Withaferin A stimulated the intrinsic apoptosis pathway in leukemia cells by phosphorylation of p38MAPK. Inhibition of p38MAPK by siRNAs resulted in reduced Bax expression and decreased apoptosis suggesting MAPK's crucial role [62]. PI3K/Akt and Erk/MAPK are inhibited by tocotrienols and β-tocopherol [63]. Colchicine upregulates the COX-2 production, Erk phosphorylation and dual specificity phosphatase (DUSP1) gene in stomach carcinogenesis [64, 65]. It also inhibits hepatocellular carcinoma cell growth by upregulating protein 12 (AKAP12) and converting beta-2 (TGF-β2) proteins [66].

2.4.5 PI3K-AKT Pathway

PI3K is a kinase which phosphorylates the plasma membrane inositols and is involved in the transduction of cellular signaling [67]. All tyrosine kinases receptors and tyrosine kinases nanoreceptors results in PI3 K activation, which releases the second messenger phosphatidylinositol (3–5)-trisphosphate (PIP3) from phosphatidylinositol 4,5-bisphosphate [68]. The pleckstrin homology domain-containing proteins are recruited to the cell membrane by PI3K-AKT pathway including AKT/PKB kinases and they induces conformational change which results in their phosphorylation by the constitutively active phosphoinositide dependent kinase 1 and 2 [69, 70]. Activated Akt translocate from cytoplasm to nucleus and triggers downstream targets that include replication, cell cycle regulation, production of growth factors, migration and angiogenesis. Akt regulates the phosphorylation and activation of mammalian rapamycin complex1 receptor, a serine/threonine kinase that plays vital roles in controlling the translation and synthesis of proteins, angiogenesis, and progression of the cell cycle.

2.4.5.1 Plant Products in the Modulation of PI3K-AKT Signaling

The in vitro study in case of food plant like plethora produces compound like polyphenols have potential to affect MAPK, AKT, and AKT signaling cascades targetting cell proliferation of cancerous cells [71]. In this case, [6]-gingerol inhibits the cell growth, activate caspase-3, and increase cancer cell MAPK and Akt activity [72]. Sulforaphane, an isothiocyanate extracted from broccoli inhibits PI3K/AKT and MAPK signaling pathways regulated by extracellular kinases that leads to cell cycle arrest and apoptosis [73]. Additionally, allicin has been shown to significantly impede cancer cell development by inhibiting the signaling of p38/MAPK [74]. Previously, emodin from aloe and cranberry myricetin have been shown to produce a negative effect on the PI3/Akt signaling pathway [75]. Dietary polyphenols like quercetin, green tea, polyphenols, and EGCG in low concentrations stimulate MAPK pathways by ERK and JNK which contributes to survival gene expression (c-fos, c-jun). HL-60 cells in the human-acute promyelocytic leukemia are inhibited by fisetin and hesperetin through the modification of several signaling networks, i.e. MAPK, NF-Fraktion, JAK/STAT, PI3K/Akt, Wnt, and mTOR pathway [76]. In pancreatic cells, kaempferol acts on the receptors of proto-oncogenic tyrosine-protein kinase (Src), Erk 1/2, and Akt, and delays development and migration [77]. Additionally, PI3K/Act pathway in pancreatic cancer cells is activated by lycopene. The inhibition of Erk and Bcl-2 signaling prevents gastric carcinogenesis

[78]. This stimulates antioxidant enzymes in cancer cells (e.g., GST, GSH, and GPx) and prevents the oxidative damage from carcinogenic agents.

2.4.6 Wnt Cell Signaling

Signaling is initiated by the wnt proteins that are secreted and further binds to seven-pass transmembrane receptors encoded by the frizzled genes [79]. This binding results in phosphorylation of the dishevelled protein whose interaction with axin inhibits essential substrates of glycogen synthase kinase 3β (GSK3β) [80]. The inactivation of GSK3β in vertebrates may result from its association with Frat-1 [81]. ⍰-catenin, Axin, and APC act as GSK3β substrates. Unphosphorylated β-catenin escapes detection by ubiquitin E3 ligase member TRCP and translocates to the nucleus where it binds with various transcription factors, i.e. TCF and LEF [82]. Discrete components in the process include casein kinases-I and II (CK-I, CK-II) which were both introduced for disheveled phosphorylation. The serine/threonine phosphatase PP2A interacts with axin and APC even though its biological position in the system remains unclear [83].

2.4.6.1 Plant Products in the Modulation of Wnt Signaling

It is indicated that genistein increased GSK-3β expression, GSK-3β binding to β-catenin, and β-catenin phosphorylation which proposed its ability to inhibit the prostate cancer through inhibition of Wnt/β-catenin signaling. It also decreased proliferation caused by Wnt 1 and also the Wnt targets. As a flavonoid, apigenin was the first identified as an inhibitor of the Wnt pathway because it down regulates the CK2 [84]. In breast, lung, and colon cancer cell lines, EGCG inhibited the Wnt signaling in a dose-dependent manner, whereas in normal cells Wnt signaling is fully active [85]. In breast cancer, treatment of EGCG (25 to 100 μM) stabilizes the HBP1 mRNA which is a suppressor of Wnt signaling. It decreases both the division and invasiveness of the breast cancer cells by HBP1 activation and subsequent suppression of Wnt signaling. Although EGCG's inhibition of Wnt is indirect, Quercetin's influence directly controls the Wnt pathway by influencing pathway molecules in several cell types. Quercetin (50 μM) decreased the β-catenin/Tcf complex after the treatment in SW480 cells (colon cancer cells [86]. Fisetin-treated melanoma cells with G1-phase arrest and suppression of Wnt/β-catenin signaling resulted in reduced cell viability. This pattern was accompanied by a decrease in the expression of Wnt proteins and their co-receptors and a corresponding increase in the development of endogenous Wnt inhibitors. Isoquercitrin (quercetin 3-O-β-d-glucopyranoside), a glycosylated analog of Quercetin, prevents the proliferation of Glioblastoma (Gbm) cells through means of control mechanisms for the Wnt/β-catenin pathway. This study has shown that 23% of β-catenin expression was found in the nuclei of control Gbm cells. However, with the treatment of isoquercitrin at the concentration of 100 μM the expression of nuclear β-catenin dropped considerably to 4%, whereas non-nuclear β-catenin rose to 77%, highlighting that isoquercitrin therapy changes β-catenin distribution in Gbm cells [87]. Genistein therapy upregulates the expression of GSK-3β and stabilizes the relationship between GSK-3β and β-catenin, contributing to phosphorylation and degradation of β-catenin. Genistein also abolished dose-dependent β-catenin/Tcf transcriptional involvement in SW480 cells [88].

2.4.7 mTOR Pathway

Mammalian target of rapamycin (mTOR) is a serine/threonine kinase, belongs to a class of kinases similar to phosphoinositide 3-kinase (PI3K). There are two types of multiprotein complexes known as mTOR complex 1 (mTORC1) and 2 (mTORC2) [89]. In general, mTORC1 regulates autonomous cell growth in reaction to the supply of nutrients and growth factors, while mTORC2 is responsible for cell division and survival. mTORC1 comprise five components known as mTOR, the catalytic subunit, regulatory-associated mTOR (Raptor) protein, mammalian lethal with Sec13 protein8 (mLST8), proline-rich Akt substrate (PRAS40); and DEP-domain-containing mTOR-interacting protein (DEPTOR) [90]. However, according to the type and position of the cell, the exact components of the mTORC1 complex can vary [91]. Raptor controls the function of mTOR positively and provides a scaffold for the recruitment of mTORC1 substrates, while PRAS40 and DEPTOR are negative regulator of mTORC1 [92, 93]. Recent studies suggest that Raptor's phosphorylation status may control the mTORC1 response [94]. The mLST8 binds to mTOR's kinase domain, which positively controls mTOR kinase activity; it appears capable of sustaining a rapamycin-sensitive relationship between Raptor and mTOR [95]. Some research indicates that mLST8 is also essential to maintain the Rictor–mTOR relationship in the mTOR2 complex, which contributes to the likelihood that mLST8 could be crucial for switching mTOR between the two forms [96]. Such a function would be dependable with the active equilibrium known to exist in mammalian cells in these complexes [97]. As already mentioned above, mTORC1 was defined by PRAS40 and DEPTOR as distinct negative regulators [98]. On the downregulation of mTORC1 activity, PRAS40 and DEPTOR recruited into the network, where they facilitate mTORC1 inhibition [99]. It has been proposed that PRAS40 controls kinase expression in mTORC1 by acting as a strong substrate-binding inhibitor [99]. The mTORC1 specifically phosphorylates PRAS40 and DEPTOR after activation, which decreases their physical interaction with mTORC1 and further enhances mTORC1 signals [99]. DEPTOR communicates adversely with mTORC1 and mTORC2, controlling their behaviors [93].

2.4.7.1 Plant Products in the Modulation of mTOR Signaling

Curcumin downregulates the protein and mRNA expression of mTOR and both the analytical and regulatory subunits of mTOR in HCT 116 cells. This indicates its antiproliferative effect by suppressing the signaling cascade of mTOR and can therefore represent a novel type of mTOR inhibitor [100]. Curcumin immediately blocks the phosphorylation of mTOR and its downstream effector proteins, i.e. p70S6 and 4E-BP1 in Rh1, Rh30, DU145, MCF-7, and Hela cell lines indicating its anticancer function mainly by suppressing mTOR signaling pathway [101]. Another polyphenol, resveratrol blocks PI3K/Akt signaling pathway and induces apoptosis in human LNCaP prostate carcinoma cells [102]. Resveratrol has also been shown to downregulate the signaling pathway for PI3K/Akt/mTOR, and together with rapamycin further improves resveratrol-induced cell death in human U251 glioma cells [103]. Resveratrol blocks the activation of the mTOR pathway in smooth muscle cells through PI3K/Akt pathway and thus suppresses the proliferation of oxLDL-induced cancer cells [104]. Resveratrol reduces mTOR and p70S6 K phosphorylation in human breast cancer cells (MDA-MB-231, MCF-7), and in conjunction with rapamycin, it also suppresses Akt's phosphorylation. The studies on pomegranate extract in the lung murine tissue sample shows that it can reduce mTOR signaling by inhibiting PI3K/Akt, pathway and its downstream molecules, i.e. p70S6 K and 4E-BP1 [105]. In phase II clinical trial, patients with high PSA (prostate serum antigen) levels post-surgery were treated with daily 8 ounce of pomegranate juice which shows statistically significant persistence of 13 months on doubling time for PSA [106]. The importance of slow PSA doubling time remains unclear [107]. Genistein inhibits MAPK pathway and Akt phosphorylation in Hela and CaSKi cells [108]. It lowers the protein expression of total and phosphorylated Akt in MCF-7 cells, indicating its potential to downregulate the PI3K/Akt signaling pathway against breast cancer [109]. In HT-29 colon cancer cells, the synthesis of genistein and indol-3-carbinol induces apoptosis and autophagy by inhibiting act and mTOR phosphorylation [110]. Synergistically, it also increase the potency of cisplatin by reducing the Akt expression in pancreatic cancer [111].

2.4.8 Phytochemicals as Modulators of Antioxidant and Anti-Inflammatory Stress

Dietary phytochemicals found in the human diet including grains, vegetables, fruits, and beverages have been well documented for their antioxidant, chemopreventive, and anticancer activities [112, 113] (Fig. 2.3). Clinical and nutritional studies strongly suggest the use of phenolic compounds to cure as well as lower the risk of different diseases including diabetes, cancer, and neurodegenerative diseases. Phenolic compounds are basically hydroxyl derivatives of aromatic carboxylic acids making them good electron/hydrogen donors to neutralize reactive oxygen species and other free radicals [114].

The antioxidant potential of phenolic compounds can be ascribed to the number of hydroxyl groups present in its chemical structure. Also, hydroxyl aromatic rings (–OH at *ortho* and *para* positions) are reported to have high antioxidant potential, e.g. flavonoids. The phenolic compounds are mainly the secondary metabolites of the plants produced in response to stress stimuli or with the attack of a pathogen in the form of a defense mechanism [115]. This property may be responsible for its similar action in mammalian systems upon the stress stimuli. The phenomena are termed as xenohormesis, demonstrating the physiological relevance to human health

Fig. 2.3 Phytochemicals as modulators of cancer progression

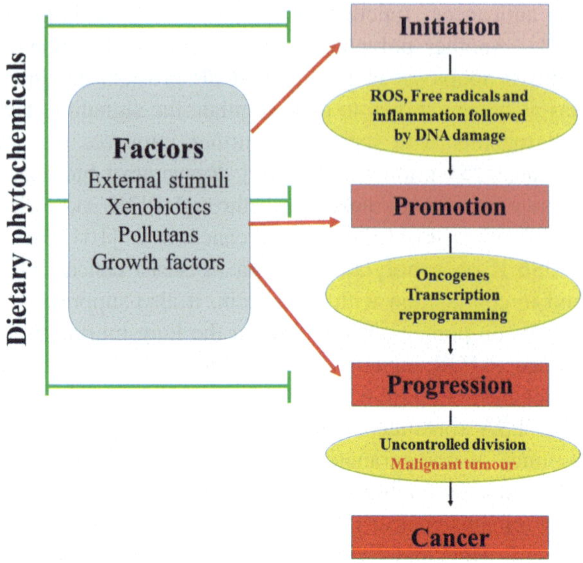

[116, 117]. Based on their chemical structures, the literature review confirmed the antioxidant and anti-inflammatory activities of phenolic compounds [118, 119].

Several factors such as physical and chemical stimuli including injury, the metabolic disorder can contribute to redox imbalance causing localized inflammations in the body. However, long-term persistent redox imbalances due to pollutants, xenobiotics, heavy metals, etc. leads to oxidative stress which can cause irreversible damage to lipids, proteins, and DNA and ultimately tumorigenesis [120]. In such conditions, phenolic compounds interrupt the auto-oxidation of biomolecules through donating electrons/hydrogens to free radicals ultimately reducing the oxidative stress. The cell has its own integrated enzymatic redox system to maintain redox homeostasis which includes catalase (CAT), superoxide dismutase (SOD), glutathione reductase (GR), and peroxidase (GPx), etc. [121]. The molecular signaling transduction pathway regulated by reactive oxygen species (ROS) mainly includes MAPK's, NFkB, and Nrf2 pathways [122]. Activation of MAPK's and NFkB pathways stimulates the upregulation of various inflammatory cytokines (e.g., TNF-α, IL-6, IL-1, etc.), thus inducing pro-inflammatory signal transduction. The other pathway involved is the Nrf2 pathway, which involves a key transcription factor, i.e., Nrf-2 (nuclear factor erythroid-related factor). This factor potentially able to modulate the expression of antioxidant enzymes, i.e. CAT, SOD, GR, GPx, GSH, etc., involved in scavenging the ROS. The enzymes directly or indirectly scavenge ROS and suppress the pro-inflammatory cytokines released as a result of ROS mediated NFkB pathway [123]. The underlying mechanism involved which is responsible for the modulation of antioxidant enzymes is ARE (antioxidant-responsive elements) mediated transcriptions via Nrf2 regulation. Translocation of Nrf2 to the nucleus stimulates ARE-mediated gene expressions. Further, the Nrf2-Keap1 pathway is responsible for maintaining ROS equilibrium in cells due to the action of xenobiotics [124]. Dietary phytochemicals such as flavonols and isoflavones interact agonistically with aryl hydrocarbon receptor (AhR) present in the cytoplasm. Consequently, activating the Nrf2 pathway mechanism to enhance antioxidant enzyme expressions to overcome oxidative stress.

These phytochemicals such as polyphenols also have been reported to induce Keap1/Nrf2 complex dissociations. In cell culture studies, quercetin, rutin induced activation of AhR in HepG2 cells; kaempferol and quercetin in Caco-2 cells (benzo (α)pyrene-induced) combined AhR/Nrf2 activation [125]. Luteolin successfully overexpressed the drug-metabolizing enzymes via., AhR/Nrf2 activation [125]. Thus, the observed multiple potential mechanisms of this plant secondary metabolite make an interesting target to overcome oxidative stress in the human body. Further, the ROS overproduction has been correlated with the enhanced release of inflammasomes via NLRP3 activation. Consequently, these inflammasomes elicit the release of cytokines in the extracellular environment which results in the activation of another signaling pathway mediated by Toll-like receptors (TLR-1) located on the cell membranes. The activation of this pathway further provoke the release of cytokines, i.e. IL-6, IL-8, TNF-α, etc. via pro-inflammatory signaling pathway which leads to inflammation [126].

The induced inflammation further may lead to the expansion of various types of degenerative diseases, i.e. Alzheimer's disease. Literature perusal suggested the neutralizing effects of dietary phytoconstituents (apigenin, procyanidin B2, and resveratrol) on production of ROS, which ultimately inhibited NLRP3 (a gene which makes cryopyrin) activation in both in vitro and in vivo experiments [127]. In addition, these phytochemicals also have been reported to inhibit the underlying mechanism/pathway via pro-inflammatory cytokines NFkB and MAPK's responsible for the activation of NLRP3 via blocking RelA/B translocations and stimulating PPARγ/SIT1 signaling pathways to alleviate inflammation [125]. Various chemopreventive phytochemicals including Gingerol from ginger (*Zingiber officinale* Roscoe.), resveratrol from grapes and red wine, curcumin from turmeric (*Curcuma longa* L.), sulforaphane, and erucin from cruciferous vegetables such as broccoli and *Eruca sativa* have been reported to target NFkB and Nrf2 pathways [128, 129]. The phenolic compounds, i.e. quinones on the one hand, act as a strong antioxidant via., donating for scavenging free radicals to overcome oxidative stress but on the other hand, during the process of losing electron/hydrogen, the molecule itself becomes a free radical in severe oxidative stress conditions [130]. The oxidized compounds start acting as pro-oxidants and can induce adverse effects on human health. This usually happens as a result of ingesting a high amount of phytochemicals in diet and usually occurs in the GI tract possibly.

2.4.9 Cell Cycle Modulators

Progression of the cell cycle is mainly regulated by cyclin-dependent kinases (CDKs) depending upon the individual phase, i.e. G1, S, G2, and M phase. Cyclin-dependent kinases consist of two subunits, i.e. regulatory and inhibitory subunit. CDK holoenzyme is produced by the association of cyclins with the catalytic subunit through non-covalent binding. The complex is activated by CDK7/cyclin H after phosphorylation of associated subunit (catalytic). For G1 phase progression, CDK4 and CDK6 bind with cyclin D; for S phase, CDK2 complex with cyclin A; for G2 phase, but transition of G1-S phase requires complex of CDK2 with cyclin E; transition from G2-M phase, requires complex of CDK1 with cyclin B. In cancer cells, regulatory units involved in the activation and progression of cell cycle, i.e. CDKs, p53 (tumor suppressor gene), pRb (retinoblastoma protein), etc. are altered by several cell survival and mitogenic signaling pathways [131, 132].

This further induces the activation of the transcription factors from E2F family. The free form of E2F depends upon the phosphorylation of retinoblastoma (Rb) protein which is regulated by CDK4/6-cyclin D1 and CDK2-cyclin E complexes. This further facilitates the transition of the G1-S phase. However, unphosphorylation of pRb and its related protein leads to inhibit cell division by seizing the E2F family [133]. Furthermore, overexpression of cyclin or CDKs, as well as their activators phosphatases, has been reported in cancer cells, e.g. the stage of the tumor initiation phase, cyclin D1 has been reported with increased expression responsible for cell proliferation. Another family specifically is known to inhibit the CDK4/6 is the INK family (p16/ink4a, p15/ink4b, etc.) helps in the inhibition of tumorigenesis. But recurrent mutations in the INK family may result in its inactivation or lack of expression leading to initiation of tumor development. Cdc25-family (protein tyrosine phosphatases) act as mitotic activators and mediates the transition of G2 to M phase, thus act as a checkpoint. It mediates the process by forming a complex with cyclin B after dephosphorylating cdc2/p34 present on tyrosine residue and gets activated. Chk/2, an upstream kinase checks the dephosphorylation of a tyrosine residue in order to inhibit the cell transition from G2 to M phase in case of tumorogenesis [134, 135]. Here, the CDC family plays a crucial role as genetic changes in cdc2/p34 kinase have been reported to its overexpression which can be related to oncogenic consequences. The above mentioned regulatory cascades thus could act as an effective therapeutic target to control tumor initiation and its progression.

Flavopiridol, a flavonoid alkaloid, has been reported to be a CDK4 and CDK7 inhibitor and is currently under investigation. This compound has shown antiproliferative activity against various types of cancers including breast, lung, bladder, and myeloid leukemia [136, 137]. Indirubin, a chemical compound used in traditional Chinese medicine known as Danggui Longhui Wan has been reported to be a CDK1 and CDK2 inhibitor [138]. Chebulagic acid isolated from *Terminalia chebula* has been reported for the induction of cell cycle arrest in G1 phase mediated anticancer activity against cancerous cell lines. It also reduced the pro-inflammatory NFkB levels in cells [139]. Wogonin from *Scutellaria radix* has been reported to decrease the expression of cyclin A, E, D1, CDK2, and CDK4 in a human colorectal cell line [140]. Mannose-binding agglutinin from bulbs of *Lycoris aurea* arrested the cell cycle in the G2/M phase via downregulating CDK/Cyclin A in the A549 cell line [141]. Fangchinoline isolated from roots of *Stephaniae tetrandrine* has shown its antiproliferative activity in breast cancer cell lines, i.e. MCF-7, MDA-MB-231 via. Cell cycle arrest in the G1 phase. It effectively enhanced levels of p21WAF1 and p27KIPI (cyclin-dependent kinase inhibitors) and decreased the expression of cyclin D and E [142, 143].

Berberine isolated from *Berberis spp.*, an isoquinoline alkaloid showed antiproliferative activity against various cancer cell lines, i.e. breast cancer, lung cancer, prostate cancer, etc. in a dose-dependent manner as well as in animal models via. Arresting cell cycle at the G0/G1 phase, inhibiting CDK2 and CDK4, reduced the activity of cyclinD1 [144, 145]. Triterpenoids from *citrus spp.* termed as limonoids possess remarkable anticancer activity against SW480 cells via arresting cell cycle at the G0/G1 phase and inhibiting CDK4 and CDK6 [146]. Polyphenols present in leaves extract of mulberry inhibited tumor growth, activated p53, blocking Rb (retinoblastoma protein) phosphorylation to reduce free E2F, arrested cell cycle at G1/S phase [147]. A flavonoid chalcone, isoliquiritigenin isolated from licorice, bean sprouts, etc. showed anticancer activity against DU145 cell line, inhibited cyclin D1 and cyclin E and activated CDC family proteins via inducing phosphorylation at TYR15 [127]. Quercetin, a flavonoid induced cell cycle arrest in G1 phase and decreased the expression of CDK2 and CDK6. It also inhibited Rb phosphorylation to inhibit cell cycle transition from the G1-S phase. Isoquercetin and rutin combined with quercetin present in seeds of buckwheat induced cell cycle arrest at G2/M phase, activate p53 gene, downregulated cyclin D1, CDK2, and CDK7. Epigallocatechin-3-gallate (EGCG) isolated from green tea showed broad range anticancer effects on human cancers, induced G0/G1 phase arrest, downregulate the expression of pro-inflammatory cytokines such as NFkB combined with activating CDKs inhibitors [148, 149].

The methanolic extract of *Maytenusroyleanus* leaves induced cell cycle arrest in G2 phase and increased the expression of CDK inhibitors [150]. Proanthocyanidins present in *Vitis vinifera* were found to be effective as an antiproliferative agent and inhibited CDK2, CDK4, and CDK6 with cell cycle arrest at the G1 phase. Cyclin D and E expressions were also reduced by the action of proanthocyanidins [151]. Another constituent resveratrol (polyphenol) from grapes was found to be effect against cancerous cell lines such as LNCap, PC-3, A431, C4-2B, and DU-145 via arresting cell cycle progression at G1 phase and reducing the expressions of cyclin/CDKs [152–155]. Genistin (flavonoid) showed a substantial antiproliferative activity among the reported flavonoids in modulating cancer cell signaling cascades. It inhibited the cell cycle progression at the G2/M phase and activated various CDKs inhibitors (CDKN1A, CDKN1C, CDKN2A, CDKN2B, CDKN2C) [156]. Curcumin from turmeric showed antiproliferative activity against immortalized umbilical vein endothelial cells via enhancing CDK1A, 1B and tumor suppressor gene (p53) [157]. From the above literature, we can conclude that the natural compounds can act as an efficient anticancer candidate in the modulation of the cell cycle.

2.4.10 Tumor Suppressor Genes Activation and DNA Repair Mechanism

Various molecular alterations have been linked with the incidence of neoplastic transformation followed by abnormal growth and differentiation of cells. Thus, underlying molecular epidemiology regarding the inactivation of important genes involved in the initiation of carcinogenesis can be analyzed through mutation spectrum analysis to study individual DNA patterns. The analysis is effective to observe changes triggered due to endogenous as well as exogenous mutagens [158]. Tumor suppression gene (p53) is involved in DNA synthesis, transcription, DNA repair, and induction of programmed cell death. Mutated tumor suppressor genes (p53) can provide effective hints regarding the etiology and molecular cascades involved in carcinogenesis. Alteration of p53 expression in normal cells due to missense mutations or frameshift mutations leads to the initiation of uncontrolled division of cells or neoplasia. Missense mutation alters the p53 expressions and activates oncogenic activity but the frameshift mutation results in loss of p53 genes [159, 160].

Further, the p53 gene is linked with various other cellular proteins forming a cascade, and the activated oncoproteins block p53 via binding to the same consequently altering its functions. Ina study, loss of p53 function rusted in the incidence of the Warburg effect via repressing the GLUT1 and GLUT4 transporters. But its activation (p53) leads to activation of the TIGAR gene and decreased glycolytic enzymes activity while enhancing the TCA cycle and ATP synthesis via oxidative phosphorylation. Activation of p53 genes in cancer cells further enhance cytochrome c activity to induce apoptosis [161]. Some studies confirmed the role of p53 protein in arresting cell growth via restore wild type p53 expression via transfection procedures due to its transactivation activity of the acidic domain at the amino-terminal region. It is also known to specifically bind DNA sequences and binds to the mdm2 gene to suppress tumor proliferation. Mdm2 gene is overexpressed by tumor cells and it can result in the inactivation of the transactivation cascade of p53 [162]. Further, p53 activates WAG1/Cipl/sdil which can deactivate cyclin-dependent kinase 2 activity, thus, arresting cell cycle through inhibition of DNA synthesis. In another study, p53 expression was elevated in irradiation treated cells and resulted in an arresting cell cycle at the G1 phase and prevented them to enter the S phase [163]. Gene p53 also induces certain repair genes such as GADD45, possibly playing a role to arrest the cell cycle in the G1 phase until the damage has been repaired [164]. As explained earlier, the mutation in the p53 gene leads to its inactivation or loss of function but there are shreds of evidence of gain of functions also.

Missense mutation occurred in conserved domains consisting of five regions of amino acids relative to particular cell types as well as tissue. But the mutant p53 can become oncogenic and is able to transform normal growing cells via interaction with RAS protein. Also, the mutant p53 is able to complex with wild type p53 as observed in fibroblast cell cultures resulting in its immortalization. In vivo studies further confirmed the oncogenic ability of the mutant p53 gene as observed in transgenic mice with overexpressed mutant p53 forms [164, 165]. About more than 30% of mice were developed with carcinomas including lymphomas, osteosarcomas, etc. Another model used was p53 knock out mouse. Mice deficient in p53 also developed various kinds of carcinomas, most prevalent were lymphomas. This model provided greater insights into the role of p53 as its loss can induce tumor formation in individuals [166]. Another gene is the retinoblastoma gene (Rb), it binds to DNA but there is sequence-specific binding. The transcription factor E2F initially discovered as an adenovirus E2 promotor activator. This factor interacts with the Rb gene and is known to regulate the expression of several genes related to the growth of cells and promote cell growth. It consists of a binding domain that binds to DNA and a transcription activating domain. Overexpression of Rb has been reported with inhibition of growth of cells and thus shows its growth-inhibiting properties via inhibiting E2F factor through forming a complex with it. Rb protein also affects other growth factors including E2F, the others are the TGF family [167].

Mutations in the Rb gene thus result in loss of its ability to suppress growth in cancer cells. Furthermore, in cancer cells such as osteosarcoma cells, bladder cancer, prostate cancer cells, introducing wild type Rb resulted in partial recovery of its growth inhibition property and was able to inhibit partially the growing cancer cells. The Rb gene occurs in two forms, i.e., phosphorylated and unphosphorylated forms. The unphosphorylated form dominates in the G1 phase and phosphorylated form in S phase, G1 and M phase. The phosphorylation of Rb protein occurs at serine and threonine residues. The growth inhibition potential of Rb proteins is inhibited by its phosphorylation as it leads to its inability to bind with E2F protein. Elevated cyclin levels further enhance the growth of cells. A complete understanding of the tumor suppressor genes could provide significant diagnostic information that can help in unraveling specific gene therapy to inhibit carcinogenesis [168, 169]. Keeping in view the effects of mutations in tumor suppressor genes, certain strategies are required to preserve the functioning of DNA.

Environmental factors such as UV radiation can damage individual nucleotide, leads to crosslinking or dimer formation of pyrimidine residues. It blocks the normal transcription of DNA. Other factors are alkylation or methylation. Due to internal thermal decomposition, urine nucleotides break off from nucleotide DNA backbone which can lead to the addition of uracil residues, which will block the transcription further. This type of error is required to be repaired to avoid lethal consequences. One type of repair system is the nucleotide excision repair system, also called a short patch repair. The uvrA and uvrB protein (as studied in *E. coli*) binds to damaged DNA fragment in an ATP dependent manner. Further, the uvrC protein removes the damaged DNA segment at 12 bp part distance. Helicase II acts on the strand and unwinds it for polymerase I to resynthesis the strand. DNA ligase seals the strands and the process of repair is completed [1]. The second type of DNA repair is mismatch repair. This type of error is repaired via the proofreading system. DNA polymerase pairs adenine with thymine and cytosine with guanine. The proteins mutS bind to DNA mismatch sequence with mutL and mutH proteins. mutH looks for a nick in the strand and helicase II unwinds the DNA. Bidirectional exonuclease removes mismatched bases and DNA polymerase binds the pairs. In the end DNA ligase seals the strand which is a mismatch free strand [170].

2.4.11 Inducing Programmed Cell Death/Apoptosis

The word apoptosis is derived from the Greek word "apo" means from and "ptosis" means falling. Apoptosis is an energy-dependent biochemical mechanism of programmed cell death. It is a genetically programmed cell death naturally occurring in all the organisms especially during the phases like embryogenesis, metamorphosis, and aging. One of the best examples of this is the human fingers differentiation observed during developmental stage of embryo, where the cells between the fingers are required to undergo apoptosis for separating the fingers. Apoptosis is a part of cell defense mechanisms like immune reactions. The apoptotic process is regulated and sequential events. Apoptotic cells become more compact, blebbing of the membrane occurs, chromatin becomes condensed and DNA is fragmented. Macrophages recognize the changed cellular surface of apoptotic cells and ultimately engulf them. One of the major changes which occur in the plasma membrane of apoptotic cells is the flipping of negatively charged phosphatidylserine from inner leaflet to the outer leaflet of the cell membrane that serves as a marker for apoptotic cells. The important feature of this process is that programmed cell death does not result in the release of cell content into the microenvironment of neighboring tissue. These cells are rapidly engulfed by the macrophages. The engulfing cells do not produce anti-inflammatory cytokines.

There is a complex mechanism behind the process of programmed cell death that is tightly regulated by an energy-derived molecular events. These pathways are both caspase-dependent as well as caspase-independent (Fig. 2.4). The caspase-dependent apoptosis is initiated either by extrinsic or intrinsic factors. There are two main types of caspase-dependent pathways viz. caspase-dependent intrinsic and caspase-dependent extrinsic pathways. Caspase-dependent extrinsic apoptotic pathway: a number of extracellular signal molecules are specialized to induce apoptosis. These extracellular signal molecules bind with cell surface receptors termed as death receptors. The FADD (Fas-associated death domain) an adaptor protein is required by the association of FasL (Ligand) and FasR (Receptor) on the surface of the cell membrane. In response to this the death-effector domain (DED) of FADD recruits and cleave procaspase-8 as active caspase-8 that is called the DISC (death-inducing signal complex). This activates caspase-3 and caspase-7, which cleaves the substrate within the cell. Nucleases are then activated, chromosomal DNA is degraded and the cell dies by apoptosis.

Apoptosis can also be induced by intercellular death signals other than exogenous signals. The defining event during the apoptotic pathway is mitochondrial matrix potential governed by outer mitochondrial membrane. After the disintegration of the outer mitochondrial membrane, cytochrome c is released into the cytosol. In the cytoplasm, cytochrome c binds to adaptor protein APAF-1 (Apoptotic Protease Activating Factor-1), a mammalian homolog of C.elegans CED-4 protein. This

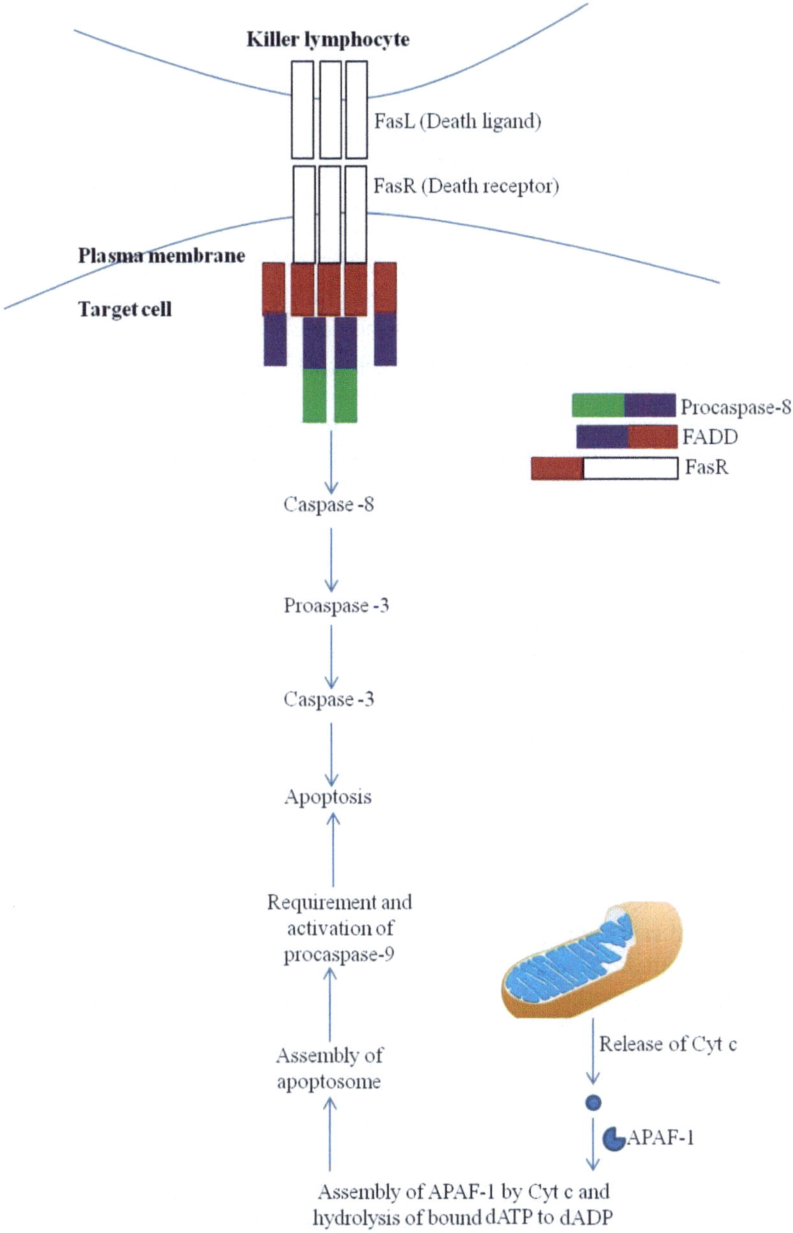

Fig. 2.4 Intrinsic and extrinsic caspase-dependent apoptotic pathways

interaction leads to hydrolysis of bound dATP to dADP and a change in the conformation of APAF-1. APAF-1 assembles into a heptameric complex, called the apoptosome. APAF-1 proteins in the apoptosome then recruit initiator procaspase-9. The incorporation of procaspase-9 triggers the auto-activating cleavage of procaspase-9 and produce caspase-9. This in turn, cleaves procaspase-3, which then cleaves the target and causes apoptosis of the cell.

2.4.11.1 Induction of Apoptosis by Phytochemicals

Most bioactive phytochemicals instigate apoptosis via an intrinsic pathway wherein the death stimulation triggers a signaling cascade activating the porosity and release of mitochondrial outer membrane components such as cytochrome c, endonuclease G, apoptosis-inducing factor (AIF), secondary mitochondrial caspase activator (SMAC), and primary apoptosis protein inhibitor (IAP) [171]. Luteolin from celery, acacetin from chrysanthemum, and 5-hydroxy-3,6,7,8,39-hexamethoxy-flavone (5-OH-HxMF) from citrus peel can disintegrate the mitochondrial membrane, release cytochrome c, activate caspases and ultimately trigger apoptosis in human cancer cells [172–175]. Human cells, on the other hand, are generally immune to apoptosis initiation by 5-OH-HxMF [174]. Since the majority of tumor cells develop relatively quickly, cellular mitogenic signals are disrupted by those which are the most active apoptosis-inducing. These results suggest that the food phytochemicals may induce programmed cell death at a limited concentrations by the generation of ROS and disruption of motor signal transductory pathways in cancer but not normal cell lines. Genistein, and EGCG causes permeabilization of outer mitochondrial membrane and inhibits cell survival signaling pathways in human hepatocellular carcinomas [176–178]. Gingerol and shogaols from ginger mediate the alterations in outer mitochondrial membrane so to disintegrate it and release cytochrome c in human cancer cells [179, 180]. Diallyl sulfide from garlic induces mitochondrial cell death through the generation of ROS and regulation of Bax/Bak but independent of Bcl-2 or Bcl-XL [181]. Allicin (from garlic) induced programmed cell death via release of AIF from mitochondria [182]. Phenethyl isothiocyanate (PEITC), an isothiocyanate from cruciferous vegetables, induce cell death via MMP disruption and release of cytochrome c from mitochondria to the cytosol in PC-3 cancer cells [183].

Acacetin from *Robinia pseudoacacia* induces antioxidant, anti-inflammatory, antiplasmodial, and antiproliferative action by inducing apoptosis and preventing cell cycle progression and substantial Fas and FasL upregulation [173, 184]. Pterostilbene a resveratrol-like dimethyl ether from vaccinium berries, found to have chemopreventive cancer activity [185]. Recently, it induces apoptosis through mitochondrial cascade and the GADD expression of Fas/FasL pathway into AGS human gastric adenocarcinoma cells [176]. Resveratrol trigger CD95 signaling dependent apoptosis in human cancer cells [186]. Coffee acid phenethyl ester (CAPE) induces apoptosis in human cancer cells through the downregulation of cyclin D1, β-catenin, c-myc, activation of Fas and p53 [187, 188]. Recent literature shown that the curcumin can induce TRAIL-related programmed cell death in malignant-glioma cells [189]. [6]-Gingerol, ginger extracted phenolic alkanone, facilitation in gastric cancer cells to TRAIL-related apoptosis with enhanced TRAIL-induced caspase-3 [190]. Treatment of TRAIL-resistant A549 cells to a high concentration of sulforaphane mediated TRAIL associated apoptosis

[191]. Luteolin is a natural flavonoid that sensitizes lung cancer cells to TNF-induced apoptosis by downregulation of NF-kB with the build-up of ROS [192]. Its co-treatment with EGCG synergistically induced apoptosis in human cancer cell lines through TRAIL [193].

2.5 Conclusion

Cancer is a deadly disease, about which despite significant advancement in knowledge, it is still a challenge. Extensive efforts are required to unravel the complete mechanism of chemopreventive agents involved in the pathophysiology and underlying intracellular signaling cascade pathways. Abnormally regulated signaling pathways often lead to anomalous cellular transformations. Phytochemicals can target these unregulated pathways to inhibit the abnormal cellular transformations or induce programmed cell death such as tumor suppressor genes (p53, Rb, cyclin-dependent kinases, and many growth factors). Phytochemicals based chemoprevention is an effective alternative to healthcare costs in the treatment of cancer. It is an inexpensive and accessible approach that can be easily acceptable by the general public and various medical practitioners. Tailored supplementation of chemopreventive phytochemicals can be a plausible approach to reduce the risk of cancer in the near future. However, more studies are required to confirm and ensure the association between the chemopreventive potential of phytochemicals and the pathophysiology of cancer.

Conflict of Interest The authors declare that there is no conflict of interest.

References

1. Daniel L (2017) Human ribosomal DNA and RNA polymerase I fate during UV-induced DNA repair. Doctoral dissertation
2. Doll R, Peto R (1981) The causes of cancer: quantitative estimates of avoidable risks of cancer in the United States today. J Natl Cancer Inst 66:1191–1308
3. Greenwald P (1996) Chemoprevention of cancer. Sci Am 275:96–99
4. Manson MM (2003) Cancer prevention: the potential for diet to modulate molecular signalling. Trends Mol Med 9:11–18
5. Milner JA, McDonald SS, Anderson DE et al (2001) Molecular targets for nutrients involved with cancer prevention. Nutr Cancer 41:1–16
6. Gescher A, Pastorino U, Plummer SM et al (1998) Suppression of tumour development by substances derived from the diet: mechanisms and clinical implications. Br J Clin Pharmacol 45:1–12
7. Kong ANT, Yu R, Hebbar V et al (2001) Signal transduction events elicited by cancer prevention compounds. Mutat Res 480:231–241
8. Agarwal R (2000) Cell signaling and regulators of cell cycle as molecular targets for prostate cancer prevention by dietary agents. Biochem Pharmacol 60:1051–1059
9. Bode AM, Dong Z (2000) Signal transduction pathways: targets for chemoprevention of skin cancer. Lancet Oncol 1:181–188

10. Manson MM, Gescher A, Hudson E et al (2000) Blocking and suppressing mechanisms of chemoprevention by dietary constituents. Toxicol Lett 112:499–505
11. Owuor ED, Kong AN (2002) Antioxidants and oxidants regulated signal transduction pathways. Biochem Pharmacol 64:765–770
12. Li Y, Sarkar FH (2002) Inhibition of nuclear factor κB activation in PC3 cells by genistein is mediated via Akt signaling pathway. Clin Cancer Res 8:2369–2377
13. Gong L, Li Y, Nedeljkovic-Kurepa A et al (2003) Inactivation of NF-κB by genistein is mediated via Akt signaling pathway in breast cancer cells. Oncogene 22:4702–4709
14. Pianetti S, Guo S, Kavanagh KT et al (2002) Green tea polyphenol epigallocatechin-3 gallate inhibits Her-2/neu signaling, proliferation, and transformed phenotype of breast cancer cells. Cancer Res 62:652–655
15. Das R, Mahabeleshwar GH, Kundu GC (2003) Osteopontin stimulates cell motility and nuclear factor κB-mediated secretion of urokinase type plasminogen activator through phosphatidylinositol 3-kinase/Akt signaling pathways in breast cancer cells. J Biol Chem 278(31):28593–28606
16. Plummer SM, Holloway KA, Manson MM et al (1999) Inhibition of cyclo-oxygenase 2 expression in colon cells by the chemopreventive agent curcumin involves inhibition of NF-κB activation via the NIK/IKK signalling complex. Oncogene 18:6013–6020
17. Bharti AC, Donato N, Singh S et al (2003) Curcumin (diferuloylmethane) down-regulates the constitutive activation of nuclear factor-κ B and IκBα kinase in human multiple myeloma cells, leading to suppression of proliferation and induction of apoptosis. Blood 101:1053–1062
18. Holmes-McNary M, Baldwin AS (2000) Chemopreventive properties of trans-resveratrol are associated with inhibition of activation of the IκB kinase. Cancer Res 60:3477–3483
19. Afaq F, Adhami VM, Ahmad N, Mukhtar H (2003) Inhibition of ultraviolet B-mediated activation of nuclear factor κB in normal human epidermal keratinocytes by green tea Constituent (-)-epigallocatechin-3-gallate. Oncogene 22:1035–1044
20. Yang F, Oz HS, Barve S et al (2001) The green tea polyphenol (-)-epigallocatechin- 3-gallate blocks nuclear factor-κB activation by inhibiting Iκ B kinase activity in the intestinal epithelial cell line IEC-6. Mol Pharm 60:528–533
21. Surh YJ, Han SS, Keum YS et al (2000) Inhibitory effects of curcumin and capsaicin on phorbol ester-induced activation of eukaryotic transcription factors, NF-κB and AP-1. Biofactors 12:107–112
22. Chun KS, Keum YS, Han SS et al (2003) Curcumin inhibits phorbol ester-induced expression of cyclooxygenase-2 in mouse skin through suppression of extracellular signal-regulated kinase activity and NF–κB activation. Carcinogenesis 24:1515–1524
23. Singh S, Aggarwal BB (1995) Activation of transcription factor NF-κ B is suppressed by curcumin (diferuloylmethane). J Biol Chem 270:24995–25000
24. Park KK, Chun KS, Lee JM et al (1998) Inhibitory effects of [6]-gingerol, a major pungent principle of ginger, on phorbol ester-induced inflammation, epidermal ornithine decarboxylase activity and skin tumor promotion in ICR mice. Cancer Lett 129:139–144
25. Surh YJ, Lee RCJ, Park KK et al (1995) Chemoprotective effects of capsaicin and diallyl sulfide against mutagenesis or tumorigenesis by vinyl carbamate and N-nitrosodimethylamine. Carcinogenesis 16:2467–2471
26. Surh YJ, Lee SS (1996) Capsaicin in hot chili pepper: carcinogen, co-carcinogen or anticarcinogen? Food Chem Toxicol 34:313–316
27. Surh YJ (2002) More than spice: capsaicin in hot chili peppers makes tumor cells commit suicide. J Natl Cancer Inst 94:1263–1265
28. Han SS, Keum YS, Seo HJ et al (2001) Capsaicin suppresses phorbol esterinduced activation of NF-κB/Rel and AP-1 transcription factors in mouse epidermis. Cancer Lett 164:119–126
29. Dong Z, Ma W, Huang C et al (1997) Inhibition of tumor promoter-induced activator protein 1 activation and cell transformation by tea polyphenols, (−)-epigallocatechin gallate, and theaflavins. Cancer Res 57:4414–4419

30. Nomura M, Ma WY, Huang C et al (2000) Inhibition of ultraviolet B-induced AP-1 activation by theaflavins from black tea. Mol Carcinog 28:148–155
31. Masuda M, Suzui M, Lim JT et al (2002) Epigallocatechin-3-gallate decreases VEGF production in head and neck and breast carcinoma cells by inhibiting EGFR-related pathways of signal transduction. J Exp Ther Oncol 2:350–359
32. Tacchini L, Dansi P, Matteucci E et al (2000) Hepatocyte growth factor signal coupling to various transcription factors depends on triggering of Met receptor and protein kinase transducers in human hepatoma cells HepG2. Exp Cell Res 256:272–281
33. Davis JN, Kucuk O, Sarkar FH (1999) Genistein inhibits NF-κB activation in prostate cancer cells. Nutr Cancer 35:167–174
34. Narayanan BA, Narayanan NK, Re GG et al (2003) Differential expression of genes induced by resveratrol in LNCaP cells: p53-mediated molecular targets. Int J Cancer 104:204–212
35. She QB, Bode AM, Ma WY et al (2001) Resveratrol-induced activation of p53 and apoptosis is mediated by extracellular-signal-regulated protein kinases and p38 kinase. Cancer Res 61:1604–1610
36. Manna SK, Mukhopadhyay A, Aggarwal BB (2000) Resveratrol suppresses TNF-induced activation of nuclear transcription factors NF-κ B, activator protein-1, and apoptosis: potential role of reactive oxygen intermediates and lipid peroxidation. J Immunol 164:6509–6519
37. Bharti AC, Aggarwal BB (2002) Nuclear factor-κ B and cancer: its role in prevention and therapy. Biochem Pharmacol 64:883–888
38. Dinkova-Kostova AT, Holtzclaw WD, Cole RN et al (2002) Direct evidence that sulfhydryl groups of Keap1 are the sensors regulating induction of phase 2 enzymes that protect against carcinogens and oxidants. Proc Natl Acad Sci U S A 99:11908–11913
39. Huang HC, Nguyen T, Pickett CB (2002) Phosphorylation of Nrf2 at Ser-40 by protein kinase C regulates antioxidant response element-mediated transcription. J Biol Chem 277:42769–42774
40. Kang KW, Park EY, Kim SG (2003) Activation of CCAAT/enhancer-binding protein β by 2′-amino-3′- methoxyflavone (PD98059) leads to the induction of glutathione S-transferase A2. Carcinogenesis 24:475–482
41. Lee JM, Hanson JM, Chu WA et al (2001) Phosphatidylinositol 3-kinase, not extracellular signal-regulated kinase, regulates activation of the antioxidant-responsive element in IMR-32 human neuroblastoma cells. J Biol Chem 276:20011–20016
42. Balogun E, Hoque M, Gong P et al (2003) Curcumin activates the haem oxygenase-1 gene via regulation of Nrf2 and the antioxidant-responsive element. Biochem J 371:887–895
43. Dickinson DA, Iles KE, Zhang H et al (2003) Curcumin alters EpRE and AP-1 binding complexes and elevates glutamate-cysteine ligase gene expression. FASEB J 17:473–475
44. Morimitsu Y, Nakagawa Y, Hayashi K et al (2002) A sulforaphane analogue that potently activates the Nrf2-dependent detoxification pathway. J Biol Chem 277:3456–3463
45. Yu R, Jiao JJ, Duh JL et al (1997) Activation of mitogen-activated protein kinases by green tea polyphenols: potential signaling pathways in the regulation of antioxidant-responsive element-mediated phase II enzyme gene expression. Carcinogenesis 18:451–456
46. Chen C, Yu R, Owuor ED et al (2000) Activation of antioxidant-response element (ARE), mitogen-activated protein kinases (MAPKs) and caspases by major green tea polyphenol components during cell survival and death. Arch Pharm Res 23:605–612
47. Thimmulappa RK, Mai KH, Srisuma H et al (2002) Identification of Nrf2-regulated genes induced by the chemopreventive agent sulforaphane by oligonucleotide microarray. Cancer Res 62:5196–5203
48. Mahmoud NN, Carothers AM, Grunberger D et al (2000) Plant phenolics decrease intestinal tumors in an animal model of familial adenomatous polyposis. Carcinogenesis 21:921–927
49. Jaiswal AS, Marlow BP, Gupta N et al (2002) β-catenin-mediated transactivation and cell–cell adhesion pathways are important in curcumin (diferuylmethane)-induced growth arrest and apoptosis in colon cancer cells. Oncogene 21:8414–8427

50. Joe AK, Liu H, Suzui M et al (2002) Resveratrol induces growth inhibition, S-phase arrest, apoptosis, and changes in biomarker expression in several human cancer cell lines. Clin Cancer Res 8:893–903
51. Blum CA, Xu M, Orner GA et al (2001) β-Catenin mutation in rat colon tumors initiated by 1,2-dimethylhydrazine and 2-amino-3- methylimidazo[4,5-f]quinoline, and the effect of postinitiation treatment with chlorophyllin and indole-3-carbinol. Carcinogenesis 22:315–320
52. Meng Q, Qi M, Chen DZ et al (2000) Suppression of breast cancer invasion and migration by indole-3-carbinol: associated with upregulation of BRCA1 and E-cadherin/catenin complexes. J Mol Med 78:155–165
53. Brack ME, Boterberg T, Depypere HT et al (2002) The citrus methoxyflavone tangeretin affects human cell–cell interactions. Adv Exp Med Biol 505:135–139
54. McEntee MF, Chiu CH, Whelan J (1999) Relationship of β-catenin and Bcl-2 expression to sulindac-induced regression of intestinal tumors in Min mice. Carcinogenesis 20:635–640
55. Dihlmann S, Siermann A, von Knebel DM (2001) The nonsteroidal anti-inflammatory drugs aspirin and indomethacin attenuate β-catenin/TCF-4 signaling. Oncogene 20:645–653
56. Mori H, Yamada Y, Hirose Y et al (2002) Chemoprevention of large bowel carcinogenesis; the role of control of cell proliferation and significance of β-catenin-accumulated crypts as a new biomarker. Eur J Cancer Prev 11:S71–S75
57. Friday BB, Adjei AA (2008) Advances in targeting the Ras/Raf/MEK/Erk mitogen-activated protein kinase cascade with MEK inhibitors for cancer therapy. Clin Cancer Res 14 (2):342–346
58. Bhui K, Tyagi S, Prakash B, Shukla Y (2010) Pineapple bromelain induces autophagy, facilitating apoptotic response in mammary carcinoma cells. Biofactors 36(6):474–482
59. Bhattacharya U, Halder B, Mukhopadhyay S, Giri AK (2009) Role of oxidation-triggered activation of JNK and p38 MAPK in black tea polyphenols induced apoptotic death of A375 cells. Cancer Sci 100(10):1971–1978
60. Jyothi D, Vanathi P, Gowri PM, Rao VRS, Rao JM, Sreedhar AS (2009) Diferuloylmethane augments the cytotoxic effects of piplartine isolated from Piper chaba. Toxicol In Vitro 23 (6):1085–1091
61. Sathya S, Sudhagar S, Priya MV, Raja RB, Muthusamy VS, Devaraj SN, Lakshmi BS (2010) 3β-Hydroxylup-20 (29)-ene-27, 28-dioic acid dimethyl ester, a novel natural product from Plumbago zeylanica inhibits the proliferation and migration of MDA-MB-231 cells. Chem Biol Interact 188(3):412–420
62. Mandal C, Dutta A, Mallick A, Chandra S, Misra L, Sangwan RS, Mandal C (2008) Withaferin a induces apoptosis by activating p38 mitogen-activated protein kinase signaling cascade in leukemic cells of lymphoid and myeloid origin through mitochondrial death cascade. Apoptosis 13(12):1450–1464
63. Sylvester PW, Ayoub NM (2013) Tocotrienols target PI3K/Akt signaling in anti-breast cancer therapy. Anti Cancer Agents Med Chem 13:1039–1047. https://doi.org/10.2174/1871520611313990116
64. Lin ZY, Kuo CH, Wu DC et al (2016) Anticancer effects of acceptable colchicine concentrations on human gastric cancer cell lines. Kaohsiung J Med Sci 32:68–73. https://doi.org/10.1016/j.kjms.2015.12.006
65. Hossan MS, Rahman S, Bashar ABMA et al (2014) Rosmarinic acid: a review of its anticancer action. World J Pharm Pharm Sci 3:57–70
66. Kuo MC, Chang SJ, Hsieh MC (2015) Colchicine significantly reduces incident cancer in gout male patients. Medicine 94:e1570. https://doi.org/10.1097/MD.0000000000001570
67. Vivanco I, Sawyers CL (2002) The phosphatidylinositol 3-kinase–AKT pathway in human cancer. Nat Rev Cancer 2(7):489–501
68. Cornélio Favarin D, Martins Teixeira M, Lemos de Andrade E, de Freitas Alves C, Lazo Chica JE, Artério Sorgi C, Faccioli LH, Paula Rogerio A (2013) Anti-inflammatory effects of ellagic acid on acute lung injury induced by acid in mice. Mediat Inflamm 2013:164202

69. Engelman JA, Luo J, Cantley LC (2006) The evolution of phosphatidylinositol 3-kinases as regulators of growth and metabolism. Nat Rev Genet 7(8):606–619
70. Yap TA, Garrett MD, Walton MI, Raynaud F, de Bono JS, Workman P (2008) Targeting the PI3K–AKT–mTOR pathway: progress, pitfalls, and promises. Curr Opin Pharmacol 8 (4):393–412
71. Wang W, Wang X, Peng L, Deng Q, Liang Y, Qing H, Jiang B (2010) CD24-dependent MAPK pathway activation is required for colorectal cancer cell proliferation. Cancer Sci 101 (1):112–119
72. Kim EC, Min JK, Kim TY, Lee SJ, Yang HO, Han S, Kim YM, Kwon YG (2005) [6]-Gingerol, a pungent ingredient of ginger, inhibits angiogenesis in vitro and in vivo. Biochem Biophys Res Commun 335(2):300–308
73. Roy SK, Srivastava RK, Shankar S (2010) Inhibition of PI3K/AKT and MAPK/ERK pathways causes activation of FOXO transcription factor, leading to cell cycle arrest and apoptosis in pancreatic cancer. J Mol Signal 5(1):10
74. Corona G, Spencer JPE, Dessi MA (2009) Extra virgin olive oil phenolics: absorption, metabolism, and biological activities in the GI tract. Toxicol Ind Health 25(4–5):285–293
75. Sun F, Zheng XY, Ye J, Wu TT, Wang JL, Chen W (2012) Potential anticancer activity of myricetin in human T24 bladder cancer cells both in vitro and in vivo. Nutr Cancer 64 (4):599–606
76. Adan A, Baran Y (2015) The pleiotropic effects of fisetin and hesperetin on human acute promyelocytic leukemia cells are mediated through apoptosis, cell cycle arrest, and alterations in signaling networks. Tumor Biol 36:8973–8984. https://doi.org/10.1007/s13277-015-3597-6
77. Lee J, Kim JH (2016) Kaempferol inhibits pancreatic cancer cell growth and migration through the blockade of EGFR-related pathway in vitro. PLoS One 11:e0155264. https://doi.org/10.1371/journal.pone.0155264
78. Kim MJ, Kim H (2015) Anticancer effect of lycopene in gastric carcinogenesis. J Cancer Prev 20:92–96. https://doi.org/10.15430/JCP.2015.20.2.92
79. He X, Saint-Jeannet JP, Wang Y, Nathans J, Dawid I, Varmus H (1997) A member of the frizzled protein family mediating axis induction by Wnt-5A. Science 275(5306):1652–1654
80. Smalley MJ, Sara E, Paterson H, Naylor S, Cook D, Jayatilake H, Fryer LG, Hutchinson L, Fry MJ, Dale TC (1999) Interaction of Axin and Dvl-2 proteins regulates Dvl-2-stimulated TCF-dependent transcription. EMBO J 18(10):2823–2835
81. Salic A, Lee E, Mayer L, Kirschner MW (2000) Control of β-catenin stability: reconstitution of the cytoplasmic steps of the wnt pathway in Xenopus egg extracts. Mol Cell 5(3):523–532
82. Hart M, Concordet JP, Lassot I, Albert I, Del los Santos R, Durand H, Perret C, Rubinfeld B, Margottin F, Benarous R, Polakis P (1999) The F-box protein β-TrCP associates with phosphorylated β-catenin and regulates its activity in the cell. Curr Biol 9(4):207–211
83. Seeling JM, Miller JR, Gil R, Moon RT, White R, Virshup DM (1999) Regulation of β-catenin signaling by the B56 subunit of protein phosphatase 2A. Science 283(5410):2089–2091
84. Landesman-Bollag E, Song DH, Romieu-Mourez R, Sussman DJ, Cardiff RD, Sonenshein GE, Seldin DC (2001) Protein kinase CK2: signaling and tumorigenesis in the mammary gland. Mol Cell Biochem 227(1–2):153–165
85. Gao ZHI, Xu Z, Hung MS, Lin YC, Wang T, Gong MIN, Zhi X, Jablon DM, You L (2009) Promoter demethylation of WIF-1 by epigallocatechin-3-gallate in lung cancer cells. Anticancer Res 29(6):2025–2030
86. Shan BE, Wang MX, Li RQ (2009) Quercetin inhibit human SW480 colon cancer growth in association with inhibition of cyclin D1 and survivin expression through Wnt/β-catenin signaling pathway. Cancer Investig 27(6):604–612
87. Amado NG, Cerqueira DM, Menezes FS, da Silva JFM, Neto VM, Abreu JG (2009) Isoquercitrin isolated from Hyptis fasciculata reduces glioblastoma cell proliferation and changes β-catenin cellular localization. Anti-Cancer Drugs 20(7):543–552

88. Park S, Choi J (2010) Inhibition of β-catenin/Tcf signaling by flavonoids. J Cell Biochem 110 (6):1376–1385
89. Laplante M, Sabatini DM (2009) mTOR signaling at a glance. J Cell Sci 122(20):3589–3594
90. Loewith R, Jacinto E, Wullschleger S, Lorberg A, Crespo JL, Bonenfant D, Oppliger W, Jenoe P, Hall MN (2002) Two TOR complexes, only one of which is rapamycin sensitive, have distinct roles in cell growth control. Mol Cell 10(3):457–468
91. Laplante M, Sabatini DM (2012) mTOR signaling in growth control and disease. Cell 149 (2):274–293
92. Schalm SS, Fingar DC, Sabatini DM, Blenis J (2003) TOS motif-mediated raptor binding regulates 4E-BP1 multisite phosphorylation and function. Curr Biol 13(10):797–806
93. Peterson TR, Laplante M, Thoreen CC, Sancak Y, Kang SA, Kuehl WM, Gray NS, Sabatini DM (2009) DEPTOR is an mTOR inhibitor frequently overexpressed in multiple myeloma cells and required for their survival. Cell 137(5):873–886
94. Gwinn DM, Shackelford DB, Egan DF, Mihaylova MM, Mery A, Vasquez DS, Turk BE, Shaw RJ (2008) AMPK phosphorylation of raptor mediates a metabolic checkpoint. Mol Cell 30(2):214–226
95. Kim DH, Sarbassov DD, Ali SM, Latek RR, Guntur KV, Erdjument-Bromage H, Tempst P, Sabatini DM (2003) GβL, a positive regulator of the rapamycin-sensitive pathway required for the nutrient-sensitive interaction between raptor and mTOR. Mol Cell 11(4):895–904
96. Guertin DA, Guntur KV, Bell GW, Thoreen CC, Sabatini DM (2006) Functional genomics identifies TOR-regulated genes that control growth and division. Curr Biol 16(10):958–970
97. Zeng Z, Sarbassov DD, Samudio IJ, Yee KW, Munsell MF, Ellen Jackson C, Giles FJ, Sabatini DM, Andreeff M, Konopleva M (2007) Rapamycin derivatives reduce mTORC2 signaling and inhibit AKT activation in AML. Blood 109(8):3509–3512
98. Vander Haar E, Lee SI, Bandhakavi S, Griffin TJ, Kim DH (2007) Insulin signalling to mTOR mediated by the Akt/PKB substrate PRAS40. Nat Cell Biol 9(3):316–323
99. Wang L, Harris TE, Roth RA, Lawrence JC (2007) PRAS40 regulates mTORC1 kinase activity by functioning as a direct inhibitor of substrate binding. J Biol Chem 282 (27):20036–20044
100. Johnson SM, Gulhati P, Arrieta I, Wang X, Uchida T, Gao T, Evers BM (2009) Curcumin inhibits proliferation of colorectal carcinoma by modulating Akt/mTOR signaling. Anticancer Res 29(8):3185–3190
101. Beevers CS, Li F, Liu L, Huang S (2006) Curcumin inhibits the mammalian target of rapamycin-mediated signaling pathways in cancer cells. Int J Cancer 119(4):757–764
102. Aziz MH, Nihal M, Fu VX, Jarrard DF, Ahmad N (2006) Resveratrol-caused apoptosis of human prostate carcinoma LNCaP cells is mediated via modulation of phosphatidylinositol 3′-kinase/Akt pathway and Bcl-2 family proteins. Mol Cancer Ther 5(5):1335–1341
103. Jiang H, Shang X, Wu H, Gautam SC, Al-Holou S, Li C, Kuo J, Zhang L, Chopp M (2009) Resveratrol downregulates PI3K/Akt/mTOR signaling pathways in human U251 glioma cells. J Exp Ther Oncol 8(1):25
104. Brito PM, Devillard R, Nègre-Salvayre A, Almeida LM, Dinis TC, Salvayre R, Augé N (2009) Resveratrol inhibits the mTOR mitogenic signaling evoked by oxidized LDL in smooth muscle cells. Atherosclerosis 205(1):126–134
105. Khan N, Afaq F, Kweon MH, Kim K, Mukhtar H (2007) Oral consumption of pomegranate fruit extract inhibits growth and progression of primary lung tumors in mice. Cancer Res 67 (7):3475–3482
106. Pantuck AJ, Leppert JT, Zomorodian N, Aronson W, Hong J, Barnard RJ, Seeram N, Liker H, Wang H, Elashoff R, Heber D (2006) Phase II study of pomegranate juice for men with rising prostate-specific antigen following surgery or radiation for prostate cancer. Clin Cancer Res 12 (13):4018–4026
107. Paller CJ, Ye X, Wozniak PJ et al (2013) A randomized phase II study of pomegranate extract for men with rising PSA following initial therapy for localized prostate cancer. Prostate Cancer Prostatic Dis 16:50–55

108. Kim SH, Kim SH, Kim YB, Jeon YT, Lee SC, Song YS (2009) Genistein inhibits cell growth by modulating various mitogen-activated protein kinases and AKT in cervical cancer cells. Ann N Y Acad Sci 1171(1):495–500
109. Anastasius N, Boston S, Lacey M, Storing N, Whitehead SA (2009) Evidence that low-dose, long-term genistein treatment inhibits oestradiol-stimulated growth in MCF-7 cells by down-regulation of the PI3-kinase/Akt signalling pathway. J Steroid Biochem Mol Biol 116 (1–2):50–55
110. Nakamura Y, Yogosawa S, Izutani Y, Watanabe H, Otsuji E, Sakai T (2009) A combination of indole-3-carbinol and genistein synergistically induces apoptosis in human colon cancer HT-29 cells by inhibiting Akt phosphorylation and progression of autophagy. Mol Cancer 8 (1):100
111. Banerjee S, Zhang Y, Wang Z, Che M, Chiao PJ, Abbruzzese JL, Sarkar FH (2007) Retracted: in vitro and in vivo molecular evidence of genistein action in augmenting the efficacy of cisplatin in pancreatic cancer. Int J Cancer 120(4):906–917
112. Kotecha R, Takami A, Espinoza JL (2016) Dietary phytochemicals and cancer chemoprevention: a review of the clinical evidence. Oncotarget 7:52517
113. Gajbhiye RL, Mahato SK, Achari A et al (2019) Cancer chemoprevention by dietary polyphenols, flavonoids, terpenoids, and saponins. In: Bioactive natural products for the management of cancer: from bench to bedside. Springer, Singapore, pp 91–109
114. Gulcin I, Beydemir S (2013) Phenolic compounds as antioxidants: carbonic anhydrase isoenzymes inhibitors. Mini Rev Med Chem 13:408–430
115. Verma N, Shukla S (2015) Impact of various factors responsible for fluctuation in plant secondary metabolites. J Appl Res Med Aromat Plants 2:105–113
116. Hooper PL, Hooper PL, Tytell M, Vígh L (2010) Xenohormesis: health benefits from an eon of plant stress response evolution. Cell Stress Chaperones 15:761–770
117. Surh YJ (2011) Xenohormesis mechanisms underlying chemopreventive effects of some dietary phytochemicals. Ann N Y Acad Sci 1229:1–6
118. Miguel MG (2011) Anthocyanins: antioxidant and/or anti-inflammatory activities. J Appl Pharm Sci 1:7–15
119. Choe KI, Kwon JH, Park KH et al (2012) The antioxidant and anti-inflammatory effects of phenolic compounds isolated from the root of Rhodiola sachalinensis A. BOR Mol 17:11484–11494
120. Ziech D, Franco R, Georgakilas AG et al (2010) The role of reactive oxygen species and oxidative stress in environmental carcinogenesis and biomarker development. Chem Biol Interact 188:334–339
121. Das P, Nutan KK, Singla-Pareek SL et al (2015) Oxidative environment and redox homeostasis in plants: dissecting out significant contribution of major cellular organelles. Front Environ Sci 2:70
122. Valavanidis A, Vlachogianni T, Fiotakis K et al (2013) Pulmonary oxidative stress, inflammation and cancer: respirable particulate matter, fibrous dusts and ozone as major causes of lung carcinogenesis through reactive oxygen species mechanisms. Int J Environ Res Public Health 10:3886–3907
123. Huang Q, Zhan L, Cao H et al (2016) Increased mitochondrial fission promotes autophagy and hepatocellular carcinoma cell survival through the ROS-modulated coordinated regulation of the NFKB and TP53 pathways. Autophagy 12:999–1014
124. Wardyn JD, Ponsford AH, Sanderson CM (2015) Dissecting molecular cross-talk between Nrf2 and NF-κB response pathways. Biochem Soc Trans 43:621–626
125. Zhang H, Tsao R (2016) Dietary polyphenols, oxidative stress and antioxidant and anti-inflammatory effects. Curr Opin Food Sci 8:33–42
126. Dubois H, Wullaert A, Lamkanfi M (2016) General strategies in inflammasome biology. In: Inflammasome signaling and bacterial infections. Springer, Cham, pp 1–22

127. Zhang B, Lai Y, Li Y et al (2018) Antineoplastic activity of isoliquiritigenin, a chalcone compound, in androgen-independent human prostate cancer cells linked to G2/M cell cycle arrest and cell apoptosis. Eur J Pharmacol 821:57–67
128. Na HK, Surh YJ (2015) Cancer chemopreventive phytochemicals targeting NF-κB and Nrf2 signaling pathways. In: Genomics, proteomics and metabolomics in nutraceuticals and functional foods. Wiley-Blackwell, Hoboken, NJ, pp 102–121
129. Dai C, Li B, Zhou Y, Li H, Xiao X, Tang S (2016) Curcumin attenuates quinocetone induced apoptosis and inflammation via the opposite modulation of Nrf2/HO-1 and NF-kB pathway in human hepatocyte L02 cells. Food Chem Toxicol 95:52–63
130. Eghbaliferiz S, Iranshahi M (2016) Prooxidant activity of polyphenols, flavonoids, anthocyanins and carotenoids: updated review of mechanisms and catalyzing metals. Phytother Res 30:1379–1391
131. Collins K, Jacks T, Pavletich NP (1997) The cell cycle and cancer. Proc Natl Acad Sci 94:2776–2778
132. Hardwick LJ, Azzarelli R, Philpott A (2018) Cell cycle-dependent phosphorylation and regulation of cellular differentiation. Biochem Soc Trans 46:1083–1091
133. Kelly BL, Wolfe KG, Roberts JM (1998) Identification of a substrate-targeting domain in cyclin E necessary for phosphorylation of the retinoblastoma protein. Proc Natl Acad Sci 95:2535–2540
134. Deshpande A, Sicinski P, Hinds PW (2005) Cyclins and cdks in development and cancer: a perspective. Oncogene 24:2909
135. Hydbring P, Malumbres M, Sicinski P (2016) Non-canonical functions of cell cycle cyclins and cyclin-dependent kinases. Nat Rev Mol Cell Biol 17:280
136. Zocchi L, Wu SC, Wu J et al (2018) The cyclin-dependent kinase inhibitor flavopiridol (alvocidib) inhibits metastasis of human osteosarcoma cells. Oncotarget 9:23505
137. Zeidner JF, Karp JE (2015) Clinical activity of alvocidib (flavopiridol) in acute myeloid leukemia. Leuk Res 39:1312–1318
138. Blažević T, Heiss EH, Atanasov AG et al (2015) Indirubin and indirubin derivatives for counteracting proliferative diseases. Evid-Based Complement Altern Med 2015:654098. https://doi.org/10.1155/2015/654098
139. Shanmuganathan S, Angayarkanni N (2018) Chebulagic acid Chebulinic acid and Gallic acid, the active principles of Triphala, inhibit TNFα induced pro-angiogenic and pro-inflammatory activities in retinal capillary endothelial cells by inhibiting p38, ERK and NFkB phosphorylation. Vasc Pharmacol 108:23–35
140. Zhao Q, Chang W, Chen R et al (2019) Anti-proliferative effect of wogonin on ovary Cancer cells involves activation of apoptosis and cell cycle arrest. Med Sci Monit 25:8465
141. Chen GL, Tian YQ, Wu JL et al (2016) Antiproliferative activities of Amaryllidaceae alkaloids from Lycoris radiata targeting DNA topoisomerase I. Sci Rep 6:38284
142. Mérarchi M, Sethi G, Fan L et al (2018) Molecular targets modulated by fangchinoline in tumor cells and preclinical models. Molecules 23:2538
143. Yang J, Hu S, Wang C et al (2019) Fangchinoline derivatives induce cell cycle arrest and apoptosis in human leukemia cell lines via suppression of the PI3K/AKT and MAPK signaling pathway. Eur J Med Chem 186:111898. https://doi.org/10.1016/j.ejmech.2019.111898
144. Barzegar E, Fouladdel S, Movahhed TK et al (2015) Effects of berberine on proliferation, cell cycle distribution and apoptosis of human breast cancer T47D and MCF7 cell lines. Iran J Basic Med Sci 18:334
145. Qin XP, Zhan XY, Chen QB et al (2018) Berberine enhances mitomycin C-induced cell cycle arrest and apoptosis of T24 bladder cancer cells. Chin J Pathophysiol 34:1025–1030
146. Lange N, Tontsa AT, Wegscheid C et al (2016) The limonoids TS3 and rubescin E induce apoptosis in human hepatoma cell lines and interfere with NF-κB signaling. Plos One 11: e0160843
147. Chan KC, Huang HP, Ho HH et al (2015) Mulberry polyphenols induce cell cycle arrest of vascular smooth muscle cells by inducing NO production and activating AMPK and p53. J Funct Food 15:604–613

148. Niedzwiecki A, Roomi MW, Kalinovsky T et al (2016) Anticancer efficacy of polyphenols and their combinations. Nutrients 8:552
149. Watson RR, Preedy VR, Zibadi S (Eds.). (2018) Polyphenols: prevention and treatment of human disease. Academic press, London
150. Shabbir M, Syed DN, Lall RK et al (2015) Potent anti-proliferative, pro-apoptotic activity of the Maytenus royleanus extract against prostate cancer cells: evidence in in-vitro and in-vivo models. Plos One 10:e0119859
151. Decean HP, Brie IC, Tatomir CB et al (2018) Targeting MAPK (p38, ERK, JNK) and inflammatory CK (GDF-15, GM-CSF) in UVB-activated human skin cells with Vitis vinifera seed extract. J Environ Pathol Toxicol Oncol 37:261–272
152. Empl MT, Albers M, Wang S et al (2015) The resveratrol tetramer r-viniferin induces a cell cycle arrest followed by apoptosis in the prostate cancer cell line LNCaP. Phytother Res 29:1640–1645
153. Zhai XX, Ding JC, Tang ZM et al (2016) Effects of resveratrol on the proliferation, apoptosis and telomerase ability of human A431 epidermoid carcinoma cells. Oncol Lett 11:3015–3018
154. Singh SK, Banerjee S, Acosta EP et al (2017) Resveratrol induces cell cycle arrest and apoptosis with docetaxel in prostate cancer cells via a p53/p21WAF1/CIP1 and p27KIP1 pathway. Oncotarget 8:17216
155. Fonseca J, Moradi F, Maddalena LA et al (2019) Resveratrol integrates metabolic and growth effects in PC3 prostate cancer cells-involvement of prolyl hydroxylase and hypoxia inducible factor-1. Oncol Lett 17:697–705
156. Fang Y, Zhang Q, Wang X et al (2016) Quantitative phosphoproteomics reveals genistein as a modulator of cell cycle and DNA damage response pathways in triple-negative breast cancer cells. Int J Oncol 48:1016–1028
157. Agarwal A, Kasinathan A, Ganesan R et al (2018) Curcumin induces apoptosis and cell cycle arrest via the activation of reactive oxygen species–independent mitochondrial apoptotic pathway in Smad4 and p53 mutated colon adenocarcinoma HT29 cells. Nutr Res 51:67–81
158. Pierotti MA, Frattini M, Molinari F et al (2016) Oncogenes. In: Holland-Frei Cancer Medicine. BC Decker, Hamilton, ON, pp 1–22
159. Rivlin N, Brosh R, Oren M et al (2011) Mutations in the p53 tumor suppressor gene: important milestones at the various steps of tumorigenesis. Genes Cancer 2:466–474
160. Muller PA, Vousden KH (2013) p53 mutations in cancer. Nat Cell Biol 15:2–8
161. Levine AJ, Puzio-Kuter AM (2010) The control of the metabolic switch in cancers by oncogenes and tumor suppressor genes. Science 330:1340–1344
162. Manfredi JJ (2010) The Mdm2–p53 relationship evolves: Mdm2 swings both ways as an oncogene and a tumor suppressor. Genes Dev 24:1580–1589
163. Tudzarova S, Mulholland P, Dey A et al (2016) p53 controls CDC7 levels to reinforce G1 cell cycle arrest upon genotoxic stress. Cell Cycle 15:2958–2972
164. Parrales A, Iwakuma T (2015) Targeting oncogenic mutant p53 for cancer therapy. Front Oncol 5:288
165. Mantovani F, Walerych D, Del Sal G (2017) Targeting mutant p53 in cancer: a long road to precision therapy. FEBS J 284:837–850
166. Liu J, Zhang C, Hu W et al (2015) Tumor suppressor p53 and its mutants in cancer metabolism. Cancer Lett 356:197–203
167. Liban TJ, Thwaites MJ, Dick FA et al (2016) Structural conservation and E2F binding specificity within the retinoblastoma pocket protein family. J Mol Biol 428:3960–3971
168. Scott MC, Sarver AL, Tomiyasu H et al (2015) Aberrant retinoblastoma (RB)-E2F transcriptional regulation defines molecular phenotypes of osteosarcoma. J Biol Chem 290:28070–28083
169. Rubin SM (2017) Inhibition of retinoblastoma protein inactivation. University of California, Santa Cruz, CA
170. Chung DC, Rustgi AK (1995) DNA mismatch repair and cancer. Gastroenterology 109:1685–1699

171. Zamzami N, Kroemer G (2001) The mitochondrion in apoptosis: how Pandora's box opens. Nat Rev Mol Cell Biol 2:67–71
172. Cheng AC, Huang TC, Lai CS et al (2005) Induction of apoptosis by luteolin through cleavage of Bcl-2 family in human leukemia HL-60 cells. Eur J Pharmacol 509:1–10
173. Pan MH, Lai CS, Hsu PC et al (2005) Acacetin induces apoptosis in human gastric carcinoma cells accompanied by activation of caspase cascades and production of reactive oxygen species. J Agric Food Chem 53:620–630
174. Pan MH, Lai YS, Lai CS et al (2007) 5- Hydroxy-3,6,7,8,39,49-hexamethoxyflavone induces apoptosis through reactive oxygen species production, growth arrest and DNA damage-inducible gene 153 expression, and caspase activation in human leukemia cells. J Agric Food Chem 55:5081–5091
175. Pan MH, Chang WL, Lin-Shiau SY et al (2001) Induction of apoptosis by garcinol and curcumin through cytochrome c release and activation of caspases in human leukemia HL-60 cells. J Agric Food Chem 49:1464–1474
176. Yeh TC, Chiang PC, Li TK et al (2007) Genistein induces apoptosis in human hepatocellular carcinomas via interaction of endoplasmic reticulum stress and mitochondrial insult. Biochem Pharmacol 73:782–792
177. Qanungo S, Das M, Haldar S, Basu A (2005) Epigallocatechin- 3-gallate induces mitochondrial membrane depolarization and caspase-dependent apoptosis in pancreatic cancer cells. Carcinogenesis 26:958–967
178. Zhao Y, Yang LF, Ye M et al (2004) Induction of apoptosis by epigallocatechin-3-gallate via mitochondrial signal transduction pathway. Prev Med 39:1172–1179
179. Wang CC, Chen LG, Lee LT et al (2003) Effects of 6- gingerol, an antioxidant from ginger, on inducing apoptosis in human leukemic HL-60 cells. In Vivo 17:641–645
180. Chen CY, Liu TZ, Liu YW et al (2007) 6-shogaol (alkanone from ginger) induces apoptotic cell death of human hepatoma p53 mutant Mahlavu subline via an oxidative stress-mediated caspase-dependent mechanism. J Agric Food Chem 55:948–954
181. Kim YA, Xiao D, Xiao H et al (2007) Mitochondria- mediated apoptosis by diallyl trisulfide in human prostate cancer cells is associated with generation of reactive oxygen species and regulated by Bax/Bak. Mol Cancer Ther 6:1599–1609
182. Park SY, Cho SJ, Kwon HC et al (2005) Caspase- independent cell death by allicin in human epithelial carcinoma cells: involvement of PKA. Cancer Lett 224:123–132
183. Xiao D, Lew KL, Zeng Y et al (2006) Phenethyl isothiocyanate-induced apoptosis in PC-3 human prostate cancer cells is mediated by reactive oxygen species-dependent disruption of the mitochondrial membrane potential. Carcinogenesis 27:2223–2234
184. Singh RP, Agrawal P, Yim D et al (2005) Acacetin inhibits cell growth and cell cycle progression, and induces apoptosis in human prostate cancer cells: Structureactivity relationship with linarin and linarin acetate. Carcinogenesis 26:845–854
185. Pan MH, Lai CS, Wang YJ et al (2006) Acacetin suppressed LPS-induced up-expression of iNOS and COX-2 in murine macrophages and TPA-induced tumor promotion in mice. Biochem Pharmacol 72:1293–1303
186. Clement MV, Hirpara JL, Chawdhury SH et al (1998) Chemopreventive agent resveratrol, a natural product derived from grapes, triggers CD95 signaling-dependent apoptosis in human tumor cells. Blood 92:996–1002
187. Watabe M, Hishikawa K, Takayanagi A et al (2004) Caffeic acid phenethyl ester induces apoptosis by inhibition of NF-kappaB and activation of Fas in human breast cancer MCF-7 cells. J Biol Chem 279:6017–6026
188. Xiang D, Wang D, He Y et al (2006) Caffeic acid phenethyl ester induces growth arrest and apoptosis of colon cancer cells via the beta-catenin/T-cell factor signaling. Anti-Cancer Drugs 17:753–762
189. Gao X, Deeb D, Jiang H et al (2005) Curcumin differentially sensitizes malignant glioma cells to TRAIL/ Apo2L-mediated apoptosis through activation of procaspases and release of cytochrome c from mitochondria. J Exp Ther Oncol 5:39–48

190. Ishiguro K, Ando T, Maeda O et al (2007) Ginger ingredients reduce viability of gastric cancer cells via distinct mechanisms. Biochem Biophys Res Commun 362:218–223
191. Jin CY, Moon DO, Lee JD et al (2007) Sulforaphane sensitizes tumor necrosis factor-related apoptosisinducing ligand-mediated apoptosis through downregulation of ERK and Akt in lung adenocarcinoma A549 cells. Carcinogenesis 28:1058–1066
192. Ju W, Wang X, Shi H et al (2007) A critical role of luteolin-induced reactive oxygen species in blockage of tumor necrosis factor-activated nuclear factor-kappaB pathway and sensitization of apoptosis in lung cancer cells. Mol Pharmacol 71:1381–1388
193. Nishikawa T, Nakajima T, Moriguchi M et al (2006) A green tea polyphenol, epigalocatechin-3-gallate, induces apoptosis of human hepatocellular carcinoma, possibly through inhibition of Bcl-2 family proteins. J Hepatol 44:1074–1082

Diverse Cancer Therapeutic Interactions: Complexities in Cancer Management

3

Prashant Tarale, Pravin Naoghare, Jyoti Tagde, Amit Bafana, and Saravanadevi Sivanesan

Abstract

Cancer is a multifactorial manifestation of diverse complexities linked with its development and progression. The environment is a source of exposure to multiple carcinogens including radiation, which causes aberration in critical regulators of cell cycle. The gene–environment interactions include highly complex downstream signaling pathways which are implicated in the process of tumorigenesis. Identification and understanding of the intricate signal mechanism is a challenging task because of the diversity of signaling networks involved in cancer. Different types of cancers exhibit a diverse set of mutations affecting a variety of signaling cascades. The better prognostic, diagnostic, and therapeutic approaches to combat cancer have been limited due to gaps in incomplete understanding of these complex molecular signaling cascades. Cancer biologists are working hard to gain insights into these intricate molecular complexities to develop efficient therapeutics for cancer. There are several anticancer drugs available for cancer treatment. However, a majority of them are associated with severe side effects and thus do not significantly contribute to patient survival. The nonspecific interaction of drugs with normal signaling pathways is primarily responsible for side effects observed. In addition, the effectiveness of therapeutic approaches becomes difficult due to individuals' genetic differences. The present

P. Tarale · S. Sivanesan
Health and Toxicity Cell, CSIR-National Environmental Engineering Research Institute (CSIR-NEERI), Nagpur, India

P. Naoghare (✉) · J. Tagde
Environmental Impact and Sustainability Division, CSIR-National Environmental Engineering Research Institute (CSIR-NEERI), Nagpur, India
e-mail: pk_naoghare@neeri.res.in

A. Bafana
Directors Research Cell, CSIR-National Environmental Engineering Research Institute (CSIR-NEERI), Nagpur, India

M. Kumar et al. (eds.), *Pharmacotherapeutic Botanicals for Cancer Chemoprevention*, https://doi.org/10.1007/978-981-15-5999-0_3

chapter entails to discuss the toxicity associated with some existing drugs used for cancer treatment. We have also discussed an integrative approach for understanding the complex interactions of drugs with various signaling networks to explore the possibility of better therapeutic options with reduced toxicity concerns. Personalized therapy has been discussed in the chapter considering its importance in effective cancer therapy.

Keywords

Cancer · Signaling complexities · Toxicity · Personalized therapy

3.1 Introduction

Cancer is known to be the leading cause of death worldwide [1, 2]. It is characterized by the uncontrolled proliferation of cells that may metastasize to the distant organs of the human body, invading tissue boundaries away from where it was originated [3]. Cancer has been divided into several types, governed by the diverse risk factors, cell type, tissue, and organ affected in the human body. At molecular level cancer arises when a key mediator of cell cycle regulation is either mutated or its normal expression is distorted [4–6]. Genes do not function in isolation but rather work in association with several other networks of signaling molecules which overall control and regulate the different stages of cell cycle [6–8]. Therefore, the mutation or disruption in single genetic entity cannot be held responsible, suggesting cancer as highly complex, multifactorial manifestations resulting in disturbance of master regulators of cell cycle [4–6, 8]. Humans are constantly exposed to multiple carcinogens present in the environment. The gene–environment interactions have been reported in the etiology of various cancer types [9]. Cell may undergo apoptosis when the damage caused by these carcinogens/xenobiotics is beyond the repair capacity of a cell. However, escaping apoptosis and survival of defective cell can prompt cancer development [8]. Immune system recognizes the tumor cells and destroyed it before it can be developed in cancerous growth. However tumor cells may develop a mechanism through alteration in certain immunological cascade that makes them un-recognizable from immune system [8]. In addition to these complex gene–environment interactions, the individual's differences in genetic susceptibility may also determine the predisposition to cancer [3, 8]. The single nucleotide polymorphisms (SNPs) in genes linked with metabolism of xenobiotics may lead to an increased risk of cancer [5]. The SNPs analysis in cancer related genetic factors might give a comprehensive list of genetic risk factors predisposing to cancer onset [5]. Recent years have witnessed a tremendous advancement with epigenetic research that greatly enhanced our understanding of molecular biology of carcinogenesis [8]. Therefore, targeting the epigenetic perturbation as a drug target for cancer treatment has become a promising area of research. Studies to unravel the mechanistic insights of cancer progression resulted in the discovery of key signaling pathways that may be targeted as cancer therapeutics [10–12]. In silico approaches at primary level of drug designing and screening are helpful to identify potential drug candidates and predicting its mechanistic interaction with target molecule. However in silico approach might not be able to complement the complex in vivo system [10–12]. Many drugs enter the clinical trials after success in the in vitro studies. However, most of them fail to make their place in the market. Currently available drugs for cancer treatment have limited therapeutic use due to their side effects reported over the prolonged treatment period. Doxorubicin is an antineoplastic drug that intercalates with DNA causing histone eviction from chromatin. As a result, epigenetic regulation and response to DNA damage are disrupted [13]. A lot of preclinical examinations are made before administrating anticancer drugs to assess its side effects [10–12]. Drug-specificity in targeting a particular cancer type without undesired side effects is of great challenge in cancer therapeutics. The ability of cancer cells to develop resistance by modulating its receptor and/or signaling networks that interact with drugs has complicated cancer management [8]. Cancer stem cells constitute the small population of entire tumor that is responsible for the maintenance and spread of the tumor [14]. The recurrence of cancer is reported in patients operated successfully with approaches like radiation and chemotherapy, strengthening the theory of cancer stem cell [9]. Recognizing and targeting the cancer stem cells may be an excellent way to overcome the cancer recurrence. Many targeted therapies target the key features specific to cancer cells like rapidly dividing ability compared to normal. However, some of the normal cells also divide rapidly and hence such drugs cause multiple side effects [10, 11]. Effective treatment of each type of cancer is limited by its own complexities at cellular level and the genetic make-up of a person. The drugs working on an individual might not be effective for others due to these complexities. This suggests that there are still unexplored factors that need to be investigated for successful cancer therapeutics. The pharmaceutical industry is focused to gain a comprehensive understanding of complex signal transduction pathways in the cancer mechanics but lacks the integrative approaches (discussed in detail later in the chapter). The undesirable and nonspecific effects of anticancer agents restrict its application to clinical practice (discussed later in the chapter) [12]. The nonspecific interaction of the drugs needs to be investigated at molecular level. The overlapping pathways in cancer and healthy cell may contribute to these side effects. Drug designing approaches should address these issues which require the use of high throughput integrated strategies [10–12].

Efficient drug discovery should only target the tumor cells without affecting healthy cells. The chapter also discusses the presently used anticancer drugs and their mechanism of action along with their toxic side effects. We have also suggested the integrative strategies that can be employed for better insight of molecular complexities. Phytochemicals have been traditionally used in Indian culture to treat a variety of diseases including cancer. We highlighted the emerging potential of phytochemicals to halt/cure the progression of cancer and combinational approaches for effective cancer treatment.

3.2 Complex Interaction of Anticancer Drugs with Signal Transduction Pathways in Cells

Anticancer drugs can trigger nonspecific abnormal cellular networks through various mechanisms such as impaired pharmacokinetics, aberrant diseases states [14], genetic polymorphisms in liver detoxifying genes [15], interactions with other drugs and receptor modulations [16]. Cellular networks are highly complex comprising of the superfluous tracks, intersecting networks, heterologous signals, and interconnections. This complex interaction boosts the strength and diversity of signaling and allows fine-tuning of desired cellular pathway with more accuracy, which may not be achieved by linear signaling cascades. Cellular signals are transduced by effecter cascades, which ultimately result in alteration of the gene expression, protein functioning, and structural organization of the cell. The canonical pharmacological approach is based on the investigation of specific physiological pathways or targets a specific protein or a metabolite. There are only a few reports available on the effect on metabolic context, which might depend on the signaling pathways specific to a certain cell type [17]. It should be considered that the complex biological mechanisms that result in a finely tuned regulation are highly adaptive to a great variety of biological situations including exposures to environmental carcinogens. Deregulation in signaling and effecter pathways may promote malignant cell growth (cancer), angiogenesis, infiltration, and metastasis. New studies have established that most of the cancer cells adopt an alternative mechanism for their survival and progression by taking advantage of feedback, cross-talk, and redundancy mechanisms [18]. Integrative biomedical and bioinformatics systems can contribute to obtain, even if approximate, a prediction of detrimental modification at metabolic or genetic level [19, 20].

3.3 Therapeutic Limitations of Anticancer Drugs

Modalities used to treat cancer are chemotherapy, neo-adjuvant therapy, radiation therapy, targeted therapy (immunotherapy), and transplantation [21–23]. Each of these therapies has their own therapeutic limitations, which may enforce complications in treatment and cure of cancer. Chronic treatment against cancer may often result in several side effects such as appetite changes, anemia, bleeding problems, constipation, diarrhea, fatigue, hair loss, infection, memory changes, mouth and sore throat or ulceration nausea and vomiting, pain, nerve damage, sexual and fertility changes, skin and nail changes, swelling (fluid retention), urination problems, and cardio-toxicity [22, 23]. Since the presently available cancer drugs have severe unwanted effects, FDA had issued a black box warning labels against the drugs with significant side effects. Some of the representative examples, which demonstrate the adverse effects of drugs due to complex cellular networks, are depicted in Fig. 3.1.

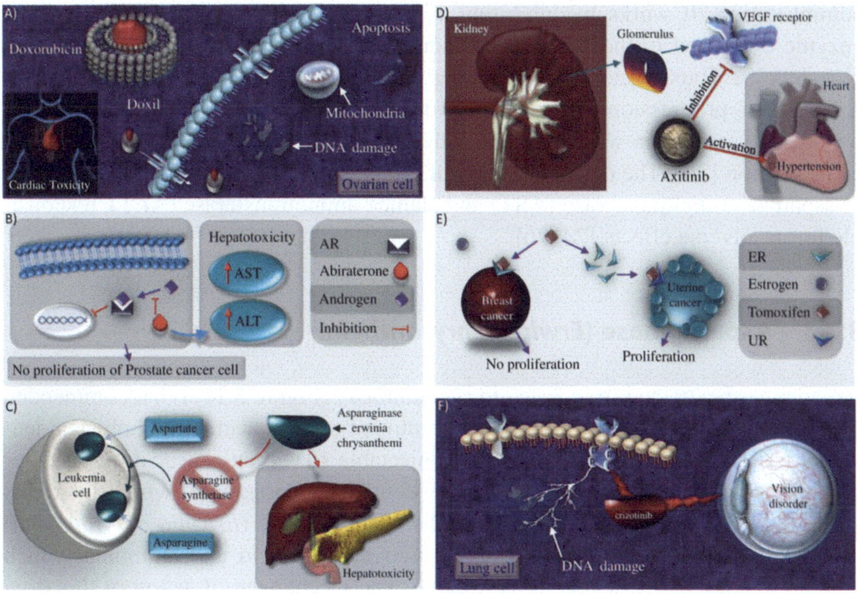

Fig. 3.1 Adverse side effects of chemotherapeutic drugs: (**a**) doxorubicin, (**b**) abiraterone, (**c**) asparaginase (Erwinia chrysanthemi), (**d**) axitinib, (**e**) tamoxifen, and (**f**) crizotinib

3.3.1 Doxorubicin

Doxorubicin (Figure 3.1a) is an antineoplastic agent used as a combination chemotherapy for the treatment of hematological malignancies (blood cancer—leukemia and lymphoma, non-Hodgkin lymphoma) [13]. It works by intercalating with DNA and stabilizing the topoisomerase II, hence preventing the double helix from getting relaxed, thus inhibiting the process of replication. This also stops the process of macromolecular biosynthesis. The incidence of cardiomyopathy has been observed in patients undergoing treatment with doxorubicin. The cytotoxic effects depend on the cumulative dose (4%- 500-550 mg/m^2; 18%- 551-600 mg/m^2; 36%- <600 mg/m^2) [13]. It increases the free radical production, decreases the expression of contractile protein and p53 mediated apoptotic protein that might contribute to cardiomyopathy and cytotoxicity. Intercalation with DNA may also cause histone eviction from a chromatin, as a result epigenome, transcriptome, and response to DNA damage are disrupted [13].

3.3.2 Abiraterone

Abiraterone (Figure 3.1b) works as a CYP17 inhibitor and prescribed in combination with prednisone for the treatment of metastatic castration resistant prostate cancer [24, 25]. Abiraterone acetate is transformed to its active form abiraterone upon administration. It works by inhibiting 17 α-hydroxylase/cyp17,20 lyse (CYP17) enzyme expressed in the prostate and adrenal tumor tissue involved in androgen biosynthesis. Thus it acts by deregulating the androgen receptor signaling which is required for progression from primary to metastatic prostate cancer. The patient undergoing treatment was reported with hepatotoxicity depending on the dose concentration [24]. The discontinuation of drug may result in an increase in activity of aspartate aminotransferase (AST) and alanine aminotransferase (ALT), responsible for its adverse effects [24–26].

3.3.3 Asparaginase (*Erwinia chrysanthemi*)

Asparaginase extracted from *Erwinia chrysanthemi* (Figure 3.1c) is a recommended candidate for a multi-chemotherapeutic treatment against acute lymphoblastic leukemia (ALL) patients [12, 27]. It is an injectable (I.V.) anticancer drug used for treating acute lymphocytic lymphoma. It contains asparaginase, which interferes with growth and spread of the cancer cells. Asparaginase (*Erwinia chrysanthemi*) catalyzes the deamidation of asparagine into aspartic acid and ammonia, which reduces the circulating levels of asparagine. The mechanism of action of this drug is thought to be based on the inability of leukemic cells to synthesize asparagine since it lacks asparagine synthetase. Administration of asparaginase results in leukemic cells specific cytotoxicity because of their dependence on an exogenous source of the amino acid asparagine for their protein metabolism and survival [12]. Five percent of patients in clinical trials have reported serious hypersensitive reaction, anaphylaxis against this drug. Asparaginase has been found to impart several side effects including an elevated level of transaminases, hepatotoxicity, hyper-bilirubinemia, etc. [12, 27].

3.3.4 Axitinib

Axitinib is found to inhibit multiple targets which include vascular endothelial growth factor receptors (VEGFRs 1, 2, and 3), platelet-derived growth factor receptor (PDGFR), and mast/stem cell growth factor receptor (SCFR; CD117) [11]. It has successfully demonstrated to repress the growth of breast cancer cell and renal cell carcinoma (RCC) in in vivo models. In clinical trials it has shown a good response in combination with gemcitabine against advanced stage pancreatic cancer. It works by binding to VEGFR inside of cancer cell, thus inhibiting the pathway that promotes the process of angiogenesis. It is considered to act by inducing autophagy as like other tyrosine kinase inhibitors like sorafenib. It is supposed to share similar kinds of target receptors along with other VEGFR-tyrosine kinase inhibitors. Forty percent of the renal carcinoma patients have reported the development of hypertension (diastolic pressure $\geq$90 mm Hg or systolic $\geq$140 mm Hg) in their first 8 and 12 weeks of axitinib treatment [11]. In another report, more than 20% of cancer patients suffered from hypertension as a result of axitinib side effects [28]. Hence, close monitoring and management of hypertension are suggested during the axitinib administration.

3.3.5 Tamoxifen

Tamoxifen is used for the treatment of endocrine receptor positive breast cancer patients in pre- and post-menopausal women. Activity of liver detoxifying enzymes; CYP2D6 and CYPP3A4 (cytochrome P450 isoforms) transform tamoxifen to its active metabolites 4-hydroxytamoxifen and 4-hydroxy-N-desmethyl tamoxifen (also known as endoxifen) that have more affinity to estrogen receptor than the parent tamoxifen [10, 29].

HER2 is a member of human epidermal growth factor receptor (HER/EGFR/ERBB) family that is abundantly expressed in cancer cells. It is reported that tamoxifen requires PAX2 for its full anticancer activity, thus at high level of PAX2 expression tamoxifen-estrogen receptor complex inhibits the expression of pro-proliferative epidermal growth factor receptor proteins, thus suppressing the tumor growth and proliferation.

Primarily tamoxifen has anti-estrogenic activity but it also has modest estrogenic property that is associated with endometrial proliferation, invasive carcinoma, and uterine sarcoma. The risk of endometrial cancer is dose and time dependent. Women treated with higher dose (40 mg/d) were reported to have more probability of developing aggressive tumors. However, the use of progestin reduces the risk of endometrial hyperplasia but its effect on endometrium and breast cancer during the course of tamoxifen is not known. Therefore, the use of progestin is not suggested for lowering the risk of cancer in women receiving tamoxifen treatment. American Cancer Society has reported that tamoxifen is a known carcinogen and can increase the risk of uterine cancer in the breast cancer patients [10, 29].

3.3.6 Crizotinib

Crizotinib (Figure 3.1f) is an approved prescription for treatment of non-small cell lung carcinoma (NSCLC). It acts as an inhibitor of anaplastic lymphoma kinase (ALK) and c-ros oncogene 1 (ROS1) [30, 31]. The NSCLC patients have a chromosomal rearrangement that creates the fusion gene between EML4 (echinoderm microtubule-associated protein-like 4) and ALK. This leads to constitutive expression of the kinase and its elevated activity culminating in carcinogenesis and malignant phenotype. This rare form of peripheral nervous system cancer occurring mostly in young children is found to have mutated ALK and could be cured by crizotinib treatment. Crizotinib is also reported to inhibit the angiogenesis in malignant tumors. The most common adverse effects observed in response to crizotinib were vision disorder (mostly vision impairment and photopsia or blurred vision) and renal abnormalities [30, 31]. This drug is under clinical trials for other cancers like neuroblastoma, anaplastic large cell lymphoma, and solid tumors. On the other hand, crizotinib has been reported to impart various side effects such as retinal abnormalities and vision disorders [31].

Why it is difficult to specifically target cancer cells while avoiding undesirable effects on the normal cells? One of the obvious reasons is the capability of cancer cells to interact with the additional cellular networks even when their potential target is impaired. For example, studies have revealed that the decline in the effectiveness of the selective insulin-like growth factor-1 receptor (IGF-1R) targeted therapy is due to the dynamic and complex association of IGF-1R with other cellular networks [32]. This in turn limits the effectiveness of IGF-1R targeted therapies by compensatory mechanism with high risk of serious side effects.

3.4 Competing the Complex Cellular Networks Towards a Rational Cancer Therapy

In recent years, several efforts have been made in elucidating the complexities in signal transduction pathways by understanding the cross-talk/feedback mechanism in the signaling biomolecules, cell-type cascades, spatial-temporal mechanisms (receptor clustering, growth factor mediated trafficking of cells, focal adhesions, filopodia/lamillipodia formations, etc.) [33], mathematical algorithms, etc. At present, many informative online resources (http://stke.sciencemag.org/, http://www.genome.jp/kegg/, http://www.reactome.org, http://www.wcrf.org/cancer_statistics/, etc.) can be efficiently used to understand the complexities in impaired cellular networks. As previously mentioned, computational biology offers the possibility to predict the onset and progression of a tumor [34]. Several analytical models have been developed to study the progression of tumor. However these models are limited by the lack of molecular information. The availability of "omics" data permits the design of more refined models that can encompass different levels of a biological system. The international initiatives, founded on Systems Biology paradigms, are attempting to elaborate a new generation of models that allows the clinicians to obtain a holistic outlook of cancer dynamics.

Currently, pharmaceutical and biotech companies have two major challenges: (a) cellular networks are still far from being entirely recognized and (b) the potential elements in the impaired cellular networks have to be validated in preclinical and clinical trials in order to establish drug targets in a cause–effect format. Therefore, many leading pharmaceutical and biotech companies have already initiated their own active drug discovery programs in order to understand the complex mechanism of cellular networks to synthesize better drug analogs with high therapeutic efficacy and low side effects. In silico methods for virtual screening provide a relevant approach to predict the interaction and impact of a newly designed chemical entity with minimum potential emergence of side effects.

In order to compete with the complex signal transduction pathways involved in cancer development, we may need to focus on the following critical aspects:

(a) It is quite rare that any single aberration in the complex cellular network can principally drive tumor growth. Thus, targeting an individual abnormal signaling component of a complex network may not be relevant to impart a significant reduction in tumor formation and progression. Effective drug targets can be identified by the better perceptive of etiology/biology of cancer and identification of multiple key targets among multiple signaling pathways implicated in the development of cancer. Some of the recently discovered targets like platelet-activating factor (PAF), chemokines and their receptors, lipoxygenases, pro-tumorogenic matrix metalloproteinases (MMPs), hyaluronic acid, guanine nucleotide exchange factors, GTPases, etc. [1, 35–38] should be considered while understanding and designing the multiple key targets against cancer. In addition, some of the less explored members in cellular pathways like mitogen activated protein kinases (MAPKs), i.e., MAP2, MAP3, etc. should also be taken into the consideration.
(b) Significant efforts should be made in understanding the protein–protein interactions, genetic/epigenetic modifications, spatio-temporal signaling, receptor clustering into micro-domains, role of scaffolds, feedback/robustness, multi-domain interactions, integrating alternative splicing into signaling pathways, cell specific signaling, and differentiation involved in the cancer development [39].
(c) Identification of appropriate subset of population for preclinical studies. Most of the cancer patients with chronic states are usually multiple-drug-resistant and are not ideal population for the study.
(d) Development of new target screening and validation methods that allow a rapid, precise, less expensive, and simultaneous analysis of multiple cellular network components while requiring a small amount of clinical specimen. In this regard, two promising technologies, i.e., high-content hyper-spectral imaging and mass spectroscopy should be extensively utilized for understanding the network complexes within the cancer cells. Integrating large-scale post-translational kinetic database and proteomic/transcriptomic datasets on signaling complexes into systems-level understanding could be one of the crucial approaches in the development of new bioinformatics tools. One of the major challenges, in integrative biomedical informatics, is the capability to mine, in an efficient way, the heterogeneous data originated by different experimental protocols and to cross-validate them. The integration of this disparate information is fundamental to investigate how molecular changes trigger the tumor emergence [40, 41].
(e) Identification and development of novel biocompatible drug delivery agents (e.g., biocompatible quantum dots, dendrimers, liposomes, etc.) that can specifically attack the target biomolecules on cancer cell without imparting any toxicity to normal cells [42, 43]. The rational design of new drugs is strongly dependent on the availability of computational tools. Immune checkpoint therapy approaches wherein immune system is activated to attack cancer cells has received a 2018 Nobel Prize in Physiology and Medicine.

(f) The development of new modalities for the precise identification of cancer subtypes in patients for efficient therapeutics: a way towards personalized therapy.
(g) Through investigation on the optimal drug combinations, neo-adjuvant/combinational therapeutic doses, and schedules essential to maximize clinical activity while minimizing toxicity [44].

3.5 Therapeutic Limitations Due to Personalized Cancer Treatment

Traditional approach in diagnosing and treatment is based on medical history of the patient, signs, symptoms and patho-physiological analysis, etc. It means that treatment was given after the appearance of the noticeable symptoms of the disease. Recent developments in medical sciences (e.g., genome-wide association studies (GWAS), human genome project, International HapMap Project, Biobanks, etc.) have enabled us to understand the role of genetic variations and disease development [45–47]. Integration of multiple heterogeneous information (molecular, cellular, clinical, and imaging) is an essential precondition to proceed with the development of personalized diagnosis and therapy [48–50]. The identification of genetic variations is fundamental to achieve this aim. In fact it is important to determine those SNPs that are critical for a specific population and for critical for all mankind. For example, cytosine to thiamine mutation (C $\rightarrow$ T) in Fig. 3.2 represents a single nucleotide polymorphism (SNP) in lung cancer patients. Usually, the carcinogens generated by the cigarette smoke, industrial pollution, etc. are excreted from the body through liver, kidney, and urine through renal elimination and biliary excretion. But, some of the carcinogenic compounds are deposited in the mucus rich environment of lungs [51]. These carcinogens (mostly lipophilic) are then metabolized and converted to hydrophilic conjugates in the liver. Finally, these hydrophilic conjugates are excreted through kidney and colon via urine and stool, respectively. SNPs in the liver detoxifying genes (phase I (cytochrome P-450) and phase II, e.g., UDP-glucuronyltransferase (GT), glutathione-S-transferase (GSH-T), etc.) of lung cancer patients can adversely alter their ability to metabolize a particular drug or susceptibility or resistance to the xenobiotic exposure. In this scenario, lung cancer patients under the same drug prescription may have different responses (Fig. 3.2), i.e., drug toxic but beneficial, drug toxic but not beneficial, drug not toxic and beneficial, and drug not toxic and not beneficial. Currently, therapeutic regimes for cancer are mostly decided based on mechanism of disease and not merely on the appearance of symptoms. More specifically, the treatment modalities are also available based on the response of an individual to a particular therapy (personalized therapy).

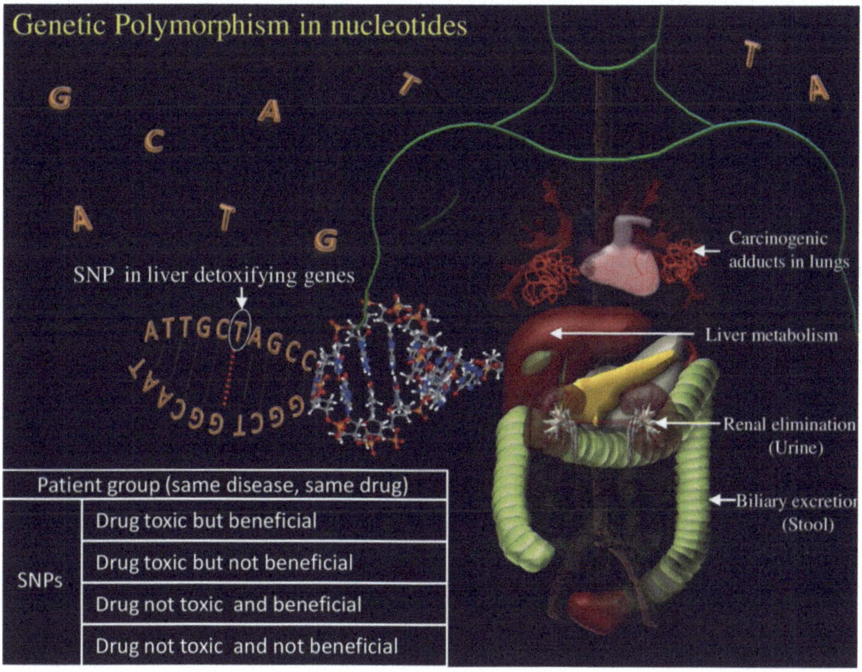

Fig. 3.2 Genetic polymorphism and drug toxicity

3.5.1 Applications of Personalized Therapy

The response of an individual to personalized therapy can be judged by analysis of various factors such as profiling genetic variations, proteins/genes, specific mutations [52]. This helps the clinicians to stratify the subtypes of the disease, disease status, drug doses/schedules, etc. at an individual level. Personalized therapy can be used to assess the risk of an individual towards a particular disease or even to a particular therapeutic approach Recent disciplines in clinical sciences (pharmacogenomics, pharmacoproteomics, and pharmacometabolomics) have a crucial role in the design of personalized therapies that can minimize the risk of adverse effects of anticancer drugs [53]. Drug responses can be measured depending on the genetic, epigenetic, proteomic, and metabolomic alterations at an individual level following this approach. It has been reported that approximately 5–10% of familial cancers (mostly breast, ovarian, bowel, melanoma, and prostate) are caused due to faulty inherited genes [51]. Some of the examples of personalized medicines include trastuzumab/herceptin (HER2 inhibitor; breast cancer), vemurafenib (B-Raf protein inhibitor; melanoma), imatinib/Gleevec (tyrosine kinase inhibitor; chronic myeloid leukemia), etc. [54–56]. Personalized medicine has enormous scope in the near future, wherein the customized polypills can be developed based on the ink-jet or fluid-jet printing mechanisms [57]. This will also allow the pharmaceutical companies to synthesize customized personalized drugs based on the theranostic analysis.

3.5.2 Lack of Personalized Therapy Approaches in Treating Cancer

A recent report according to the Union for International Cancer Control (UICC) entails that out of the 0.175 million children diagnosed with cancer every year, 0.090 million children die due to disease [58]. Another surprising report according to International Society of Paediatric Oncology (SIOP) and International Confederation of Childhood Cancer Parent Organizations (ICCCPO) depicts that 80% children with high-socioeconomic status and adequate facilities survive, whereas 80% children with the low-socioeconomic settings with inadequate treatment facilities do not survive [59]. These data highlight the major bottlenecks (lack of personalized therapy, lack of drugs without severe side effect and cancer stage, and the economic status of patients) associated with cancer therapeutics.

Personalized medicine market in the USA has already crossed $200 billion with a projected annual elevation of 11% [60]. On the other hand, the world personalized medicine market has rised with a compound annual growth rate over 4% during 2012–2016 [61].

At present only a few hospitals and academic institutions in the world are implementing personalized therapy into practice. Thus, more efforts are needed to boost personalized therapy options all around the world. Recently, MD Anderson Cancer Center (University of Texas, USA) has made considerable progress in designing novel personalized and targeted therapies against cancer [62]. Some of their newly identified potential targets are already in the clinical trials. These include AKT inhibitor (MK-2206) [63] for advanced breast tumor in the patients with PIK3CA mutation and/or PTEN loss, combinatorial therapeutics (BATTLE-2 trials) [64] for advanced refractory non-small cell lung cancer (NSCLC), PI3K inhibitors for in gliomas and studies on refractory metastatic colorectal cancer [65]. In addition, researchers at MD Anderson had also developed eight new therapeutic assays to identify undesirable mutations in the tumor cells that can be efficiently applied in personalized therapy. The assay targets include guanine nucleotide binding protein G(q) subunit alpha (GNAQ), PTEN, p53, v-Ki-ras2 Kirsten rat sarcoma viral oncogene homolog (KRAS), hepatocyte growth factor receptor (HGFR), isocitrate dehydrogenases 1 and 2 (IDH1 and IDH2), and platelet-derived growth factor-alpha (PDGFRA) [66–68]. Recent developments in cancer research have identified new promising therapeutic cancer targets as detailed in Table 3.1. Memorial Sloan Kettering researchers (MSK, USA) have also developed a genetic sequencing test called MSK-IMPACT, which screens for the mutations in more than 400 genes that have a role in cancer. This test is presently offered to all patients with advanced cancers. The results may help to categorize patients into clinical trials for drugs that are based on the unique genetic pattern of their tumors. MSK currently has about 30 clinical trials under progress that designate patients to targeted therapy depending on MSK-IMPACT test. These developments can give new dimensions to our basic

Table 3.1 New therapeutic anticancer targets

New therapeutic anticancer targets	References
VCP (valosin-containing protein)/p97 complex	[67]
Lin 28 oncomirs (*miR-17, miR-21, miR-125b, miR-155, miR-569*) snoRNAs and host genes	[68]
JAMs (junctional adhesion molecules)	[88]
JAB1/CSN5 (c-Jun activation domain-binding protein-1/constitutive photomorphogenic-9 signalosome)	[89]
DDR (discoidin domain receptor) tyrosine kinases	[90]
Id1 (inhibitor of DNA-binding protein1)	[91]
Tumor stem cells	[92]
Telomerase	[42]
Dysadherin	[93]
Lymphangiogenesis	[94]
CD47–CRT (calreticulin)	[95]
Stat3-Syk (signal transducer and activator of transcription 3-spleen) tyrosine kinase	[96]
Hypoxia-induced autophagy	[97]
Rho family of GTPase	[98]
CDK4/D1	[99]
Glutamine transporter, glutaminase, glutamate dehydrogenase, aspartate transaminase	[100]
Topoisomerases	[101]
Histone deacetylase inhibitors	[102]

understanding about cancer development and also help researchers in designing new therapeutic modalities against cancer. Such efforts could lay the necessary foundation for the future of personalized therapy based approaches. This could also help to recognize the role of genetic variation in the development of drug resistance.

3.6 Possible Strategies to Overcome Drug Resistance Pathways

The combinational treatment has a long proved to be effective in cancer chemotherapy. The first successful attempt was in 1963 for the treatment of childhood leukemia with "VAMP," which combined four drugs (amethopterin, vincristine, 6-mercaptopurine, prednisone) [69]. Various dietary phytochemicals have shown the potential to be used in combination with conventional anticancer drugs. Recently, combination of cisplatin with emetine (alkaloid found in *Ipecacuanha* species) demonstrated synergism action against the ovarian cancer cell line A2780. Combined treatment of cisplatin along with phytochemicals (emetine) may be a better way to treat ovarian cancer patients with minimal side effects [70]. Kaempferol is a natural flavonoid that occurs in natural sources like apples, tea, broccoli onions, leeks, citrus fruits, grapes, and red wines and known to have potential anticancer properties in many cancer cell lines [71]. Kaempferol in combination with doxorubicin or cisplatin has shown an effective synergistic cytotoxic effect in HCT-15 and MDA MB 231 cell lines [72].

There are also evidences that seaweed compounds such as Eckol, a phlorotannin from brown seaweed could be used in combination with anticancer drugs particularly temozolomide (TMZ) against malignant gliomas [73].

Efforts should be made to overcome the therapeutic limitations associated with the drug resistance considering a few potential key aspects: (a) Establishing the link between the MDR inducing molecules and the established alterations (e.g., mutant p53 expression) in cancer cell can help in generating the organ specific fingerprints that can be used to design better prognostic markers for drug resistance. (b) Problems with receptor modulations due to interpersonal genetic variations. This problem can be settled by the introduction of personalized therapy in immunotherapy models. (c) Anticancer vaccines may not work for all the patients with similar cancer types due to the differences in their immunogenic potentials. Some patients may not express the targeted antigens. (d) Since the immunogenic efficiencies are compromised at an advanced stage of cancer, immunotherapy should be used at an initial stage of cancer development to obtain a meaningful immunological response. (e) Identification of appropriate drug combinations with high efficacy at low doses and low toxicity. (f) Proper amalgamation of combinatorial therapies to overcome the therapeutic limitations. (g) Elucidation of non-characterized MDR inducing molecules in cancer cells, including members of MDR-1 and MRP super-families. h) Development of novel inhibitors against MDR inducers.

3.7 Emerging Role of Phytochemicals in Cancer Prevention

The dietary factors like vegetables, fruits, and whole grains are the major sources of natural bioactive compounds abundant in dietary phytochemicals. In the USA, National Cancer Institute (NCI) promotes a "Five-A-Day" program to encourage people to consume a minimum of five servings of fruits and vegetables in a day to reduce the risk of cancer and other chronic diseases. The beneficial mechanism for synergistic action of bioactive compounds maybe attributed to regulating more than one signal transduction pathways or increasing the drug availability [74]. Thousands of phytochemicals with great chemical diversity are naturally produced by plants with function highly related to their chemical backbone and functional groups, which might be used to aid in cancer therapy for better results. Alkaloids are among the major class of phytochemicals containing at least one nitrogen atom, obtained basically from medicinal herbal plants [75]. The alkaloids are reported to have wide pharmacological activities including anti-inflammatory, anti-bacterial, psychotropic, stimulant, and anticancer activity [76, 77]. Carcinogens are transformed into DNA damaging reactive form through the activity of metabolic enzymes; phytochemicals reported to impact the detoxification of carcinogens via promoting the activity of phase I and phase II metabolic enzymes including cytochrome p450 and antioxidant enzymes [78–80]. Cancer invasion is very detrimental which allows cancer cells to invade surrounding tissues, enters in circulation, and establishes a secondary tumor at distant sites in the body through a process called metastasis [81]. Phytochemicals like curcumin, resveratrol, and sulforaphane were reported to significantly demonstrate anti-invasion property via diminishing the activity of MMPs required for cell migration [82, 83]. miRNA is known to play a vital role in cancer progression via regulating the expression of various ontogenic genes like Myc, Ras, and tumor suppressor p53 gene [84]. Resveratrol has been found to decrease the expression of oncogenic miRNAs such as miR-21, miR-155, miR-196a, miR-17, and miR-92a-2 in colon tumor cells [85]. The present data imply that regular consumption of vegetables, fruits, and whole grain can reduce the risk of cancer development. Certain phytochemicals are known to have great potential to prevent cancer via anti-inflammatory, anti-metastasis, and pro-apoptosis effects. However these phytochemicals need to be further investigated for their cellular interaction with signal transduction pathways. Even though limitation exists and it is difficult to characterize the dietary phytochemicals at the clinical level but investigating their interaction with signaling targets may help to understand their anticancer mechanism so that a safe, effective, and affordable cancer therapy could be adopted for human beings [86, 87].

3.8 Conclusion

Cancer is mechanistically highly complex manifestation and hence multidimensional approaches are needed to understand these signaling complexities. In this regard, elucidation of natural macrophage activating factors and synthetic immune-modulators which can trigger the production of tumoricidal macrophages would be highly promising. This approach could immensely lower down the toxicity issues and enhance the natural defense system of the body to a greater extent. Combinational therapy with phytochemicals is another very promising area that needs to be focused for effective therapeutic intervention.

Conflict of Interest The authors declare that there is no conflict of interest.

References

1. Li F, Park SJ, Ling D, Park W, Han JY, Na K, Char K (2013) Hyaluronic acid-conjugated graphene oxide/photosensitizer nanohybrids for cancer targeted photodynamic therapy. J Mater Chem B 1(12):1678–1686
2. Bray F, Ferlay J, Soerjomataram I, Siegel RL, Torre LA, Jemal A (2018) Global cancer statistics 2018: GLOBOCAN estimates of incidence and mortality worldwide for 36 cancers in 185 countries. CA Cancer J Clin 68(6):394–424
3. Shaw AT, Kim DW, Nakagawa K, Seto T, Crinó L, Ahn MJ, Wu YL (2013) Crizotinib versus chemotherapy in advanced ALK-positive lung cancer. N Engl J Med 368(25):2385–2394
4. Wu X, Zhao H, Suk R, Christiani DC (2004) Genetic susceptibility to tobacco-related cancer. Oncogene 23(38):6500
5. McLean MH, El-Omar EM (2014) Genetics of gastric cancer. Nat Rev Gastroenterol Hepatol 11(11):664

6. Eramo A, Lotti F, Sette G, Pilozzi E, Biffoni M, Di Virgilio A, De Maria R (2008) Identification and expansion of the tumorigenic lung cancer stem cell population. Cell Death Differ 15 (3):504
7. Qi WX, He AN, Shen Z, Yao Y (2013) Incidence and risk of hypertension with a novel multitargeted kinase inhibitor axitinib in cancer patients: a systematic review and meta-analysis. Br J Clin Pharmacol 76(3):348–357
8. Campos-Parra AD, López-Urrutia E, Orozco Moreno LT, López-Camarillo C, Meza-Menchaca T, Figueroa González G, Pérez-Plasencia C (2018) Long non-coding RNAs as new master regulators of resistance to systemic treatments in breast cancer. Int J Mol Sci 19 (9):2711
9. Stratton MR, Campbell PJ, Futreal PA (2009) The cancer genome. Nature 458(7239):719–724
10. Raj TS, Chitrathara K (2015) Screening and prevention of carcinoma endometrium in uterine Cancer. Springer, New Delhi, pp 33–44
11. Hussain I, Tabassum N (2012) Axitinib-a new potent and selective VEGFR inhibitor. J Pharm Res 5(11):5157–5159
12. Keating GM (2013) Asparaginase Erwinia chrysanthemi (Erwinaze®): A Guide to Its Use in Acute Lymphoblastic Leukemia in the USA. BioDrugs 27(4):413–418
13. Hanf A, Oelze M, Manea A, Li H, Münzel T, Daiber A (2019) The anti-cancer drug doxorubicin induces substantial epigenetic changes in cultured cardiomyocytes. Chem Biol Interact 313:108834
14. Evers R, Piquette-Miller M, Polli JW, Russel FG, Sprowl JA, Tohyama K, International Transporter Consortium (2018) Disease-associated changes in drug transporters may impact the pharmacokinetics and/or toxicity of drugs: a white paper from the international transporter consortium. Clin Pharmacol Ther 104(5):900–915
15. Kapahtia S, Hazam RK, Asim M, Karra VK, Chowdhury SJ, Das BC, Kar P (2018) Role of glutathione S transferase M1 and T1 gene polymorphism in hepatitis B related liver diseases and cryptogenic cirrhosis. J Clin Exp Hepatol 8(2):169–172
16. Das DN (2015) Elucidating mechanisms of benzo[a]pyrene mediated apoptotic and autophagic cell death and its prevention with phytotherapeutics. Doctoral dissertation
17. Rodic S, Vincent MD (2018) Reactive oxygen species (ROS) are a key determinant of cancer's metabolic phenotype. Int J Cancer 142(3):440–448
18. Liao H, Ji F, Helleday T, Ying S (2018) Mechanisms for stalled replication fork stabilization: new targets for synthetic lethality strategies in cancer treatments. EMBO Rep 19(9):e46263
19. Doshi J, Dutta S, Bose K (2019) Analysing Cancer Signalling pathways: a structural bioinformatics approach in unravelling cancer signaling pathways: a multidisciplinary approach. Springer, Singapore, pp 271–293
20. Vilaça P, Maia P, Giesteira H, Rocha I, Rocha M (2018) Analyzing and designing cell factories with OptFlux in metabolic network reconstruction and modeling. Humana Press, New York, pp 37–76
21. Storey KB (ed) (2005) Functional metabolism: regulation and adaptation. Wiley, Hoboken
22. Pégorier S, Stengel D, Durand H, Croset M, Ninio E (2006) Oxidized phospholipid: POVPC binds to platelet-activating-factor receptor on human macrophages: implications in atherosclerosis. Atherosclerosis 188(2):433–443
23. Lu P, Weaver VM, Werb Z (2012) The extracellular matrix: a dynamic niche in cancer progression. J Cell Biol 196(4):395–406
24. Yadav HK, Srouji AE, Mohammed A, Dibi M (2018) Advancements in anticancer drug delivery. Front Anti-Cancer Drug Discov 9:287
25. Schweizer MT, Antonarakis ES (2012) Abiraterone and other novel androgen-directed strategies for the treatment of prostate cancer: a new era of hormonal therapies is born. Ther Adv Urol 4(4):167–178
26. Alrawashdeh H, Madi L, Elhada AH, Ahmed A, Serheed D (2018) A case of probable oxybutynin-induced increase in liver enzymes. Ther Clin Risk Manag 14:1657

27. Duncan R (2003) The dawning era of polymer therapeutics. Nat Rev Drug Discov 2 (5):347–360
28. Bianchi L, Rossi L, Tomao F, Papa A, Zoratto F, Tomao S (2013) Thyroid dysfunction and tyrosine kinase inhibitors in renal cell carcinoma. Endocr Relat Cancer 20(5):R233–R245
29. Kalampokas T, Sofoudis C, Anastasopoulos C, Boutas I, Melloy S, Kondi-Pafiti A, Salakos N (2013) Effect of tamoxifen on postmenopausal endometrium. Eur J Gynaecol Oncol 34 (4):325–328
30. Migliorino MR, Santo A, Romano G, Cortinovis D, Galetta D, Alabiso O, Giuffrida D (2017) Economic burden of patients affected by non-small cell lung cancer (NSCLC): the LIFE study. J Cancer Res Clin Oncol 143(5):783–791
31. Ho WL, Wong H, Yau T (2013) The ophthalmological complications of targeted agents in cancer therapy: what do we need to know as ophthalmologists? Acta Ophthalmol 91 (7):604–609
32. Ma H, Zhang T, Shen H, Cao H, Du J (2014) The adverse events profile of anti-IGF-1R monoclonal antibodies in cancer therapy. Br J Clin Pharmacol 77(6):917–928
33. Charney DS, Nestler EJ (eds) (2017) Charney and Nestler's neurobiology of mental illness. Oxford University Press, New York
34. Overby CL, Tarczy-Hornoch P (2013) Personalized medicine: challenges and opportunities for translational bioinformatics. J Pers Med 10(5):453–462
35. Heldin P, Kolliopoulos C, Lin CY, Heldin CH (2020) Involvement of hyaluronan and CD44 in cancer and viral infections. Cell Signal 65:109427
36. Li C, Zhao Y, Yang D, Yu Y, Guo H, Zhao Z, Yin X (2015) Inhibitory effects of kaempferol on the invasion of human breast carcinoma cells by downregulating the expression and activity of matrix metalloproteinase-9. Biochem Cell Biol 93(1):16–27
37. Lazer G, Katzav S (2011) Guanine nucleotide exchange factors for RhoGTPases: good therapeutic targets for cancer therapy? Cell Signal 23(6):969–979
38. Tsoupras AB, Iatrou C, Frangia C, Demopoulos CA (2009) The implication of platelet activating factor in cancer growth and metastasis: potent beneficial role of PAF-inhibitors and antioxidants. Infect Disord Drug Targets 9(4):390–399
39. Hongo K, Kazama S, Sunami E, Kitayama J, Watanabe T (2013) Perianal adenocarcinoma associated with anal fistula: a report of 11 cases in a single institution focusing on treatment and literature review. Hepato-Gastroenterology 60(124):720–726
40. Hayes-Jordan A, Green HL, Lin H, Owusu-Agyemang P, Fitzgerald N, Arunkumar R, Ludwig J (2014) Complete cytoreduction and HIPEC improves survival in desmoplastic small round cell tumor. Ann Surg Oncol 21(1):220–224
41. Pavlovic M, Balint B (2015) Bioengineering and cancer stem cell concept. Springer International Publishing, Cham
42. Bashash D, Ghaffari SH, Mirzaee R, Alimoghaddam K, Ghavamzadeh A (2013) Telomerase inhibition by non-nucleosidic compound BIBR1532 causes rapid cell death in pre-B acute lymphoblastic leukemia cells. Leuk Lymphoma 54(3):561–568
43. Finsterer J, Ohnsorge P (2013) Influence of mitochondrion-toxic agents on the cardiovascular system. Regul Toxicol Pharmacol 67(3):434–445
44. Poch Martell M, Atenafu EG, Minden MD, Schuh AC, Yee KW, Schimmer AD, Brandwein JM (2013) Treatment of elderly patients with acute lymphoblastic leukaemia using a paediatric-based protocol. Br J Haematol 163(4):458–464
45. Chen Z, Tao S, Gao Y, Zhang J, Hu Y, Mo L, Qin X (2013) Genome-wide association study of sex hormones gonadotropins and sex hormone-binding protein in Chinese men. J Med Genet 50(12):794–801
46. Tibayrenc M (2007) Human genetic diversity and the epidemiology of parasitic and other transmissible diseases. Adv Parasitol 64:377–462
47. Khan MS, Khan IA, Barh D (eds) (2016) Applied molecular biotechnology: the next generation of genetic engineering. CRC Press, Boca Raton

48. Arrigo P, Cardo PP, Izzotti A (2010) Proteomic-based screening of miRNAs metabolic pathway targeting. Int J Comput Biol Drug Des 3(2):164–173
49. Arrigo P, Scartezzini P, Romano P (2002) AgeWa: an integrated approach for antisense experiment design. IEEE Trans Nanobioscience 99(4):167–171
50. Naoghare PK, Kim MJ, Song JM (2008) Uniform threshold intensity distribution-based quantitative multivariate imaging cytometry. Anal Chem 80(14):5407–5417
51. Wang H, Khor TO, Yang Q, Huang Y, Wu TY, Saw CL, Kong ANT (2012) Pharmacokinetics and pharmacodynamics of phase II drug metabolizing/antioxidant enzymes gene response by anticancer agent sulforaphane in rat lymphocytes. Mol Pharm 9(10):2819–2827
52. Roubille C, Pelletier JP, Martel-Pelletier J (2013) New and emerging treatments for osteoarthritis management: will the dream come true with personalized medicine? Expert Opin Pharmacother 14(15):2059–2077
53. Everett JR, Loo RL, Pullen FS (2013) Pharmacometabonomics and personalized medicine. Ann Clin Biochem 50(6):523–545
54. Lee J, Ou SHI (2013) Towards the goal of personalized medicine in gastric cancer—time to move beyond HER2 inhibition part I: targeting receptor tyrosine kinase gene amplification. Discov Med 15(85):333–341
55. Pellowska M, Merk D, Schubert-Zsilavecz M (2013) Advances in personalized medicine–medicinal chemistry and pharmacology of vemurafenib and ivacaftor die Pharmazie. Int J Pharm Sci 68(7):484–491
56. Angelini S, Ravegnini G, Fletcher JA, Maffei F, Hrelia P (2013) Clinical relevance of pharmacogenetics in gastrointestinal stromal tumor treatment in the era of personalized therapy. Pharmacogenomics 14(8):941–956
57. Alfano KM (2013) US Patent Application No 13/842414
58. Rodriguez-Galindo C, Friedrich P, Alcasabas P, Antillon F, Banavali S, Castillo L, Quah TC (2015) Toward the cure of all children with cancer through collaborative efforts: pediatric oncology as a global challenge. J Clin Oncol 33(27):3065
59. Kempe F (2011) Berlin 1961: Kennedy Khrushchev and the most dangerous place on earth Penguin. Penguin Books, London
60. Balfour-Lynn IM (2014) Personalised medicine in cystic fibrosis is unaffordable. Paediatr Respir Rev 15:2–5
61. Lambin P, Leijenaar RT, Deist TM, Peerlings J, De Jong EE, Van Timmeren J, van Wijk Y (2017) Radiomics: the bridge between medical imaging and personalized medicine. Nat Rev Clin Oncol 14(12):749
62. Rozovski U, Hazan-Halevy I, Keating MJ, Estrov Z (2014) Personalized medicine in CLL: current status and future perspectives. Cancer Lett 352(1):4–14
63. Sangai T, Akcakanat A, Chen H, Tarco E, Wu Y, Do KA, Meric-Bernstam F (2012) Biomarkers of response to Akt inhibitor MK-2206 in breast cancer. Int J Cancer Clin Res 18 (20):5816–5828
64. Kim ES, Herbst RS, Wistuba II, Lee J, Blumenschein GR, Tsao A, Alden CM (2011) The BATTLE trial: personalizing therapy for lung cancer. Cancer Discov 1(1):44–53
65. Camacho LH, Garcia S, Panchal AM, Lim J, Hong DS, Ng C, Kurzrock R (2010) Exploratory study of hepatic arterial infusion oxaliplatin with systemic 5-fluorouracil/bevacizumab in patients with refractory solid tumor and extensive liver metastases. Clin Colorectal Cancer 9 (5):311–314
66. Sánchez NS, Mills GB, Mills Shaw KR (2017) Precision oncology: neither a silver bullet nor a dream. Pharmacogenomics 18(16):1525–1539
67. Jiang N, Shen Y, Fei X, Sheng K, Sun P, Qiu Y, Mi J (2013) Valosin-containing protein regulates the proteasome-mediated degradation of DNA-PKcs in glioma cells. Cell Death Dis 4(5):e64
68. Banno K, Iida M, Yanokura M, Kisu I, Iwata T, Tominaga E, Aoki D (2014) MicroRNA in cervical cancer: OncomiRs and tumor suppressor miRs in diagnosis and treatment. Sci World J 2014:1

69. DeVita VT, Chu E (2008) A history of cancer chemotherapy. Cancer Res 68(21):8643–8653
70. Alam MN, Yu JQ, Beale P, Huq F (2020) Cisplatin in combination with emetine and patulin showed dose and sequence dependent synergism against ovarian cancer. Synergy 10:100060
71. Liu RH (2003) Health benefits of fruit and vegetables are from additive and synergistic combinations of phytochemicals. Am J Clin Nutr 78(3):517S–520S
72. Kalyani CH, Narasu ML, Devi YP (2017) Synergistic growth inhibitory effect of flavonol–kaempferol and conventional chemotherapeutic drugs on cancer cells. Int J Pharm Pharm Sci 9:123–127
73. Ferreira J, Ramos AA, Almeida T, Azqueta A, Rocha E (2018) Drug resistance in glioblastoma and cytotoxicity of seaweed compounds alone and in combination with anticancer drugs: a mini review. Phytomedicine 48:84–93
74. Yang J, Liu RH (2009) Synergistic effect of apple extracts and quercetin 3-β-D-glucoside combination on antiproliferative activity in MCF-7 human breast cancer cells in vitro. J Agric Food Chem 57(18):8581–8586
75. Michael JP (2008) Indolizidine, quinolizidine alkaloids. Nat Prod Rep 25(1):139–165
76. Bahramsoltani R, Farzaei MH, Farahani MS, Rahimi R (2015) Phytochemical constituents as future antidepressants: a comprehensive review. Rev Neurosci 26(6):699–719
77. Robertson J, Stevens K (2014) Pyrrolizidine alkaloids. Nat Prod Rep 31(12):1721–1788
78. Wang Z, Sun Y, Gao B, Lu Y, Fang R, Gao Y, Chen H (2014) Two co-existing germline mutations P53 V157D and PMS2 R20Q promote tumorigenesis in a familial cancer syndrome. Cancer Lett 342(1):36–42
79. Bowen-Forbes CS, Zhang Y, Nair MG (2010) Anthocyanin content antioxidant anti-inflammatory and anticancer properties of blackberry and raspberry fruits. J Food Compost Anal 23(6):554–560
80. Davenport DM, Wargovich MJ (2005) Modulation of cytochrome P450 enzymes by organosulfur compounds from garlic. Food Cosmet Toxicol 43(12):1753–1762
81. Yodkeeree S, Chaiwangyen W, Garbisa S, Limtrakul P (2009) Curcumin demethoxycurcumin and bisdemethoxycurcumin differentially inhibit cancer cell invasion through the down-regulation of MMPs and uPA. J Nutr Biochem 20(2):87–95
82. Eastham LL, Howard CM, Balachandran P, Pasco DS, Claudio PP (2018) Eating green: shining light on the use of dietary phytochemicals as a modern approach in the prevention and treatment of head and neck cancers. Curr Top Med Chem 18(3):182–191
83. Özenver N, Efferth T (2020) Small molecule inhibitors and stimulators of inducible nitric oxide synthase in cancer cells from natural origin (phytochemicals marine compounds antibiotics). Biochem Pharmacol:113792
84. Croce CM, Calin GA (2005) miRNAs cancer and stem cell division. Cell 176(1):6–7
85. Tili E, Michaille JJ, Alder H, Volinia S, Delmas D, Latruffe N, Croce CM (2010) Resveratrol modulates the levels of microRNAs targeting genes encoding tumor-suppressors and effectors of TGFβ signaling pathway in SW480 cells. Biochem Pharmacol 80(12):2057–2065
86. Miller PE, Snyder DC (2012) Phytochemicals and cancer risk: a review of the epidemiological evidence. Nutr Clin Pract 27(5):599–612
87. HemaIswarya S, Doble M (2006) Potential synergism of natural products in the treatment of cancer. Phytother Res 20(4):239–249
88. Goetsch L, Haeuw JF, Beau-Larvor C, Gonzalez A, Zanna L, Malissard M, Corvaia N (2013) A novel role for junctional adhesion molecule-a in tumor proliferation: modulation by an anti-JAM-A monoclonal antibody. Int J Cancer 132(6):1463–1474
89. Pan Y, Zhang Q, Tian L, Wang X, Fan X, Zhang H, Yang H (2012) Jab1/CSN5 negatively regulates p27 and plays a role in the pathogenesis of nasopharyngeal carcinoma. Cancer Res 72(7):1890–1900
90. Kim HG, Tan L, Weisberg EL, Liu F, Canning P, Choi HG, Mandinova A (2013) Discovery of a potent and selective DDR1 receptor tyrosine kinase inhibitor. ACS Chem Biol 8 (10):2145–2150

91. Mern DS, Hasskarl J, Burwinkel B (2010) Inhibition of id proteins by a peptide aptamer induces cell-cycle arrest and apoptosis in ovarian cancer cells. Br J Cancer 103(8):1237–1244
92. J Prud'homme G (2012) Cancer stem cells and novel targets for antitumor strategies. Curr Pharm Des 18(19):2838–2849
93. Yonar D, Süloğlu AK, Selmanoğlu GÜLDENİZ, Sünnetçioğlu MM (2019) An Electron paramagnetic resonance (EPR) spin labeling study in HT-29 Colon adenocarcinoma cells after Hypericin-mediated photodynamic therapy. BMC Mol Cell Biol 20(1):16
94. Han L, Su W, Huang J, Zhou J, Qiu S, Liang D (2014) Doxycycline inhibits inflammation-induced lymphangiogenesis in mouse cornea by multiple mechanisms. PLoS One 9(9): e108931
95. Wemeau M, Kepp O, Tesniere A, Panaretakis T, Flament C, De Botton S, Chaput N (2010) Calreticulin exposure on malignant blasts predicts a cellular anticancer immune response in patients with acute myeloid leukemia. Cell Death Dis 1(12):e104–e104
96. Fm U (2012) Stat3-Syk molecular complex as a target for anti-cancer nanomedicines. J Nanomed Nanotechol 3:e110
97. Mazure NM, Pouysségur J (2010) Hypoxia-induced autophagy: cell death or cell survival? Curr Opin Cell Biol 22(2):177–180
98. Guan R, Xu X, Chen M, Hu H, Ge H, Wen S, Pi R (2013) Advances in the studies of roles of rho/rho-kinase in diseases and the development of its inhibitors. Eur J Med Chem 70:613–622
99. Wang HM (2013) An effective approach for transient thermal analysis in a functionally graded hollow cylinder. Int J Heat Mass Transf 67:499–505
100. Jin L, Alesi GN, Kang S (2016) Glutaminolysis as a target for cancer therapy. Oncogene 35 (28):3619–3625
101. Delgado JL, Hsieh CM, Chan NL, Hiasa H (2018) Topoisomerases as anticancer targets. Biochem J 475(2):373–398
102. Cappellacci L, Perinelli DR, Maggi F, Grifantini M, Petrelli R (2019) Recent progress in histone deacetylase inhibitors as anticancer agents. Curr Med Chem. https://doi.org/10.2174/0929867325666181016163110

Natural Products as Chemosensitizers for Adjunct Therapy in Cancer Management

4

Suchisnigdha Datta, Priyanka Saha, Subhabrata Dey, and Dona Sinha

Abstract

Conventional cancer treatment including surgery, chemotherapy, and radiotherapy has their respective limitations and complications. A large number of cancers are relatively resistant and highly refractory to cytotoxic chemotherapy or radiotherapy. One of the main reasons behind the cancer-related death is the development of multidrug resistance which may hinder an effective anticancer treatment. Many of the first-order antineoplastic drugs like anthracyclines, platins, taxanes, cyclophosphamides show resistance and/or adverse side effects after initial success in chemotherapy. Other than individual's genetic variation which is the basis of intrinsic resistance, an array of epigenetic dysregulation is acquired parallel to the cancer progression. Genetic or epigenetic alterations influence various cellular processes involved in drug metabolism such as modulation in drug uptake and efflux, activation of drug-detoxifying machinery, drug target alteration and compartmentalization, activation of error-prone DNA repair, and death evasion. Systemic cytotoxicity due to chemotherapy/radiotherapy refrains improvement of patient survival. Restoration of chemosensitivity by chemical modulators was extensively explored but trials did not indicate towards absolute efficacy. Many natural compounds possess the property of suppression or even reversal of drug resistance. Most of these natural chemosensitizers are phytochemicals e.g. polyphenols, quinones, alkaloids, carotenoids, and steroids. Natural chemosensitizers are promising adjuvants as they selectively alter the drug–transporter interaction by their unique conformational features. In general, these

S. Datta · P. Saha · D. Sinha (✉)
Department of Receptor Biology and Tumor Metastasis, Chittaranjan National Cancer Institute, Kolkata, India
e-mail: dona.sinha@cnci.org.in

S. Dey
Department of Clinical and Translational Research, Chittaranjan National Cancer Institute, Kolkata, India

M. Kumar et al. (eds.), *Pharmacotherapeutic Botanicals for Cancer Chemoprevention*, https://doi.org/10.1007/978-981-15-5999-0_4

molecules can increase the uptake and retention of chemotherapeutic drugs in tumor cells, induce pro-apoptotic signals, cell-cycle arrest and DNA damage, or can even control the changed drug targets expression. When combined, these mechanisms enhance the cytotoxicity of anticancer drugs by increasing their response and promoting either a synergistic or an additive effect even in resistant cells. So, the future direction towards a successful chemoprevention needs a combinatorial approach between natural chemosensitizers and conventional anticancer chemotherapeutics which would aim for drug-dose minimization by enhanced drug-efficacy on the one hand and reduced side effects on the other hand by the virtue of their low toxicity, pleiotropy, and immune-modulatory properties.

Keywords

Cancer · Phytochemicals · Chemosensitizers · Combinatorial approach

Abbreviations

ABC	ATP-binding cassette
AIF	Apoptosis inducing factor
Akt	Protein kinase B
ALDH	Aldehyde dehydrogenase
AMPK	5′ adenosine monophosphate-activated protein kinase
Bax	Bcl-2 associated X protein
Bcl-XL	B-cell lymphoma-extra-large
BCR-Abl	Breakpoint cluster region-abelson murine leukemia viral oncogene homolog 1
BCRP	Breast cancer resistance protein
Bim	Bcl-2-like protein 11
BRCA	Breast cancer 1
CKIs	CDK inhibitors
CSC	Cancer stem cells
CYP	Cytochromes
DMEs	Drug metabolizing enzymes
DPD	Dihydropyrimidine dehydrogenase
EGCG	Epigallocatechin-3-gallate
EGFR	Epidermal growth factor receptor
EMT	Epithelial-mesenchymal transition
ERK	Extracellular signal regulated kinase
GSK3β	Glycogen synthase kinase 3 beta
GST	Glutathione S-transferase
HCC	Hepatocellular carcinoma
HIF-1	Hypoxia-inducible factor 1
HO-1	Hemeoxygenase 1

HSP	Heat shock protein
IAPs	Inhibitors of apoptosis protein
JNK	c-Jun N-terminal kinase
Keap-1	Kelch-like ECH-associated protein 1
MAPK	Mitogen-activated protein kinase
Mcl-1	Myeloid cell leukemia 1
MDR	Multidrug resistance
MGMT O6	Methylguanine DNA methyl transferase
MRP	Multidrug resistance-associated protein
NHEJ	Non-homologous end joining
NF-κB	Nuclear factor kappa-light-chain-enhancer of activated B cells
Nrf-2	Nuclear factor erythroid 2-related factor 2
PARP	Poly [ADP-ribose] polymerase
P-gp	P- glycoprotein
PI3K	Phosphoinositide 3-kinase
RFC	Reduced folate carrier
ROS	Reactive oxygen species
SLC	Solute carrier
SNPs	Single nucleotide polymorphisms
STAT3	Signal transducer and activator of transcription *3*
TSCC	Tongue squamous cell carcinoma
UCP-2	Uncoupling protein-2
XRCC1	X-ray repair cross-complementing protein 1

4.1 Introduction

An alarming rise in the global cancer burden is the cumulative outcome of consistently increasing occurrence and limitations of existing anticancer treatment regimens. Conventional cancer treatment is frequently confronted with high frequency of poor prognosis which ultimately terminates into death [1]. Therefore, the researchers are focused on solving this enigma with effective anticancer therapy with lesser side effects.

4.1.1 Conventional Strategies of Cancer Management

The anticancer treatments extensively prescribed and practiced are called conventional treatments which broadly include—surgery, radiotherapy, and chemotherapy.

Surgery Surgery is an appropriate treatment where it almost kills 100% of cancer cells and it is operated by zero order kinetics. It is the oldest form of cancer therapy but still forms an ultimate and main treatment for solid tumors. For an effective result, oncologists review different therapeutic options, history of the malignancy, proper analysis of biopsy examination and how all these factors may be integrated into a well-organized and appropriate treatment algorithm [2]. The grade and stage of the cancer determine the strategy of treatment which often comprises of chemotherapy and/or radiotherapy along with surgery. However, surgery can provide long-term remission only in early stages when metastasis has not been initiated. Moreover, surgery is not free of its mutilating effects since it involves morbidity, loss of function of removed organ and other cosmetic issues [3].

Radiotherapy Radiotherapy uses high-energy ionizing radiation for targeting the cancer mass which is often applied in combination with surgery or chemotherapy. Radiation causes double strand breaks of DNA and can also affect cellular processes like cell cycle and apoptosis. The treatment is determined by the radio sensitivity of the tumor cells and the tolerance power of the surrounding tissues [4]. Some of the side effects of radiation therapy include reduced sensation in the exposed tissue and skin problems in the treated area like itching and soreness [5].

Chemotherapy Since 1930s, chemotherapy is used in treating cancer where one or more anticancer drugs are used for the treatment. Chemotherapy uses synthetic or semi-synthetic chemical drugs which either inhibit cellular proliferation or promote cytotoxicity but in a non-specific and non-localized manner. These drugs target rapidly dividing cancer cells more than the normal resting cells [6]. Neoadjuvant chemotherapy is used to shrink the primary tumor and is given before local treatment like surgery. Adjuvant chemotherapy is used to remove any residual cancer cells after any local treatment like surgery and radiotherapy thereby reducing the risk of further metastasis [7]. Chemotherapeutic agents can interfere with cell division and can initiate apoptosis. They are anti-mitotic drugs and can affect rapidly growing normal cells of bone marrow, digestive tract, and hair follicles. Most common side effects may include immune-suppression, hair loss, and mucositis. These may cause diseases like rheumatoid arthritis, multiple sclerosis, systemic lupus erythematosus, and many more [8].

4.1.2 Limitations and Complications of Chemotherapy

Chemotherapy, being the systemic anticancer therapy can address widespread cancers from any anatomic location. Normal cells with high proliferation rate endure unavoidable damage due to non-specific targeting by chemotherapeutic drugs. This gives rise to most of the adverse effects like immune suppression, mucositis, alopecia, and thrombocytopenia [9] along with gastro-intestinal problems like nausea, anorexia, bowel issues, oral ulcers, etc. According to the nature or mode of action, chemotherapeutic drugs are of several types: alkylating agents, antimetabolites, topoisomerase poisons, anti-microtubule agents, receptor blockers, kinase inhibitors, anticancer antibiotics, and many others which can be both cell cycle dependent or independent [9]. Oral or intravenous are the two most common routes of administration. Based on the mode of action of the chemical formulation, the cytotoxic outbreak takes place in an acute or chronic manner (Table 4.1)

Table 4.1 Modality and toxicity of chemotherapeutic drugs

Types of chemotherapy depending on modality	Chemotherapeutic drugs	Mechanism of acquiring resistance	Cell-cycle specificity	Side effects	Ref
Alkylating agents	cyclophosphamide, cisplatin	↑detoxification, ↑DNA repair	Independent	DNA damage leading to bone marrow depletion and acute leukemia	[10]
Antimetabolites	methotrexate, fluorouracil, gemcitabine	↑thymidylate synthase, ↑DHFR activity, ↑EMT, ↑survival factors	Dependent		[11]
Topoisomerase poisons	doxorubicin, etoposide	↑efflux, ↑autophagy topoisomerase mutation, ↑NF-kB	Independent	Allergic inflammation, tachycardia	
Anti-microtubule agents	vincristine, vinblastine, paclitaxel, docetaxel	↑efflux, ↑autophagy, ↑microtubule machinery, β-tubulin mutation, ↑cancer stemness	Dependent	Peripheral neuropathy	
Kinase inhibitors	imatinib, trastuzumab	↑ growth factors, PI3K mutation	Independent	Skin and heart problems	
Anticancer antibiotics	doxorubicin, bleomycin	↓Topo-II, ↑MRP, ↑Wnt,	Independent	Long-term cardiopathy	

[12]. After these side effects, next comes the challenge of resistance or non-responsiveness towards a successful chemotherapy.

The mechanisms responsible for failure cancer chemotherapy may be pharmacological, physiological, and/or cellular mechanisms. The pharmacological mechanisms of chemotherapy failure revolve around optimal dosing as the drug remains ineffective when under-dosed and toxic when overdosed. Chemotherapy may fail due to various physiological mechanisms which may include improper distribution of the chemotherapeutic agents due to existence of the blood–brain barrier and/or blood–testicular barrier and poor vasculature due to angiogenesis [13]. Those tumor cells which are not present in contact with the blood vessels face hypoxic conditions due to the unavailability of new blood vessels to the entire tumor. This leads to issues with drug delivery to the tumor. The cellular mechanisms which may lead to chemotherapy resistance include drug efflux by deregulated ATP-binding cassette [ABC] transporters, reduced drug uptake and metabolism, drug sequestration, altered drug targets, deregulated DNA repair, and reduced apoptosis [14].

4.2 Chemoresistance

Cancer chemoresistance is the inherent or adaptive property of the cancerous cells to lose sensitivity against chemotherapeutic drugs and to grow aggressively in presence of the drugs. It is the survival strategy of malignant cells which leads to untreatable disease relapse and metastasis. Most of the cancer cells possess not only an intrinsic nature of survival against the therapeutic toxicity [innate resistance] but are also constantly acquire spontaneous resistance after exposure with the help of their ever-changing genetic and epigenetic escape strategies [acquired resistance]. As cancer is a dynamic disease, intratumoral heterogeneity, which shapes the tumor microenvironment, lies at the root of resistance. Individual cells of a same cancer case show differential drug-sensitivity due to genetically distinct tumor-cell subpopulations [spatial heterogeneity] or changes in the molecular make-up with the course of disease progression [temporal heterogeneity] [9].

4.2.1 Mechanisms of Chemoresistance

Like the multifactorial disease cancer, its treatment is also dependent on multiple genetic components. Host and tumor genetic alterations which are responsible for cancer formation are also determinants of efficiency of anticancer treatments. Mutations or chromosomal rearrangements like deletion and amplification of genes [15] involved in DNA synthesis and repair, cellular energy production, genotoxic metabolism, growth signaling, immune phenotyping, i.e. which are directly associated with cellular proliferation are most susceptible for growing innate antineoplastic resistance, e.g. organic-anion-transporting polypeptide, glutathione S-transferase [GST], thymidylate synthase, dihydropyrimidine dehydrogenase, multidrug resistance-associated protein [MRP], uridine 5′-diphospho-glucuronosyltransferase, cytochromes P[CYP] 450, topoisomerase, epidermal growth factor receptor [EGFR], ki-ras2 Kirsten rat sarcoma viral oncogene homolog, breakpoint cluster region-Abelson murine leukemia viral oncogene homolog 1 [BCR-Abl], breast cancer 1[BRCA], excision repair 1- endonuclease non-catalytic subunit, p53, Fc fragment of IgG receptor IIIa.

There are several mechanisms behind acquiring the nongenetic alterations which can downregulate the sensitivity of the neoplastic cells against the antineoplastic drugs. Epigenetic changes which can directly alter the expression include DNA methylation at the promoter region and/or histone modification by deacetylation, which causes gene silencing and hypermethylation, which promotes overexpression of the genes. These modifications can affect the drug efficacy at multiple levels starting from drug–plasma membrane interaction, drug metabolism, drug targeting and sequestration, and finally to DNA damage response and death evasion [16]. Recent advancement in this field confirms the role of small RNA structures like small interfering [si]RNA and micro [mi]RNA in gene silencing by destabilization or poor translation of mRNAs which finally restricts the chemosensitivity in different cancers [17].

Depending on the mode of acquisition, acquired chemoresistance can be divided into pre-target, on-target, post-target, and off-target resistance (Fig. 4.1). Pre-target resistance affects the active drug-availability to the cells [18]. The decreased uptake of most anticancer agents occurs through tight control of transporters of plasma membrane, mediated via solute carrier family, reduced folate carrier, and altered lipid bilayer [16]. In resistant tumor cells the most common phenomenon is an increased expression of multidrug resistance [MDR] and MRP efflux pumps. Hypoxia-inducible factor 1 [HIF-1] is highly expressed in cancer cells which induces efflux of drugs from the cancer cells by the increased expression of P-glycoprotein [P-gp] and the MDR1 gene [19]. Once in the intracellular compartment, enzymatic inactivation of the functional drugs is the next step of acquiring cellular resistance. This drug metabolism involves phase I [by oxidoreduction] and II detoxifying enzymes [conjugation of hydrophilic groups]. Pharmacokinetic mechanism of drug resistance involves overexpression of drug metabolizing enzymes [DMEs], namely: aldehyde dehydrogenase [ALDH], GST, dihydropyrimidine dehydrogenase, NAD [P]H quinone dehydrogenase 1, CYP450 which are mainly responsible for inactivation, degradation, and premature clearance of anticancer drugs. Lysosomal sequestration of hydrophobic and weak base chemotherapeutic agents, enhanced protein trafficking and secretion by ABC transporters is associated with poor drug response [16]. Structural or functional modifications in the target molecules of the drug may stop the drug from acting and cause irreversible resistance.

On-target resistance impedes the targeting and damage causing ability of the drug. One of the basic tendencies of cytotoxic drugs, i.e. killing of the highly dividing cells is tackled by the enhanced and even error-prone repair of the DNA lesion by various repair pathways including non-homologous end joining [NHEJ] [15]. Other intracellular pathways that can cause chemoresistance include death receptor pathway, nuclear factor kappa-light-chain-enhancer of activated B cells

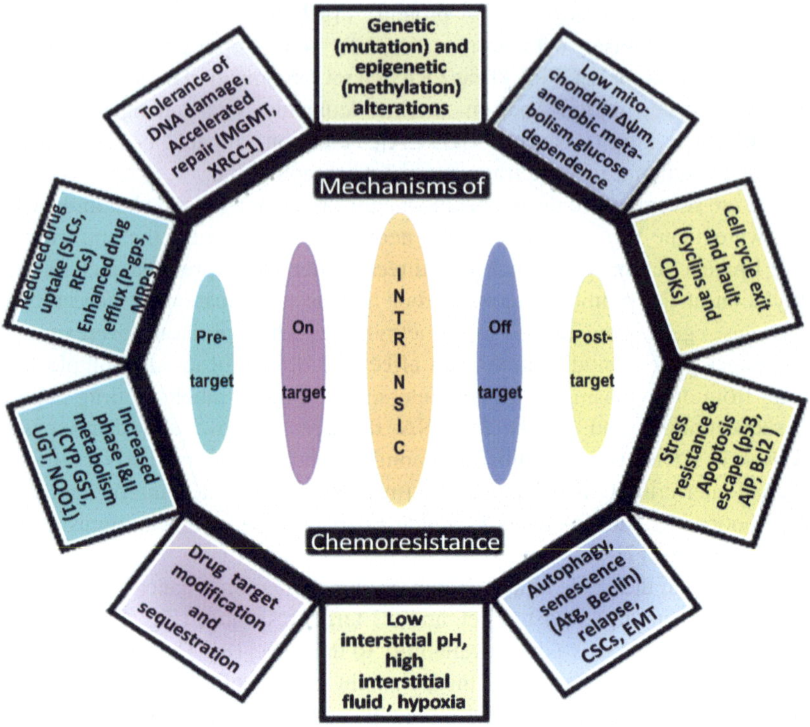

Fig. 4.1 Mechanisms of acquiring chemoresistance: Pre-target, post-target, on-target, and off-target acquired resistance and intrinsic resistance development against anticancer chemotherapy

[NF-κB] pathway, phosphoinositide 3-kinase[PI3K]/protein kinase B [Akt] pathway, activation of anti-apoptotic proteins, and inhibitors of apoptosis protein [IAPs] [20]. Insufficient activation of apoptotic pathways and escape from cell cycle checkpoints limits the ultimate purpose, i.e. cancer-killing ability of the chemotherapeutic drug. During exposure to chemotherapeutic agents mitochondrial uncoupling protein-2[UCP-2] a major controller of mitochondrial reactive oxygen species [ROS] induces chemoresistance via inhibition of ROS induced apoptosis [21].

Post-target resistance involves all the general or drug-specific mechanisms which are acquired by the cancer cell to evade the drug action. Changes in interstitial fluid flow can cause efflux of fluid from the tumor mass into its surroundings. Cell adhesion mediated drug resistance can cause activation of anti-apoptotic signals by integrin mediated ligand receptor binding mainly integrin $\alpha_4\beta_1$. In leukemia cells adhesion mediated by β1 integrin causes an increase in degradation of Bcl-2-like protein 11 [Bim] and leads to drug resistance. pH plays another important role in providing chemoresistance. Solid tumors have low interstitial pH due to accumulation of CO_2 and carbonic acid that makes the environment acidic as a result of glycolysis. Drugs like doxorubicin, mitoxantrone, vincristine, and vinblastine which are weakly basic drugs undergo protonation in acidic pH, but the cell membrane is negatively charged which cause restrictions to the cellular influx of positively charged drugs [22]. In acidic extracellular pH, the level of heat shock protein [HSP]27 increases which can cause chemoresistance to cisplatin [23].

Lastly, off-target resistance includes all those indirect mechanisms which may lead to the inefficiency of the drug or adaptive response of the cells. Enhanced absorption and metabolic turnover of glucose and glutamine in an anaerobic direction is a feature of highly proliferating cancer cells. Higher expression of proteins involved in transportation and metabolism of glucose and glutamine is associated with chemoresistance and poor prognosis [24]. Destabilization of mitochondrial membrane potential [$\Delta\Psi M$], oxidative phosphorylation deregulation, and mutation or depletion of mitochondrial DNA are common features of drug resistant cases [25].

4.2.2 Multidrug Resistance

MDR is a process in which neoplastic cells which were initially responding to a single anticancer drug subsequently develop resistance to an array of architecturally and mechanistically unrelated drugs which may have multiple molecular targets. Acquired resistance develops simultaneously against almost every anticancer drug and may develop by multiple mechanisms. In ABC, subfamily B, member 1 [ABCB1/P-gp/MDR1] was responsible for failure of cancer management of liver, kidney, colon, lymphoma, and leukemia. P-gp is a 170 kDa surface glycoprotein which acts as energy dependent ABC transporter and functions as a drug efflux pump. In humans, there exists two members of P-gp gene family, known as MDR1 [widely distributed] and MDR3 [limited distribution]. At first, P-gp identifies the cytotoxic drug, followed by ATP binding and hydrolysis which provides the energy for efflux of the drug. Thus, they reduce the effectiveness of drugs like anticancer, antibiotic, and immunosuppressant [26]. In later years more members joined this group like ABCC1/MRP1 [caused chemoresistance in prostate, lung, and breast cancer], ABCG2/BCRP [evoked resistance in breast cancer and leukemia] and at present it has increased to a family of 49 ABC transporter genes arranged into seven subfamilies designated from A to G. A pivotal role of ABC transporters is prevention of entry of chemotherapeutic drugs into cancer cells by diffusion or active uptake [27]. Some of the important MDR-ABC transporters and their substrates [28] have been enlisted in Table 4.2.

Table 4.2 Important MDR-ABC transporters and their substrates

Important MDR-ABC transporters	Anticancer drugs [Substrates]
ABCB1/MDR1/P-gp	Paclitaxel, colchicine, vincristine
ABCC10/ MRP7	Paclitaxel, vincristine, gemcitabine
ABCG2/BCRP/MXR	Mitoxantrone, SN-38, doxorubicin
ABCC1/MRP1	Doxorubicin, topotecan, vincristine

A plethora of different mechanisms which may initiate and promote MDR [27, 29] are as follows:

1. Abnormal metabolism, absorption, and distribution of chemotherapeutic drugs to tumor site which may vary in different types of cancer.
2. Tumor tissues may be inaccessible to the drugs due to enhanced hydrostatic pressure and altered vasculature in tissues.
3. Heterogeneity of cancer inflicts different molecular signatures in subpopulation of cancer cells and affects sensitivity towards chemotherapy.
4. Retarded uptake of drugs, downregulation of drug influx due to changed surface receptors/carriers, aggravated drug efflux by drug transporters, increased elimination of drugs due to detoxification, inhibition of drugs via glutathione [GSH] mediated reduction, and sequestration of chemotherapeutic drugs in lysosomes or other intracellular organelles and intercellular vesicles.
5. Modification in lipid metabolism [ceramide pathway], deregulation of cell cycle, increased capacity of DNA repair in cancer cells, inability to undergo apoptosis mediated through robust expression of antiapoptotic genes in cancer cells caused by chromosomal abnormalities, and changed drug targets such as topoisomerase II.

MDR mechanisms have been divided into two major groups—pump resistance and non-pump resistance. Degradation of drug by lysosome, phase I and phase II enzymes mediated drug inactivation, antiapoptotic and antioxidant defence mechanisms, comprises some of the non-pump resistance mechanisms. Antiapoptotic defence mechanism is a major hurdle against anticancer drug-induced apoptosis. These mechanisms can lead to MDR either individually or synergistically. This forms the basis of the inefficiency of the treatment regimens that comprise of many chemotherapeutic agents with varied unrelated targets. To circumvent MDR many synthetic biological-response modifiers have been used, but most of them remain clinically unsuccessful due to severe side effects. Certain natural compounds with multiple bioactive moieties may be good reversal multi-functional agents but not as specific regulators of target proteins. These agents can antagonistically regulate the key factors contributing to MDR [30].

4.3 Strategies to Overcome Chemoresistance with Different Generations of Chemosensitizers

The riddle of chemoresistance has been attempted to be solved by researchers with different generations of chemosensitizers which were supposed to increase efficacy of the chemotherapeutic drugs in cancer cells which became chemoresistant. The chemosensitizers have been classified into different generations based on relative affinity, toxicity, and specificity.

4.3.1 First-Generation Chemosensitizers

First-generation chemosensitizers included the calcium channel blockers verapamil, immunosuppressive drugs such as cyclosporine A, and the antimalarial drug quinine. However, they were not effective due to non-specific toxicity and less affinity to the ABC transporter which in turn resulted in requirement of high doses for in vivo efficacy [14].

4.3.2 Second-Generation Chemosensitizers

These were synthetic analogues of first-generation chemosensitizers like dexverapamil and PSC833 [valspodar, modified from cyclosporine A]. They were successful in reversal of MDR in vitro, but imparted in vivo toxicity and inflicted drug–drug interaction via CYP450 suppression in clinical trials [14].

4.3.3 Third-Generation Chemosensitizers

The third-generation chemosensitizers, R1010933 [laniquidar], LY335979 [zosuquidar], GF120918 [elacridar], VX-710 [biricodar], and XR9576 [tariquidar], exhibited affinity for P-gp, low toxicity, and better functionality. However, clinical trials failed due to non-specific interactions with various types of ABC transporters [14].

4.4 Phytochemicals as a Respite against Drug Resistance

The hurdles encountered with three generations of chemosensitizers prompted the use of natural products as fourth-generation chemosensitizers [14, 31]. The multi-functional targeted approach and low toxicity of the natural products made them potent partners for adjunct therapy during chemotherapy.

4.4.1 Classes of Phytochemicals Active against Drug Resistance

A wide spectrum of phytochemicals belonging to alkaloids, flavonoids, phenylpropanoids, saponins, esters, phenols, terpenoids, ketones have been reported with MDR reversal efficacy [32]. A recent systematic review reported that the principal compounds which acted as chemosensitizers were phenolic derivatives and flavonoids [33]. It is not possible to cover all the compounds in a single chapter. So, we have restricted ourselves to discuss about some important phytochemicals with chemosensitizing property. The chemistry and sources of these natural compounds (Table 4.3) have been discussed in the following section.

Table 4.3 Chemistry and sources of phytochemicals used for chemosensitization

Phytochemical	Structure	IUPAC nomenclature	Source	Ref
Polyphenols				
Curcumin [Diferuloyl methane, flavonoid]	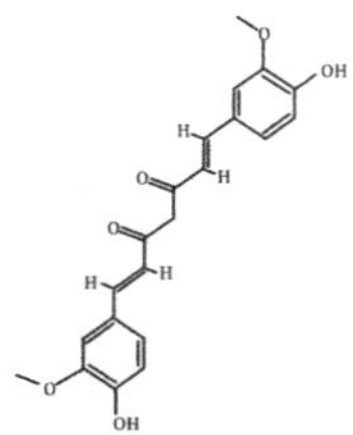	[1*E*,6*E*]-1,7-bis[4-hydroxy-3-methoxyphenyl] hepta-1,6-diene-3,5-dione	Rhizome of *Curcuma longa*	[34] [35]
Hispidin	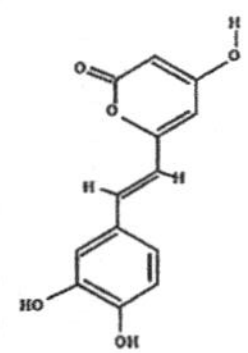	6-[[*E*]-2-[3,4-dihydroxyphenyl] ethenyl]-4-hydroxypyran-2-one	*Phellinus linteus*	[36] [37]
Resveratrol [Stilbenoid]	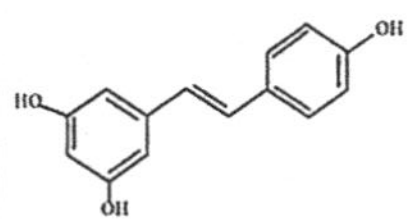	5-[[*E*]-2-[4-hydroxyphenyl] Ethenyl] benzene-1,3-diol	Grapes, peanuts, pistachios, grapes, red and white wine, blueberries, cranberries	[38] [39]
Scutellarin [Flavonolignan]	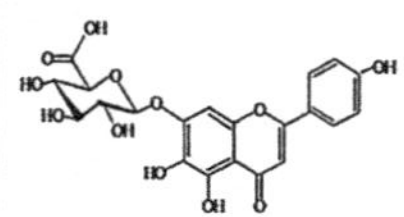	[2*S*,3*S*,4*S*,5*R*,6*S*]-6-[5,6-dihydroxy-2-[4-hydroxyphenyl]-4-oxochromen-7-yl] oxy-3,4,5-trihydroxyoxane-2-carboxylic acid	*Scutellaria barbata*	[40] [41]

Fisetin [flavone]	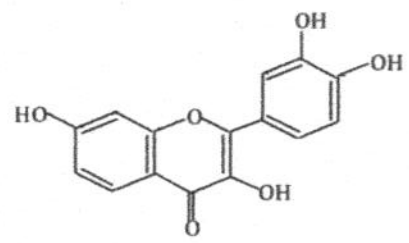	2-[3,4-dihydroxyphenyl]-3,7-dihydroxychromen-4-one	Fruits, vegetables, like strawberries, apples, persimmons, grapes, onions, cucumbers nuts, and wine	[42] [43]
Rutin [Flavonol]	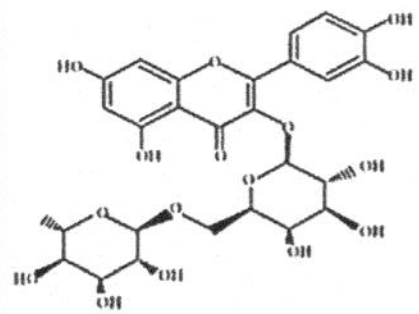	2-[3,4-dihydroxyphenyl]-5,7-dihydroxy-3-[[2*S*,3*R*,4*S*,5*S*,6*R*]-3,4,5-trihydroxy-6-[[[2*R*,3*R*,4*R*,5*R*,6*S*]-3,4,5-trihydroxy-6-methyloxan-2-yl]oxymethyl]oxan-2-yl]oxychromen-4-one	Apple, buckwheat, apricots, cherries, grapes, grapefruit, plums, and oranges	[44] [45]
Quercetin [Flavanol]	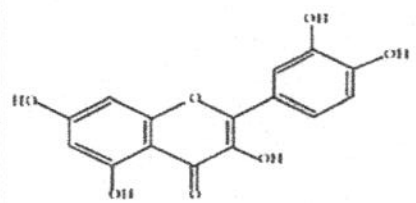	2-[3,4-dihydroxyphenyl]-3,5,7-trihydroxychromen-4-one	Leafy vegetables and fruits, such as capers, lovage, dill, cilantro, onions, various berries [choke-berries, cranberries and lingonberries] and apples	[46] [47]
Baicalein [flavone]	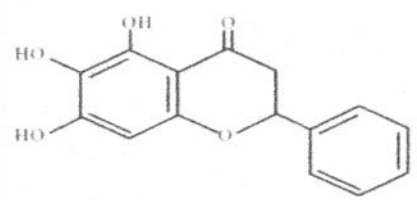	5,6,7-trihydroxy-2-phenylchromen-4-one	Roots of *Scutellariae radix, Scutellaria baicalensis, Scutellaria lateriflora,* and *Oroxylum indicum*	[48] [49]
Silibinin [Flavonolignan]	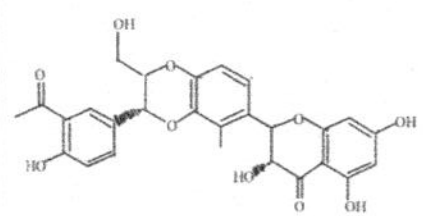	[2R,3R]-3,5,7-trihydroxy-2-[[2R,3R]-3-[4-hydroxy-3-methoxyphenyl]-2-[hydroxymethyl]-2,3-dihydro-1,4-benzodioxin-6-yl]-2,3-dihydrochromen-4-one	*Silybum marianum*	[50]

(continued)

Table 4.3 (continued)

Phytochemical	Structure	IUPAC nomenclature	Source	Ref
Myricetin [flavonoid]	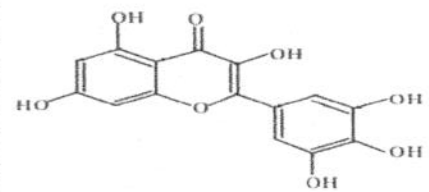	3,5,7-trihydroxy-2-[3,4,5-trihydroxyphenyl] chromen-4-one	Vegetables, fruits, nuts, berries, tea, and red wine	[51]
Mangiferin [Xanthonoid]	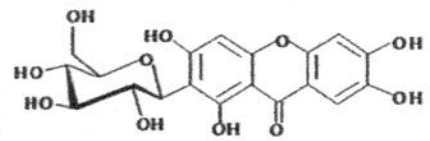	1,3,6,7-tetrahydroxy-2-[[2S,3R,4R,5S,6R]-3,4,5-trihydroxy-6-[hydroxymethyl]oxan-2-yl] xanthen-9-one	*Magnifera indica*	[52]
Xanthohumol [Prenylflavonoid]	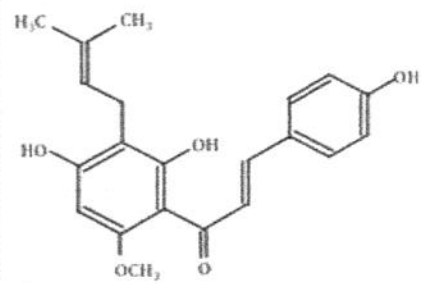	[E]-1-[2,4-dihydroxy-6-methoxy-3-[3-methylbut-2-enyl] phenyl]-3-[4-hydroxyphenyl] prop-2-en-1-one	Hop plant, *Humulus lupulus L.*	[53] [54]
EGCG [Gallate ester]	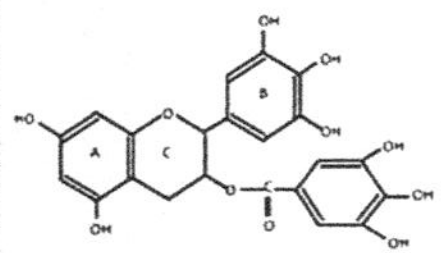	[[2*R*,3*R*]-5,7-dihydroxy-2-[3,4,5-trihydroxyphenyl]-3,4-dihydro-2*H*-chromen-3-yl] 3,4,5-trihydroxybenzoate	Tea leaves	[55] [56]
Chlorogenic acid [Ester]	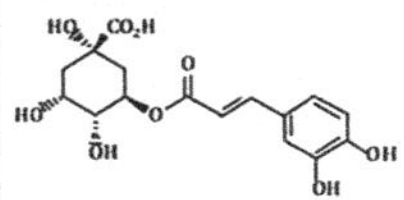	[1*S*,3*R*,4*R*,5*R*]-3-[[*E*]-3-[3,4-dihydroxyphenyl] prop-2-enoyl] oxy-1,4,5-trihydroxycyclohexane-1-carboxylic acid	Apples, artichoke, betel, burdock, carrots, coffee beans, eggplants, grapes, kiwi fruit, pears, plums, potatoes, tomatoes, coffee	[57] [58]
Nobiletin [Methoxyflavone]	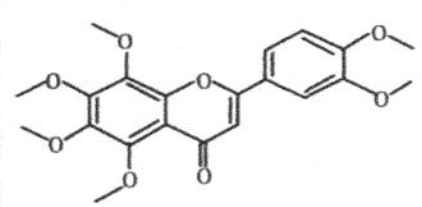	2-[3,4-dimethoxyphenyl]-5,6,7,8-tetramethoxychromen-4-one	Citrus fruits, peels of oranges [*Citrus sinensis*]	[59] [60]

Naringenin [Flavanone]	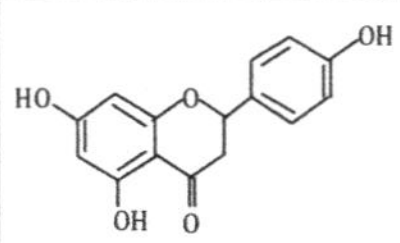	5,7-dihydroxy-2-[4-hydroxyphenyl]-2,3-dihydrochromen-4-one	Grapefruit [*Citrus paradisi*], oranges, tomato [*Solanum lycopersicum*], *Thymus vulgaris*	[61] [62]
Coumarins				
Psoralen [Furanocoumarin]	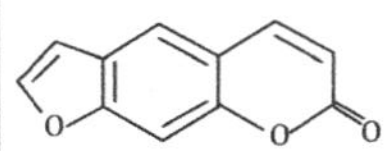	Furo[3,2-g] chromen-7-one	*Psoralea corylifolia*	[63] [64]
Cnidiadin [Furanocoumarin]	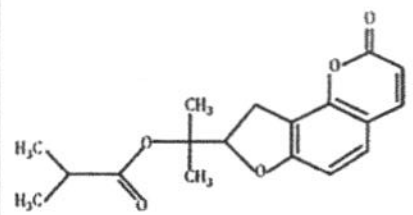	2-[[8S]-2-oxo-8,9-dihydrofuro[2,3-h] chromen-8-yl] propan-2-yl 2-methylpropanoate	*Tordylium apulum*	[65] [66]
Conferone [Sesquiterpene coumarin ether]	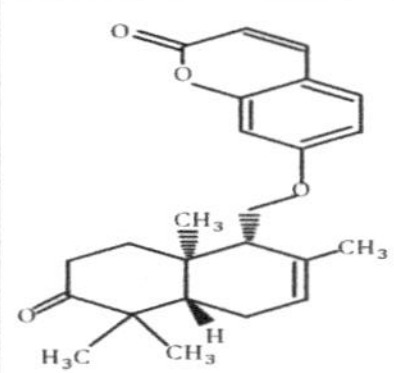	7-[[2,5,5,8*a*-tetramethyl-6-oxo-4,4*a*,7,8-tetrahydro-1*H*-naphthalen-1-yl] methoxy] chromen-2-one	Roots of *Ferula schtschurowskiana* and other species of *Ferula*	[67] [68]
Alkaloids				
Neferine	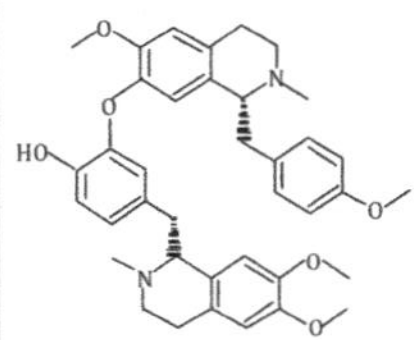	4-[[[1*R*]-6,7-dimethoxy-2-methyl-3,4-dihydro-1*H*-isoquinolin-1-yl]methyl]-2-[[[1*R*]-6-methoxy-1-[[4-methoxyphenyl]methyl]-2-methyl-3,4-dihydro-1*H*-isoquinolin-7-yl]oxy]phenol	Seed-embryos of *Nelumbo nucifera*	[69] [70]

(continued)

Table 4.3 (continued)

Phytochemical	Structure	IUPAC nomenclature	Source	Ref
Isoliensinine [Bis-benzyl isoquinoline alkaloids]	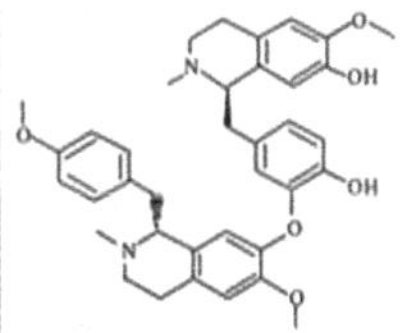	[1R]-1-[[4-hydroxy-3-[[[1R]-6-methoxy-1-[[4-methoxyphenyl]methyl]-2-methyl-3,4-dihydro-1H-isoquinolin-7-yl]oxy]phenyl]methyl]-6-methoxy-2-methyl-3,4-dihydro-1H-isoquinolin-7-ol		[71] [69]
Cernumidine [Guanidinic alkaloid]	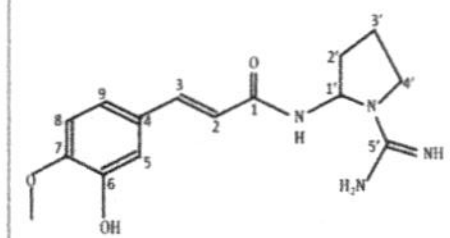	[*E*]-*N*-[1-carbamimidoylpyrrolidin-2-yl]-3-[3-hydroxy-4-methoxyphenyl]prop-2-enamide	Leaves of *Solanum cernuum Vell.* [Solanaceae]	[72] [73]
Dendrobine [Dendrobine-type alkaloid]	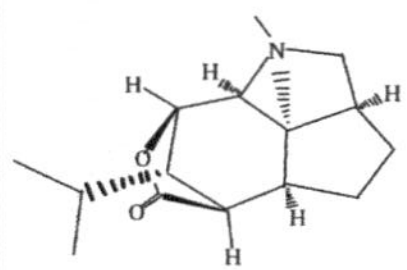	[1*S*,4*S*,7*S*,8*R*,11*R*,12*R*,13*S*]-2,12-dimethyl-13-propan-2-yl-10-oxa-2 azatetracyclo[5.4.1.$1^{8,11}$.$0^{4,12}$]tridecan-9-one	*Dendrobium nobile*	[74] [75]
Honokiol [biphenyl-type neolignane]	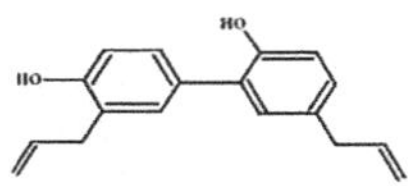	2-[4-hydroxy-3-prop-2-enylphenyl]-4-prop-2-enylphenol	Bark of *Magnolia species*	[76] [77]
Crebanine	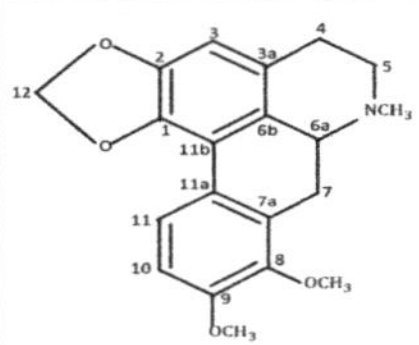	[12*R*]-15,16-dimethoxy-11-methyl-3,5-dioxa-11-azapentacyclo[10.7.1.$0^{2,6}$.$0^{8,20}$.$0^{14,19}$]icosa-1[20],2[6],7,14[19],15,17-hexaene	Tuber of *Stephania venosa*	[78] [79]

O-methylbulbocapnine [Aporphine alkaloids]	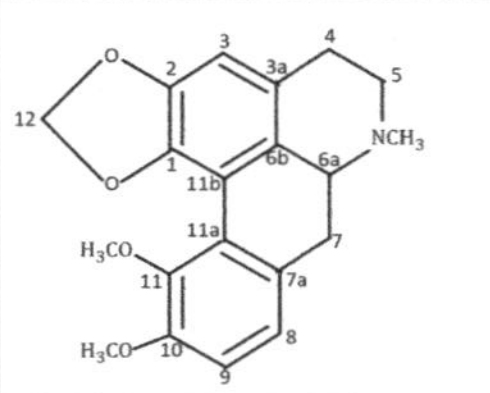	[12*S*]-17,18-dimethoxy-11-methyl-3,5-dioxa-11-azapentacyclo[10.7.1.$0^{2,6}$.$0^{8,20}$.$0^{14,19}$]icosa-1[20],2[6],7,14[19],15,17-hexaene		[80] [78]
Isostemofoline	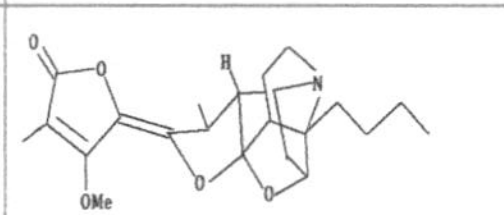	[5*E*]-5-[[1*S*,4*S*,5*S*,6*S*,8*S*,9*S*,13*R*]-9-butyl-4-methyl-2,14-dioxa-10-azapentacyclo[6.5.1.$0^{1,5}$.$0^{6,10}$.$0^{9,13}$]tetradecan-3-ylidene]-4-methoxy-3-methylfuran-2-one	*Stemona curtisii* and *Stichoneuron halabalesis*	[81]
11Z -didehydrostemofoline	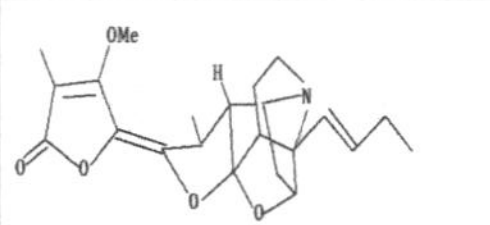	[5Z]-5-[[1S,4R]-9-[[E]-but-1-enyl]-4-methyl-2,14-dioxa-10 azapentacyclo [6.5.1.01,5.06,10.09,13] tetradecan-3-ylidene]-4-methoxy-3-methylfuran-2-one		
11E-didehydrostemofoline [Stemona alkaloids]	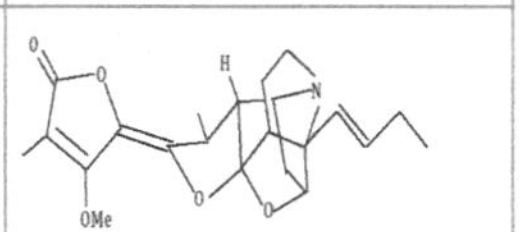			
Stemofoline [Stemona alkaloids]	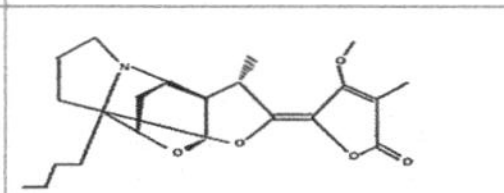	[5E]-5-[9-butyl-4-methyl-2,14-dioxa-10-azapentacyclo [6.5.1.01,5.06,10.09,13]tetradecan-3-ylidene]-4-methoxy-3-methylfuran-2-one	Roots of *Stemona aphylla* and *S. burkillii*	[81] [82]
Capsaicin [Capsinoid]	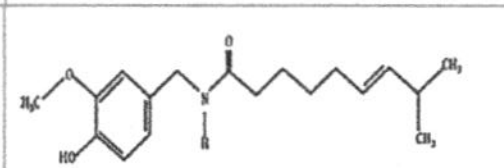	[E]-N-[[4-hydroxy-3-methoxyphenyl] methyl]-8-methylnon-6-enamide	*Capsicum frutescens*	[83] [84]
Piperine [Piperidine alkaloid]	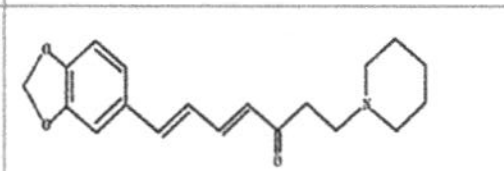	[2E,4E]-5-[1,3-benzodioxol-5-yl]-1-piperidin-1-ylpenta-2,4-dien-1-one	*Piper nigrum*	[85] [86]

(continued)

Table 4.3 (continued)

Phytochemical	Structure	IUPAC nomenclature	Source	Ref
Piperlongumine	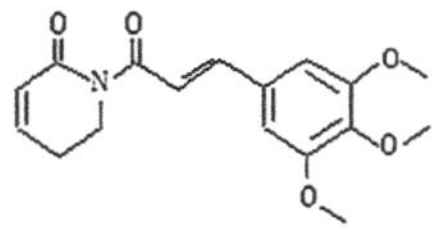	1-[[*E*]-3-[3,4,5-trimethoxyphenyl] prop-2-enoyl]-2,3-dihydropyridin-6-one	*Piper longum*	[87] [86]
Tetrandrine [Bis-benzylisoquino-line alkaloid]	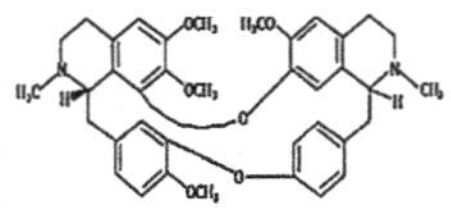	[1S,14S]-9,20,21,25-tetramethoxy-15,30-dimethyl-7,23-dioxa-15,30-diazaheptacyclo [22.6.2.23,6.18,12.114,18.027,31.022,33]hexatriaconta-3 [36],4,6[35],8,10,12[34],18,20,22[33],24,26,31-dodecaene	Root of *Stephania tetrandra S. Moore*	[88] [89]
Isotetrandrine	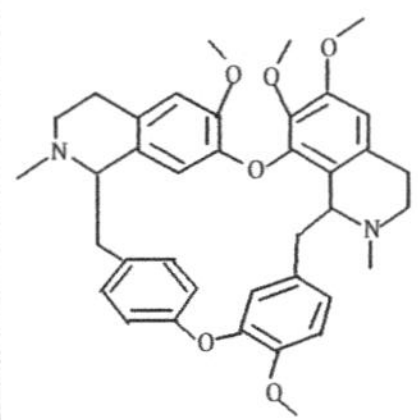	9,20,21,25-tetramethoxy-15,30-dimethyl-7,23-dioxa-15,30-diazaheptacyclo [$22.6.2.2^{3,6}.1^{8,12}.1^{14,18}.0^{27,31}.0^{22,33}$]hexatriaconta-3 [36],4,6[35],8,10,12[34],18,20,22[33],24,26,31-dodecaene	*Caulis mahoniae*	[90] [91]
Terpenoids				
Borneol [Monoterpenoid]	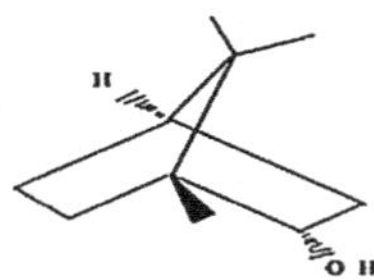	1,7,7-trimethylbicyclo[2.2.1]heptan-2-ol	Essential oil of medicinal plants like valerian, chamomile, and lavender	[92] [93]

Thymoquinone [Monoterpenoid]	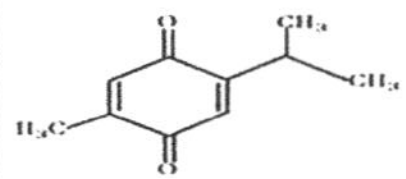	2-methyl-5-propan-2-ylcyclohexa-2,5-diene-1,4-dione	*Nigella sativa*	[94] [95]
Neoandrographolide [Diterpene glycoside]	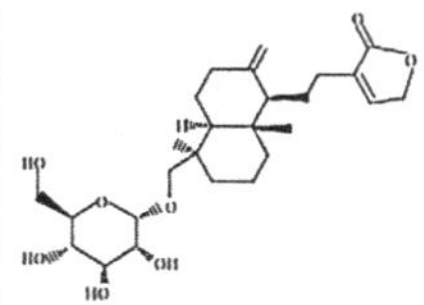	4-[2-[[1R,4aS,5R,8aS]-5,8a-dimethyl-2-methylidene-5-[[[2R,3R,4S,5S,6R]-3,4,5-trihydroxy-6-[hydroxymethyl] oxan-2-yl]oxymethyl]-3,4,4a,6,7,8-hexahydro-1H-naphthalen-1-yl]ethyl]-2H-furan-5-one	*Andrographis paniculata*	[96] [97]
Saikosaponin D [Triterpenoid saponin]	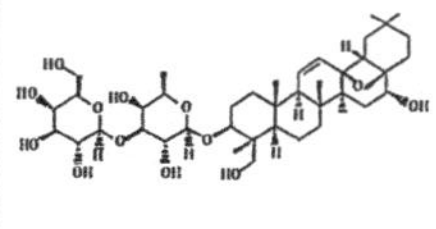	[2*S*,3*R*,4*S*,5*S*,6*R*]-2-[[2*R*,3*R*,4*S*,5*S*,6*R*]-3,5-dihydroxy-2-[[[1*S*,2*R*,4*S*,5*R*,8*R*,9*R*,10*S*,13*S*,14*R*,17*S*,18*R*]-2-hydroxy-9-[hydroxymethyl]-4,5,9,13,20,20-hexamethyl-24-oxahexacyclo[15.5.2.$0^{1,18}$.$0^{4,17}$.$0^{5,14}$.$0^{8,13}$]tetracos-15-en-10-yl]oxy]-6-methyloxan-4-yl]oxy-6-[hydroxymethyl] oxane-3,4,5-triol	*Bupleurum chinense DC*	[98] [99]
β-Carotene [carotenoid [Tetraterpenoid]]	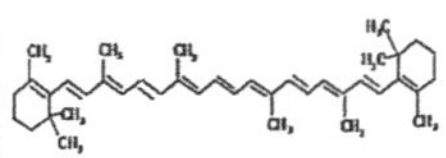	1,3,3-trimethyl-2-[[1*E*,3*E*,5*E*,7*E*,9*E*,11*E*,13*E*,15*E*,17*E*]-3,7,12,16-tetramethyl-18-[2,6,6-trimethylcyclohexen-1-yl]octadeca-1,3,5,7,9,11,13,15,17-nonaenyl]cyclohexene	• carrots, sweet potatoes, dark leafy greens-kale and spinach, romaine lettuce, squash, cantaloupe, red and yellow peppers, apricots, etc.	• [100] [101]
Alantolactone [Sesquiterpene lactone]	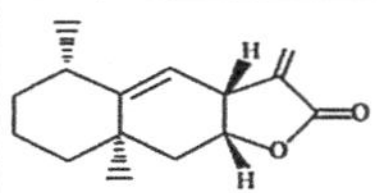	[3aR,5S,8aR,9aR]-5,8a-dimethyl-3-methylidene-5,6,7,8,9,9a-hexahydro-3aH-benzo[f][1]benzofuran-2-one	*Inula helenium*	[102] [103]

(continued)

Table 4.3 (continued)

Phytochemical	Structure	IUPAC nomenclature	Source	Ref
Marine phytochemicals				
Parguerenes I	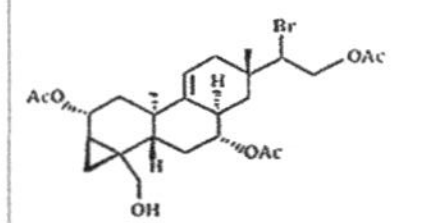	15-bromoparguer-9[11]-ene-2,7,16,19- tetrol-2,7,16-triacetate	Red alga *Laurencia filiformis*	[104]
Parguerenes II [Bromoditerpenes]	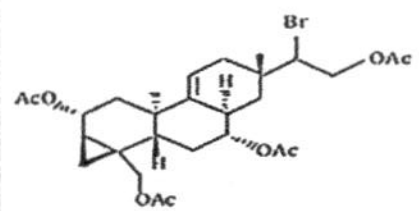	Parguerene II- 15-bromoparguer9[11]-ene-2,7,16,19-tetrol-2,7,16,19-tetraacetate		
Others				
Gambogic acid [Xanthonoid]	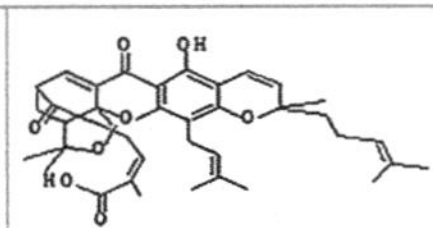	[Z]-4-[[1S,2S,8R,17S,19R]-12-hydroxy-8,21,21-trimethyl-5-[3-methylbut-2-enyl]-8-[4-methylpent-3-enyl]-14,18-dioxo-3,7,20-trioxahexacyclo [15.4.1.02,15.02,19.04,13.06,11]docosa-4[13],5, 9,11,15-pentaen-19-yl]-2-methylbut-2-enoic acid	Gamboge, a dry resin secreted from the *Garcinia hanburyi*	[105] [106]
Sulforaphane [Organosulfur compound]	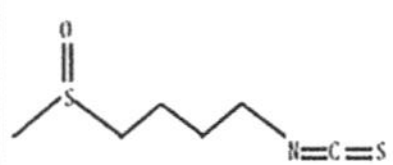	1-isothiocyanato-4-methylsulfinylbutane	Broccoli	[107] [108]
Cryptotanshinone [Tanshinones]	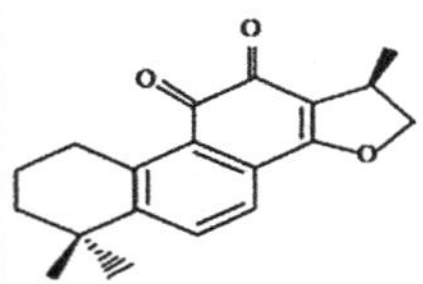	[1*R*]-1,6,6-trimethyl-2,7,8,9-tetrahydro-1*H*-naphtho [1,2-g][1]benzofuran-10,11-dione	Roots of *Salvia miltiorrhiza Bunge* [Danshen]	[109] [110]

Anethole [aromatic unsaturated ether]	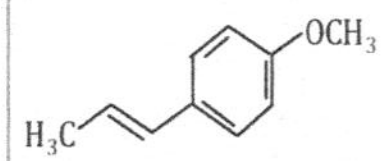	1-methoxy-4-[[*E*]-prop-1-enyl] benzene	Anise oil, fennel oil, and camphor	[111] [112]
Digitonin [steroidal saponin]	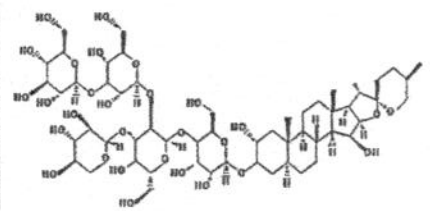	[2S,3R,4S,5S,6R]-2-[[2S,3R,4S,5S,6R]-2-[[2S,3R,4S,5R,6R]-2-[[2R,3R,4R,5R,6R]-6-[[1R,2S,3S,4R,5'R,6R,7S,8R,9S,12S,13S,15R,16R,18S]-3,15-dihydroxy-5′,7,9,13-tetramethylspiro [5-oxapentacyclo[10.8.0.02,9.04,8.013,18]icosane-6,2′-oxane]-16-yl]oxy-4,5-dihydroxy-2-[hydroxymethyl] oxan-3-yl]oxy-5-hydroxy-6-[hydroxymethyl]-4-[[2S,3R,4S,5R]-3,4,5-trihydroxyoxan-2-yl]oxyoxan-3-yl] oxy-3,5-dihydroxy-6-[hydroxymethyl]oxan-4-yl]oxy-6-[hydroxymethyl]oxane-3,4,5-triol	Foxglove plant, *Digitalis purpurea*	[113, 114]

4.4.1.1 Polyphenols

Curcumin [diferuloylmethane], the principal constituent of the spice turmeric is isolated from the rhizomatous part of *Curcuma longa* plant, belonging to the Zingiberaceae [ginger] family. It is endowed with a wide variety of pharmacological properties—antioxidant, antibacterial, antiviral, antifungal, anti- inflammatory, hepatoprotective, and also anti-tumor activity. It is used as a traditional medicinal herb in pan Asia [34]. Hispidin [member of phenols and 2-pyranones] a polyphenol, isolated mainly from a medicinal mushroom, *Phellinus linteus* is widely used in Africa and east Asia, has been associated with anticancer, antioxidant, and DNA-damage protection properties [36]. Resveratrol [trans-3, 4′, 5-trihydroxystilbene] is principally derived from peanuts, grapes, pistachios, red and white wine, cranberries, blueberries, and other fruits. It is present in extracts of more than seventy plant species and has been attributed with a wide array of pro-health effects including anticancer properties. The presence of the 4-hydroxyl group, m-hydroquinone moieties, of the phenol ring mediates the pharmacological properties of resveratrol [38]. Scutellarin[4,5,6-trihydroxyflavone-7-glucuronide] is a flavonoid from traditional Chinese medicinal herb, *Scutellaria barbata* have been found to impart synergism with chemotherapeutic drugs against resistant cancer cells [41]. It has been intertwined with many other health promoting properties including ROS-scavenging, anti-inflammatory, and cardio-vascular protective function [115]. Fisetin [3,3,4,7-tetrahydroxyflavone], another flavonoid is available in fruits, vegetables, like apples, grapes, strawberries, cucumbers, onions, nuts, and wine. Fisetin also exhibited anti-oxidative, anti-inflammatory, cytotoxic, pro-apoptotic, anti-tumorigenic, and neuro-protective efficacy in a number of studies [42, 116]. Rutin, a glycoside belonging to flavanol group of flavonoids is found in apricots, apple, plums, cherries, grapes, and oranges. Rutin is formed by combination of disaccharide rutinose with a flavonolic aglycone quercetin. It has been reported with antioxidant, chemopreventive, anti-tumor, neuroprotective, and cardioprotective properties [45]. Quercetin is secondary polyphenolic metabolite that belongs to the flavonol class and is largely found in various vegetables and fruits, such as dill, capers, cilantro, apples, various berries [e.g., chokeberries, blueberries, and cranberries,], and onions. It has a C6-C3-C6 carbon framework with a benzo-[γ]-pyrone skeletal structure and A and B benzene rings, connected by a C three-carbon pyrone ring. Quercetin is known as pentahydroxy flavonol because it possesses five hydroxyl groups on its flavonol skeletal framework at 3, 3′, 4′, 5, and 7 carbons. The pharmacological characteristics of quercetin including anticancer effect are attributed to the extent of substitution of the different functional moieties on the flavonol molecule [117]. Baicalein is a flavone, extracted from the roots of *Scutellariae radix, Scutellaria baicalensis, Scutellaria lateriflora, and Oroxylum indicum,* the traditional Chinese medicinal herbs. Baicalein has a di-orthohydroxyl group on its ring-A [48]. It was abundantly used in ancient China, and has been attributed with antibacterial, antiviral, anti-inflammatory, antioxidant, anti-tumor, neuroprotective, and cardioprotective efficacy [118]. Silibinin [flavonolignan], a flavonoid isolated from *Silybum marianum,* was used as a folk medicine and has been found effective against liver disorders, poisoning by death cap mushroom [*Amanita phalloides*] as well as elicited antioxidant and anticancer activity [50]. Myricetin is a flavonoid produced by plants of Myricaceae, Anacardiaceae, Polygonaceae, Pinaceae, and Primulaceae families. It is abundantly found in tea leaves, berries, and wines. It has both free and glycosidically bound forms. Due to similarity with quercetin it is also known as hydroxy quercetin. It is endowed with pharmacological properties—anti-inflammatory, analgesic, antitumor, hepatoprotective, and antidiabetic effects [51]. Mangiferin [xanthonoid] is a polyphenolic C-glycosylxanthone structure compound, found in many herbal sources, including mango tree [*Mangifera indica*]. Mangiferin has a wide pharmacological profile—antibacterial, antifungal, anthelmintic, antiparasitic, antitumor, and many other health beneficiary properties [52]. Xanthohumol [prenylated flavonoid], a chalcone extracted from *Humulus lupulus*, elicited growth inhibitory effects against cancers of colon, prostate, ovarian, breast, and white blood cells [53]. The prime green tea catechin, [−]-epigallocatechin-3-gallate (EGCG) has been evidenced with antioxidant effects, chemopreventive role, cardio-protective efficacy, and others [55]. Chlorogenic acid, a phenolic derivative of the hydroxycinnamic acid, comprises of a caffeic acid and a quinic acid. It is found in apples, carrots, betel, burdock, coffee seeds, eggplants, grapes, potatoes kiwi fruit, plums, tomatoes, pears, etc. It has been evidenced with anti-oxidant, anti-inflammatory, anticancer, antidiuretic, and many other health promoting properties [57]. Nobiletin [polymethoxylated flavone] is found in *Citrus depressa* and *Citrus sinensis* [oranges]. It was associated with a wide range of pro-health effects including anti-inflammatory, anti-tumor, neuroprotective, and chemopreventive effects [119]. Naringenin [4′,5,7-trihydroxyflavanone] is the aglycone form of naringin [naringenin-7-rhamnoglucoside] which is found as narirutin [naringenin- 7-O-rutinoside] or naringenin-glucoside [naringenin-7-O-glucoside], depending on the sugar motive. It is found in citrus fruits like grapefruit [*Citrus paradisi*], oranges, and in tomato [*Solanum lycopersicum*]. Its structure imparts lipophilic ability for easy cellular absorption and facilitates reversal of MDR [61].

4.4.1.2 Coumarins

Psoralen [furanocoumarin] is obtained from Chinese medicinal herb, *Psoralea corylifolia*, and has been reported with antineoplastic properties against various cancers including leukemia [63].Conferone was the first among sesquiterpene coumarins to be isolated from *Ferula schtschurowskiana* [67]. Sesquiterpene coumarins are found in plants of the family *Ferula* [Apiaceae] with a geographical distribution from the Mediterranean Sea to Central Asia and in some plants of the families- Asteraceae and Rutaceae. Cnidiadin [furanocoumarin] obtained from dried leaves of traditional Chinese medicine and Greek spice, *Tordylium apulum* [65] has been evidenced with MDR reversal ability in experimental models.

4.4.1.3 Alkaloids

Neferine and isoliensinine are bis-benzylisoquinoline alkaloid which are extracted from embryo of lotus *Nelumbo nucifera.* It consists of an isoquinoline attached to a benzyl group. Neferine along with chemosensitizing properties also exerted anti-angiotensive, anti-arrhythmic, anti-thrombotic, cholinesterase inhibition, and anti-diabetes effects [120]. Isoliensinine exerted anticancer, anti-diabetic, and anti-pulmonary fibrosis potential [69]. Cernumidine is guanidinic alkaloid in nature which is isolated from *Solanum cernuum* leaves that belongs to the family Solanaceae. It is a Brazilian herb which wildly grows in Rio de Janeiro and Minas Gerais. It may be used to treat many diseases like liver damage, skin infections, and chemosensitizing effects [121]. Dendrobine is an alkaloid isolated from *Dendrobium nobile,* belonging to the Orchidaceae family. It enhanced immunity and exhibited anticancer activities against ovarian cancer, promyelocytic leukemia, and sarcoma [74]. Piperine, a piperidine alkaloid extracted from a common spice pepper *[Piper nigrum]* was found to exhibit immunomodulatory, anti-carcinogenic, stimulatory, hepatoprotective, and anti-inflammatory effects [122]. Honokiol, a traditional Chinese herbal medicine, is a biphenyl neolignan isolated from the bark, seed, and cone of *Magnolia species.* It elicited apoptotis, cell cycle arrest and autophagy against cancer cells [123]. Crebanine and O-methylbulbocapnine are aporphine alkaloid in nature, isolated from the tube of *Stephania venosa* [78]. It was attributed with antiarrhythmic, antimicrobial, anticancer, and multidrug resistance reversal activities [124]. Isostemofoline, a stemona alkaloid, extracted from *Stemona curtisii* grown in Malaysia has been evidenced with MDR reversal property [81]. Stemofoline, another stemona alkaloid isolated from *Stemona* of family Stemonaceae, is also native to southeastern Asia and is used for its biological and medicinal properties. It has been reported to have anti-carcinogenic, anti-inflammatory, and anti-proliferative activities [125].Capsaicin, a capsinoid alkaloid, is derived from the genus *Capsicum* belonging to the family Vanilloid. It consists of a benzene ring associated with a long hydrophobic carbon tail with a polar amide moiety and has the capacity of reversing MDR [126]. Piperlongumine, an amide alkaloid isolated from a medicinal plant *Piper longum* L. [long piper], acts as a potential anticancer agent. The structure consists of two Michael acceptors – 2,3- and 7,8- unsaturated bonds. It also has proapoptotic, anti-invasive, and antiangiogenic properties [127]. Isotetrandrine, an isoquinoline alkaloid, isolated from *Caulis mahoniae* exhibited MDR reversal characteristic property [128]. It ameliorated oxidative stress in liver cancer cells by upregulating antioxidant defence mechanism [129]. Tetrandrine is a bis-benzylisoquinoline alkaloid, found as an active constituent in the roots of *Stephania tetrandra and S. moore.* Tetrandrine exhibited immunomodulatory, proapoptotic, anti-hepato fibrogenetic activity and also has the ability of reversing MDR [130].

4.4.1.4 Terpenoids

Borneol is a terpenoid available in the essential oils of many medicinal plants, such as lavender and chamomile and has diverse uses in food and drug industries. Borneol contains anti-inflammatory and chemosensitizing properties [92]. Thymoquinone is a monoterpenoid in nature and isolated from *Nigella sativa.* It also exhibits anti-inflammatory and anti-proliferative properties of tumor cells and angiogenesis [95]. Neoandrographolide, a diterpene glycoside is extracted from the traditional medicinal herb *Andrographis paniculata.* It was known to have chemosensitizing, anti-inflammatory, and immunosuppressive properties [97].Saikosaponin D, another terpenoid saponin is a derivative of *Bupleurum chinense* DC. It was evidenced with a potential of reversing MDR, anti-inflammatory, anti-tumor, anti-infectious, and other pharmacological properties [98]. β-carotene is a tetraterpenoid extracted from carrots, spinach, romaine lettuce, sweet potatoes, squash, dark leafy greens-kale, cantaloupe, red and yellow peppers, which have the potential of reversing MDR [100]. Alantolactone is a sesquiterpene lactone which is a natural compound extracted from *Inula helenium,* which has the properties of inhibiting inflammatory process and inducing apoptosis in tumor cells [102].

4.4.1.5 Marine Phytochemicals

Parguerenes I and II are extracted from a red alga, *Laurencia filiformis* from southern Australian collection of marine algae. It is brominated diterpene in nature and has various pharmacological properties like antitumor, antimicrobial, and antiviral [104].

4.4.1.6 Others

Gamboic acid, a xanthonoid, extracted from a dry resin [gamboge] from the tree *Garcinia hanburyi,* found in Southeast part of Asia. It possess anticancer, pro-apoptotic, and chemosensitizing properties [105]. Sulforaphane is an isothiocyanate and isolated from cleaved product of glucoraphanin. It has been extracted from broccoli, cabbage, and cauliflower of cruciferous family. It was found with potent anticancer, anti-inflammatory, and chemosensitizing property against in many cancers [107]. Cryptotanshinone is obtained from the roots of herbal medicine *Salvia miltiorrhiza* Bge which belong to lamiaceae plants of China. It exhibited many pharmacological effects like anti-inflammatory, anti-oxidative, and anti-tumorigenic [109]. Anethole, an aromatic unsaturated ether is obtained from anise oil, fennel oil, and camphor. It was associated with anti-inflammatory, anti-carcinogenic, and anti-oxidative properties [111]. Digitonin, a steroidal saponin is extracted from the foxglove plant *Digitalis purpurea*. It has been found as a reversible membrane permeabilization agent in mammalian cells [114].

4.4.2 Mechanisms Underpinning Chemosensitization by Different Groups of Phytochemicals

A myriad of phytochemicals are found to resensitize resistant malignant cells towards anticancer drugs by varied mechanisms. Different classes of herbal compounds have been investigated in preclinical models for their role in adjunct therapy with chemotherapy (Table 4.4). The major mechanisms by which the phytochemicals imparted sensitization of chemotherapeutic drugs in non-responding cancer cells have been discussed in the following section.

Table 4.4 Mechanisms underpinning sensitization of chemotherapeutic drugs by phytochemicals

Phyto-chemical	Resistant drug	Model	Effect	Mechanism	Ref
Polyphenols					
Curcumin	Cisplatin	In vitro: Leiomyosarcoma cells [LMC] [isolated from Wistar rats]- MRC-5 In vivo: Female Wistar rats	Apoptosis, drug dose minimization, impairment of mitochondrial function, in vitro sensitization of cisplatin, cisplatin induced nephrotoxicity in vivo	↑S phase arrest, ↑mitochondrial damage, no change in MDA	[34]
	Doxorubicin	Breast cancer cells -MCF-7/DOX, MDA-MD-231/DOX [doxorubicin resistant]	Reversal of doxorubicin resistance	↓ATPase activity of ABCB4	[131]
	Irinotecan	Colon cancer cells-LoVo [drug sensitive], LoVo/CPT-11 [irinotecan resistant]	Reduction of cancer stemness, induction apoptosis, sensitization of irinotecan	↓CD44, ↓CD24, ↓CD133, ↓EpCAM, ↓Bcl-2, ↑Bax, ↑cleaved caspase3, ↑caspase 9	[132]
	Carboplatin	Lung cancer cells- A549	Inhibition of tumor cell growth, migration and invasion, induction of apoptosis, chemosensitization of carboplatin	↓MMP-2, ↓MMP-9, ↑caspase-3, ↑caspase-9, ↓Bcl-2, ↓NF-κB, ↓Akt/IKKα pathway, ↑ERK1/2	[133]
	Cytarabine	Primary AML cells	Sensitization of cytarabine, synergistic effect in combination and drug dose minimization	↓MDR1, ↓LRP, ↓BCRP and ↓FLT3	[134]
	Cisplatin	A549, A549/DDP [cisplatin resistant]	Inhibition of cell proliferation, induction of apoptosis, synergism with cisplatin	↓Fanconi anemia [FA]/BRCA pathway, ↓ FA complementation group D2 [$FANCD_2$] mono ubiquitination and nuclear foci	[135]

	Cisplatin	Human colon cancer cells- HCT116 [mismatch repair deficient], HCT116R, HCT116 + ch3 [cisplatin resistant]	Disintegration of colonospheres, apoptosis, reduction of cancer stemness, sensitization of cisplatin	↓CD133, ↓CD44, ↓ALDH1	[136]
EGCG	Etoposide	Lung cancer cells- A549, NCI-H23, normal lung cells- BEAS-2B	Chemosensitization towards etoposide	↓Nrf2, ↑Keap1, ↓MRP-1, ↑p53, ↓pERK1/2, ↓PARP1, ↓XRCC1	[137]
	Doxorubicin	In vitro: Human multidrug resistant carcinoma cell line [KB-A1] and a drug-sensitive cell line [KB-3-1] In vivo: Nude mice transplanted with KBA-1 cells	Chemosensitization towards doxorubicin	Structural alteration of P-gp, ↓P-gp activity	[138]
Hispidin	Gemcitabine	Human pancreatic cancer cell lines- BxPC-3 and AsPC1 and BxPC-3 $CD44^{+}$CSC	Apoptosis, reduction of cancer stemness, sensitization towards gemcitabine	↓Bcl-2, ↓NF-κB, ↑cleaved caspase 3, ↓CD44, ↓Sox-2, ↓Nanog	[36]
Resveratrol	Cisplatin	Breast cancer cell line- MCF-7R [cisplatin resistant]	Sensitization of cisplatin, induction of apoptosis	↑p53ser20, ↓p53ser15, ↓p53ser46, ↓RAD51, ↓Bcl-2, ↑Bax, ↑CK1, ↑CHK2, ↑AMPK, ↓ATMser1981	[38]
	5-FU	Colon cancer cell lines- DLD-1, SW480 and COLO201, DLD-1/5FU [5-FU-resistant], DLD-1/OXA [OXA-resistant]	Induction of apoptosis and sensitization of 5-FU	↓PI3K/Akt, ↓MAPK/ERK1/2, ↑miR34a, ↓E2F3, ↓Sirt 1	[139]
	5-FU	Human colorectal carcinoma cell lines- HT-29 and SW-620	Reduction of cell survivability, imbalance in cellular antioxidant activities	↑ROS, ↑lipid peroxidase, ↓STAT3, ↓AKT	[140]

(continued)

Table 4.4 (continued)

Phyto-chemical	Resistant drug	Model	Effect	Mechanism	Ref
Scutellarin	Cisplatin	Ovarian cancer cells OVCAR-3 and SKOV-3	Sensitization of cisplatin, DNA damage, induction of apoptosis	↑platinum DNA adducts, ↑Bax, ↓Bcl-2, ↓PARP, ↑cleaved caspase 3	[41]
Fisetin	Etoposide	Osteosarcoma cell lines -MG-63, Saos-2, and U2OS	Inhibition of cell proliferation, cell cycle arrest	↑G2 phase cell cycle arrest, ↓ cyclin B1, ↓cyclin E1	[141]
	Oxaliplatin Irinotecan	In vitro: LoVo [parent cell], OR [Oxaliplatin-resistant LoVo cell], CPT11 [Irinotecan-resistant LoVo cell] In vivo: Male nude mice inoculated with the above colon cancer cells	Induction of apoptosis	↑cleaved caspase 3, ↑caspase 8, ↑cyt C, ↓Bcl-2, ↓pIGFIR, ↓PI3K/Akt	[116]
	Sorafenib	In vitro: HeLa In vivo: Balb/c females injected with HeLa cells	Induction of apoptosis	↑cleaved caspase 3, ↑cleaved caspase 8, ↑cleaved PARP, Bax/Bcl-2, ↑DR-5	[142]
	Sorafenib	In vitro: Melanoma cells-A375 In vivo: Nude mice injected with A375 cells or SK-MEL-28 cells	Reduction of BRAF-mutated tumor cells, induction of apoptosis	↑cleaved caspase 3, ↑Bax, ↑Bak, ↑cleaved PARP,↓Bcl-2, ↓ Mcl-1, ↓PI3K/Akt/mTOR, ↓MEK, ↓ERK	[42]
Nobiletin	Adriamycin	A549	Enhanced apoptosis, chemosensitization towards adriamycin	↓MYCN, ↓MRP1, ↓Akt, ↓GSK3β, and ↓β-catenin. ↑caspase3, ↑PARP cleavage	[143]
Rutin	Cyclo-phosphamide, methotrexate	Breast cancer cell lines-MDA-MB-231, MCF-7, primary human mammary fibroblasts	Apoptosis, cell cycle arrest, sensitization of cyclo-phosphamide and methotrexate	↑G0/G1 and G2/M cell cycle arrest, ↓P-gp, ↓BCRP	[144]

Baicalein	Cisplatin	In vitro: A549, A549/CDDP In vivo: A549 xenograft in Balb/c-nu mice	Suppression of EMT, sensitization of cisplatin	↑E-cadherin, ↓vimentin, ↓PI3K/Akt/ NF-κB	[118]
Silibinin	Doxorubicin Paclitaxel	Breast cancer cells-MDA-MB-435, MCF-7 [drug sensitive], MDA-MB-435/Dox [doxorubicin resistant], MCF-7/Pac [paclitaxel resistant]	Inhibition of cell growth, apoptosis	↓pSTAT3, ↓pERK, ↓pAkt, ↑cleaved caspase-3, ↓survivin	[145]
	Cisplatin, 5-FU	In vitro: Esophageal squamous cell carcinoma cells-KYSE270, KYSE510, T.tn and NE2-hTERT [normal esophageal cell line] In vivo: KYSE270 xenograft in nude mice; Human Esophageal cancer tumor and adjacent normal tissues	Induction of apoptosis	↓Bcl2, ↑Bax, ↑cleaved caspase 3, ↑pAMPK	[146]
Chlorogenic acid	5-FU	Human HCC cell lines-HepG2, Hep3B	Inhibition of cell proliferation, chemo-sensitization of 5-FU	↓ERK1/2, ↑ROS	[147]
Naringenin	DNA damaging drugs	Human colorectal cancer cell lines -SW1116 and SW837, human breast cancer cell lines -HTB26, HTB132, and normal human fibroblast cells-CRL1554	Apoptosis, cell-cycle arrest, potentiation of DNA damaging drugs	↓PI3K/Akt signaling pathway, ↓NF-κB, ↓Cdk4, ↓Cdk6, ↓Cdk7, ↑p18, ↑p19, ↑p21, ↑Caspases-3, 7, 8 and 9, ↑Bak, ↑AIF and ↑Bax., ↓Bcl2, ↓x-IAP and ↓c-IAP-2	[148]
Myricetin	Cisplatin	Ovarian cancer cell lines-OVCAR-3 and A2780/CP70 [Cisplatin resistant] and normal ovarian cells IOSE-364	Induction of apoptosis	↑p53, ↑p21, ↓c-myc, ↓Bcl2, ↓Bcl-xL, ↑Bad, ↑Bax	[149]

(continued)

Table 4.4 (continued)

Phyto-chemical	Resistant drug	Model	Effect	Mechanism	Ref
	5-FU	In vitro: Human esophageal squamous carcinoma cell line-EC9706 In vivo: Balb/c nude mice injected with EC9706cells	Apoptosis, chemosensitization to 5-FU	↓CyclinD1, ↓Survivin, ↓Bcl-2, ↑Caspase-3, ↑p53	[150]
Mangiferin	Doxorubicin	MCF-7	Reduced cell viability and evoked sensitization towards doxorubicin	↓P-gp	[52]
Xanthohumol	Adriamycin, Cisplatin	In vitro: ALL cell lines -L1210 mouse, ALL- PO, RS4, Nalm-6, 697 In vivo: B6D2F1 female mice injected with L1210 cells	Growth arrest, apoptosis, Reduction of extravasation and tissue invasiveness, Chemosensitization	↓Cyclin D1, ↑PARP, ↓FAK, ↓AKT, ↓NF-κB, ↑intracellular calcein accumulation, no over expression of MRP-1, P-gp	[53]
Coumarins					
Psoralen	Adriamycin	MCF-7[drug sensitive] and MCF-7/Adr [adriamycin resistant]	Reversal of MDR	↓P-gp activity	[63]
Conferone	Cisplatin	Human urinary bladder grade II carcinoma cells	Increased cytotoxicity of cisplatin	↑DNA damage	[151]
	Vinblastine	MDCK-MDR1,	Reversal of MDR1 multidrug resistance, alteration in bioavailability of P-gp substrates	↓P-gp, ↑drug uptake	[67]
Cnidiadin	Vinblastine, vincristine	MDCK-MDR1 [resistant canine kidney epithelial cells], KB/VCR cells [vincristine resistant human papilloma cells]	Reversal of MDR1 multidrug resistance, alteration in bioavailability of P-gp substrates	↓P-gp, ↑drug uptake	[65]

Alkaloids					
Cernumidine	Cisplatin	Urinary bladder carcinoma cells-T24	Inhibition of cell proliferation and migration, mitochondrial membrane depolarization, induction of apoptosis, sensitization towards cisplatin	↓EGFR, ↓pERK1/2, ↓MMP-9, ↓MMP-2, ↑Bcl-2	[121]
Dendrobine	Cisplatin	In vitro: A549 cells In vivo: Balb/c female mice inoculated with A549 cells	Induction of apoptosis	↑cytosolic cyt C and ↑AIF, ↑cleaved PARP, ↓pro caspase 3, ↑Bax, ↑Bim, ↑pJNK, ↑p38	[74]
Piperlongumine	Oxaliplatin	In vitro: Gastric cancer cells-SGC-7901, BGC-823, AGS and HCT116 In vivo: BALB/c nu/nu female mice injected with HCT116	Induction of apoptosis, sensitization of oxaliplatin	↑pJNK, ↑p-p38, ↑ROS, ↓Trx1	[152]
Piperine	Paclitaxel	KB-V1, KB-3-1	Sensitization of paclitaxel, apoptosis	↑Bax/Bcl-2, ↑cleaved PARP, ↑caspase 3, ↓pAkt, ↓Mcl-1	[153]
	Doxorubicin and mitoxantrone	Resistant cancer cell lines -MCF-7/DOX and A549/DDP	Sensitization towards doxorubicin and mitoxantrone	↓ABCB1, ↓ABCC1, ↓ABCG2	[154]
Honokiol	Cisplatin	Oral squamous cell carcinoma cells-SAS, OECM-1, GNM	Reduction of stemness, migration, invasion, colony formation capability, and potentiation of cisplatin	↓ALDH1, ↓CD44, ↓IL-6, ↓pSTAT 3	[155]
Neferine	Oxaliplatin	HCC cells -HepG2, Bel-7402, human normal liver cell line -L02	Cytotoxicity and apoptosis, inhibition of EMT-induced migration and invasion, Sensitization towards oxaliplatin	↑E-cadherin, ↓Vimentin, ↓snail, ↓N-cadherin	[120]

(continued)

Table 4.4 (continued)

Phyto-chemical	Resistant drug	Model	Effect	Mechanism	Ref
Stemofoline	Vinblastine, paclitaxel, doxorubicin	Cervical carcinoma cells-KB-V1, KB-3-1	Sensitization towards vinblastine, paclitaxel, and doxorubicin	↓P-gp	[125]
Isotetrandine	Doxorubicin	MCF-7/Dox	Reversal of doxorubicin resistance	↓P-gp activity	[91]
Tetrandrine	Doxorubicin	In vitro: Epidermoid carcinoma-KB and KBv200 cells In vivo: Nude mice injected with KBv200 cells	Sensitization of doxorubicin, apoptosis, inhibition of drug efflux	↓IκB-α, ↓Bcl-2,↓ P-gp activity	[156]
Saikosaponin D	Doxorubicin	MCF-7, MCF-7/Adr	Sensitization towards doxorubicin	↓P-gp drug efflux, ↓P-gp, ↓MDR-1	[98]
Terpenoids					
Alantolactone	Oxaliplatin	Human pancreatic cancer cell lines -MIA PaCa-2, PANC-1	Accumulation of autophagosomes due to impaired autophagic degradation, inhibition of cell proliferation, chemosensitization to oxaliplatin	↓Cathepsin B/D, ↓transcription factor EB	[102]
β-Carotene	Paclitaxel, etoposide, doxorubicin, 5FU, mitoxantrone	HeLaS3 and NCI-H460 [sensitive cancer cells] and KB-vin and NCI-H460/MX20 [resistant cancer cells]	Sensitization towards chemotherapeutic drugs	↓P-gp efflux function, ↑P-gp ATPase activity, ↓BRCP efflux function	[100]

Borneol	Selenocystine	Hepatocellular carcinoma cells -HepG2	Induction of apoptosis, enhancement of cellular drug uptake, potentiation of selenocystine	↓Caspase3/7/8/9, ↓mitochondrial membrane potential, ↑Bad, ↑Bax, ↓Bcl-XL, ↓Bcl-2,↓Mcl-1, ↑truncation of bid, ↑cleaved PARP, ↑ROS, ↑ DNA damage, ↑p53 Ser15, ↑p-histone H2A.X, ↓PARP, ↓pAKT, ↓pERK, ↑phospo p38	[92]
Neoandrographolide	Etoposide	Human T lymphocyte cell line-Jurkat cells	Sensitization of etoposide	↓XIAP	[97]
Marine phytochemicals					
Parguerenes I and II	Vinblastine, doxorubicin, and paclitaxel	Human colon cancer cell line-SW620, SW620 Ad300, and human lung cancer cell line-NCI-H460, H460 MX20	Reversal of drug resistance	↓P-gp transport activity, ↓MRP1	[157]
Others					
Sulforaphane	Cisplatin	Human colangiocarcinoma cells -HuCCT-1 and TFK-1	Apoptosis, DNA damage response, and sensitization towards cisplatin	↑cleaved caspase3, ↑cleaved PARP, ↑Bax, ↓Bcl-2, XIAP, H2A.X, ↑pATM, ↑pATR, pChk1, ↑pChk2, ↑p-p53	[107]
	Cisplatin	Human osteosarcoma cells-OS-732 and MG-63	Growth inhibition, sensitization of cisplatin	↑p53, ↑p27, ↑p21,↑Bax, ↓cyclin D, E	[158]
Cryptotanshinone	Paclitaxel	Human TSCC cell lines CAL 27 and SCC 9	Inhibition of cell migration and proliferation, apoptosis	↓p-STAT3,↓ Bcl-2, ↓CDK2, ↓snail, ↓MMP2 ↑E-cadherin, ↑P53, ↑P21, ↑β-catenin	[159]

(continued)

Table 4.4 (continued)

Phyto-chemical	Resistant drug	Model	Effect	Mechanism	Ref
	Cisplatin.	A2780	Enhanced DNA damage, inhibition of migration and invasion, chemosensitization to cisplatin	↑γH2AX, ↓MMP2/9 [dose-dependent manner], ↑cleaved caspase3/9	[109]
Gambogic acid	Adriamycin	In vitro: A549, NCI-H460 and normal lung fibroblasts-MRC-5, normal mammary epithelial -MCF-10A In vivo: A549 xenograft in Balb/c nude mice [105]	Synergism with adriamycin, induction of apoptosis	↑cleaved caspase3, 8, 9, ↑cleaved PARP, ↓Bcl-2, ↓Bcl-xl ↑Bax, ↓XIAP, ↓survivin, ↑Fas, ↓P-gp activity, ↓nuclear p65, ↑IκBα	[105]
	Cisplatin	A549, NCI-H460	Potentiation of cisplatin-induced apoptosis	↑caspase3, ↑caspase8, ↑Fas, ↓Bcl-2, ↓Bcl-xl, ↑Bax, ↑caspase9, ↑PARP, ↓survivin, ↓NF-kB, ↓MAPK/HO-1, ↑ROS	[160]
Multiple phytochemicals					
Neferine Isoliensinine	Cisplatin	Colorectal adenocarcinoma cells-HCT-15	Disruption of mitochondrial trans-membrane potential [ΔΨM], apoptosis and sensitization of cisplatin	↑intracellular Ca2+, ↑ROS, ↓Bcl-2, ↑Bax, ↑cleaved PARP, ↑caspase 3, ↑caspase 9, ↑cyt c ↓PI3K/Akt	[69]
Crebanine, O-methylbulbo-capnine	Cisplatin	A2780 [Cisplatin sensitive] and SKOV3 [cisplatin resistant]	Apoptosis and sensitization of cisplatin	↑Caspase 3 and 8, ↑cleaved PARP, ↓BcL-xL, ↓cIAP-2, ↓survivin, ↓pAkt, ↓NF-κ Bp65, ↓IL-6	[161]
Capsaicin Piperine	Doxorubicin	Colon cancer cells-Caco-2 and T-lymphoblastic leukemic cells CEM/ADR 5000	Synergism in sensitization of doxorubicin	↓P-gp activity	[162]

EGCG Tannic acid Curcumin Digitonin	Doxorubicin	Caco-2 and CEM/ADR 5000	Synergism in sensitization of doxorubicin in combination of doxorubicin +polyphenol and even better in doxorubicin+ polyphenol+ digitonin	↓P-gp activity	[163]
Isostemofoline, 11Z-didehydrostemofoline, 11E-didehydrostemofoline	Doxorubicin and paclitaxel	K562, K562/Adr, human fibroblasts, peripheral blood mononuclear cells, and red blood cells	Sensitization towards paclitaxel and doxorubicin, minimum toxicity towards normal cells	↓P-gp function	[81]
EGCG Sulforaphane	Paclitaxel	SKOV3-ip1 [paclitaxel-sensitive] and SKOV3TR-ip2 [resistant]	Induction of G2/M cell cycle arrest, apoptosis and potentiation of paclitaxel	↑DNA damage, ↑γH2AX, ↓telomerase reverse transcriptase [hTERT], ↓Bcl-2	[164]
Anethole Curcumin	Cisplatin and oxaliplatin	A2780 [parent], A2780cisR [cisplatin-resistant] and A2780ZD0473R [ZD0473-resistant] cells	Sensitization of chemo-therapeutic drugs only with prior addition of a single phytochemical	——	[111]
Quercetin Thymoquinone	Cisplatin and oxaliplatin	A2780 [parent] and A2780cisR [Cis-resistant]	Sensitization of chemo-therapeutic drugs only with prior addition of a single phytochemical, ineffective for chemo-sensitization during concurrent addition	↑cellular platinum accumulation, ↑platinum DNA adducts	[95]

4.4.2.1 Counteraction of Drug Efflux Pumps

The most important reason behind MDR is the hyperfunctionality of membrane bound ABC drug transporter pumps of P-gp and MRPs. Though three consecutive generations of pump inhibitors have arrived so far, their clinical outcome has not been very promising. Considering the combinatorial chemistry, the structure-activity relationship specificity and potency have definitely increased but the cross-reactivity and unwanted toxicity to the non-cancerous cells still remains as the main issue. Currently the natural phytochemical based pump-inhibition has emerged as the fourth generation transporter-inhibitor [31]. Polyphenols EGCG or curcumin in combination with doxorubicin and digitonin suppressed P-gp activity and increased the chemotherapeutic efficacy of these drugs against Caco-2 and CEM/ADR 5000 [163]. Doxorubicin resistance was reversed by curcumin by inhibiting ATPase activity of drug efflux protein ABCB4 in resistant breast cancer cells [131]. Curcumin downregulated MDR genes and showed synergism with cytarabine which in turn caused drug dose minimization against primary leukemic cells [134]. Binding of EGCG with carboxyl-terminal DNA binding domain of P-gp, retarded its activity along with sensitization of resistant KB-A1 cells and xenograft models towards doxorubicin [138]. A flavonoid, Rutin, sensitized breast carcinoma cells towards cyclophosphamide and methotrexate by counteracting P-gp and breast cancer resistance protein [BCRP] mediated drug efflux pumps [144]. Magniferin another flavonoid reversed resistance of MCF-7 towards doxorubicin by inhibiting P-gp [52]. Psoralen, a furanocoumarin exhibited reversal of adriamycin resistance by suppressing P-gp induced drug efflux in resistant breast cancer cells [63]. Conferone exhibited reversal of MDR-1 mediated resistance against vinblastine in MDR1-transfected Madin-Darby canine kidney [MDCK-MDR1] cells [67]. Cnidiadin sensitized resistant cells [MDCK-MDR1 cells and vincristine resistant KB human oral epidermoid carcinoma cell line] overexpressing P-gp to vinca alkaloids by competitive inhibition of the drugs by P-gp [65]. Stemofoline, belonging to the group of stemona alkaloids proved effective in modulating P-gp and improving the responsiveness of the chemo-resistant cervical cancer cells KB-V1 towards vinblastine, paclitaxel, and doxorubicin [125]. Piperine was reported to inhibit ABCB1, ABCC1, and ABCG2 molecules which, respectively, encoded P-gp, MRP1, and BCRP drug efflux associated proteins and increased the responsiveness of doxorubicin and mitoxantrone in resistant breast and lung carcinoma cells [154]. Tetrandrine, a bis-benzylisoquinoline alkaloid, imparted reversal of P-gp influenced paclitaxel resistance against human epidermoid carcinoma [156]. Alkaloids, capsaicin, and piperine exhibited synergism in inhibiting P-gp function and chemosensitizing Caco-2 and CEM/ADR 5000 towards doxorubicin [162]. A triterpenoid saponin, saikosaponin D caused reversal of doxorubicin resistance of breast cancer cells by inhibiting P-gp and MDR-1 mediated drug efflux system [98]. β-carotene, a carotenoid caused reversal of MDR by increasing the activity of ATP-dependent P-gp and inhibiting the drug efflux potentiality of P-gp and BRCP [100]. Marine phytochemicals parguerenes I and II are bromoditerpenes reversed resistance of drugs—vinblastine, doxorubicin, and paclitaxel in human colon cancer cell line SW620, SW620-Ad300, and human lung cancer cell line—

NCI-H460, H460 MX20 by inhibiting P-gp transport activity and MRP1 mediated drug efflux [157]. Stemona alkaloids caused reversal of P-gp mediated adriamycin resistance in K562 cells [81]. Gambogic acid, a xanthonoid, was found to synergize adriamycin against A549 cells and A549 xenograft by inducing apoptosis through inhibition of NFκB and P-gp [90].

4.4.2.2 Cell Cycle Arrest

Many cytotoxic drugs cause DNA damage which induce death via cell cycle arrest. Resistant cancer cells harbor deregulated cell-cycle checkpoint proteins as a result of which they fail to recognize and repair drug-induced damage and cell cycle progresses in an uninterrupted manner. It further makes the cancer cells more unresponsive. On the other hand, growth arrested quiescent or even senescent cells can re-enter into the cell cycle process to cause relapse. Cyclin dependent kinases [CDKs] and their regulatory cyclins which comprise cell cycle machinery are ideal targets [165]. The first generation CDK inhibitors [CKIs] were mostly G1 or G2 blocker, but had non-specific toxicities. Whereas some novel small molecular derivatives are found to cause G2/M, S/G2 arrest efficiently. Flavonoids, alkaloids, purine, indoles, pyrimidines, thiazole, staurosporines are some types of phytochemicals which have been identified as CKIs [18]. Curcumin enhanced S phase cell cycle arrest, aggravated mitochondrial damage, induced apoptosis, and sensitized leiomyosarcoma cells towards cisplatin but could not inhibit cisplatin-induced nephrotoxicity in female Wistar rats [34]. A flavonoid, fisetin, inflicted G2 phase cell cycle arrest by inhibiting cyclins B1 and E1 and sensitized the osteosarcoma cells -Saos-2 and U2OS towards etoposide [141]. Myricetin caused reduction in cyclin D1 and Bcl-2 level and sensitized human esophageal squamous cancer cells to 5-FU [150]. Sulforaphane sensitized cisplatin by inducing G1 phase arrest and upregulating p53-p21 mediated apoptosis in osteosarcoma cells [158]. EGCG and sulforaphane potentiated paclitaxel against paclitaxel resistant ovarian cancer SKOV3TR-ip2 cells by inducing G2/M phase arrest and by promoting apoptosis [164].

4.4.2.3 p53 Deregulation

p53 controls most if not all processes associated with chemoresistance. As it regulates the transcription of genes responsible for detecting DNA-damage and repairing, cell cycle surveillance, extrinsic as well as intrinsic apoptosis, autophagy and senescence, drug efflux, intracellular metabolism and so on, its mutation or deregulation directly affects the responsiveness of any chemotherapeutic drugs. Tp53 mutations which are most common and crucial for cancer development are equally responsible for cancer chemoresistance [166]. Mutated p53 may remain responsible for not detecting DNA lesions and corresponding cell cycle arrest, for anti-apoptotic induction, for upregulating drug efflux and metabolism and many other events. Targeting the mutated p53 approach explores inactivation and destabilization of mutated faulty p53 and restores the wild type functionality in cancer cells. Resveratrol induced phosphorylation of p53ser20 which inhibited RAD51, aggravated apoptosis, and sensitized resistant MCF-7_R towards cisplatin

[38]. Myricetin induced p53/c-myc mediated apoptosis in cisplatin resistant ovarian cancer cells [149].

4.4.2.4 Apoptosis and Mitochondrial Membrane Depolarization

Upon the exposure to therapeutic drugs most cancer cells evolve to over express anti-apoptotic proteins to escape apoptotic death signaling. Initially drug development was aimed to design small molecular inhibitors for different anti-apoptotic Bcl-2 family proteins. Though the first one of this kind was against protein Bcl-2 but trials failed largely because its inhibitory efficacy could not cover the entire set of anti-apoptotic molecules robustly. So, targeting the other potent anti-apoptotic oncogenes myeloid leukemia cell differentiation protein [Mcl-1], B-cell lymphoma-extra large [Bcl-XL], etc. will give room to the Bcl-2 inhibitors for effective chemosensitization. A report exhibited that curcumin could sensitize cisplatin resistant lung cancer cells by inhibiting Fanconi anaemia [FA]/BRCA pathway and inducing apoptosis [135]. A flavonoid scutellarin sensitized resistant ovarian cancer cells towards cisplatin by increasing platinum DNA adducts and inducing apoptosis [41]. Naringenin potentiated the sensitivity towards DNA damaging drugs through reduced expression of PI3K/Akt pathway and induced caspase 3, p21, Bcl-2 associated X protein [Bax], and apoptosis inducing factor [AIF] [148]. Fisetin sensitized irinotecan and oxaliplatin resistant colon cancer cells towards apoptosis [116] and imparted more efficacy of sorafenib via death receptor [DR]-5/caspase3/caspase 8 mediated apoptosis in cervical cancer [142]. Another flavonoid, silibinin, sensitized doxorubicin and paclitaxel in non-responsive breast cancer cells by inducing apoptosis in a signal transducer and activator of transcription *3* [STAT3]/ extracellular signal-regulated kinase [ERK]/Akt mediated pathway [145]. Silibinin also sensitized cisplatin and 5-FU against resistant esophageal cancer cell lines by regulating 5′ adenosine monophosphate-activated protein kinase [AMPK] and inducing apoptosis [146]. Dendrobine, an alkaloid chemosensitized A549 cells towards cisplatin by augmenting c-Jun N-terminal kinase [JNK]/p38 stress pathway and Bax/Bim mediated apoptosis [74]. Piperlongumine sensitized oxaliplatin against gastric cancer cell lines and in vivo xenograft model by p38/JNK mediated apoptosis [152]. Piperine, an alkaloid from common spice pepper, sensitized paclitaxel induced apoptosis by downregulating pAkt and Mcl-1 in Hela cells [153]. Isoquinoline derived aporphine alkaloids, O-methylbulbocapnine sensitized resistant ovarian cancer cells SKOV3 towards cisplatin by promoting interleukin-6 [IL-6]/Akt/NF-κB mediated apoptosis [161]. A monoterpenoid borneol was observed to potentiate selenocystine induced apoptosis in hepatocellular carcinoma, HepG2 cells by propagating cellular uptake of selenocystine, increasing ROS induced DNA damage and by blocking Akt and ERK pathways [92]. Neoandrographolide was found to sensitize etoposide by suppressing the overexpression of X chromosome-linked inhibitor of apoptosis protein in S-Jurkat cells [97]. Gambogic acid also potentiated cisplatin by downregulating NF-κB, mitogen-activated protein kinase [MAPK]/Heme oxygenase [HO]-1 pathway and inducing apoptosis in lung carcinoma cells [160].

As most of the chemotherapeutic drugs show their cytotoxicity by induction of apoptosis and mitochondria plays a major role in this process, mitochondrial dysfunction is intricately associated with death-escape and chemoresistance. Use of mitochondriotropic agents that can directly act on mitochondrial permeability transition pore complex can target the anti-apoptotic molecules [e.g., Bcl-2]. Exploiting the higher mitochondrial membrane potential of cancerous cells, lipophilic and cationic cytotoxic agents can selectively kill more resistant cancer cells. Inhibition of ROS scavenging will maintain a more oxidative environment inside the mitochondria that can also impart selective lethality to the cancer cells [167]. Flavonoids apigenin and fisetin as well as alkaloid honokiol proved effective in increasing the intracellular accumulation of doxorubicin by differentially modulating GSH content and ROS generation inside the resistant MES-SA/Dx5 cells [168]. Lotus derived alkaloids, neferine and isoliensinine increased efficacy of cisplatin in HCT-15 cells by increasing intracellular Ca^{+2} concentration, disintegration of mitochondrial membrane potential and induction of apoptosis [69].

4.4.2.5 Autophagy

Autophagy which recycles damaged subcellular components also repairs the damage caused by the antineoplastic drugs giving the cancer cells stress-free pro-survival advantage. Autophagic and apoptotic machineries share many molecules in common. The same autophagy which helps the cancer cells to evolve as chemoresistant by inhibiting apoptotic signaling, can also induce apoptotic cell death in presence of excessive stress. Moreover, extensive autophagy may also lead cell killing even in cancer. ROS-mediated cell killing is also prevented by autophagy in cancer as ROS signaling triggers autophagy, which in turn facilitates clearance of ROS-leaking damaged organelles [169]. Therefore, autophagy ultimately limits oxidative-stress induced cellular damage and death. A case specific up/down regulation of autophagy may reverse the resistance property of cancer. So, phytochemicals which are natural antioxidants are excellent candidates for autophagy-managed chemosensitization [170]. Alantolactone, a sesquiterpene lactone, induced chemosensitivity towards oxaliplatin in human pancreatic cancer cell lines by suppression of lysosomal autophagy and induction of apoptosis [102].

4.4.2.6 Suppression of Cancer Stemness, EMT, Migration, and Invasion

Cancer stem cells [CSC] are a small subpopulation of primary cancer cells highly heterogenous in nature and capable of extended self-renewal and are progenitor of all differentiated cancer cell lineages [171]. Due to their low growth rate with the inherent quiescence, high DNA-damage repair property, enhanced drug efflux, detoxification and ALDH activity, upregulation of epithelial–mesenchymal transition [EMT], Wnt, Notch, Hedgehog signaling are the few prominent features of CSC which give them the intrinsic resistance property. This inertness is responsible for the relapse of the disease even after surgical or therapeutic removal of the tumor mass. Targeted eradication of CSCs can make the anti-therapies more effective and also prevent recurrence [172]. Curcumin reduced the cancer stemness made the resistant colon cancer cells susceptible to irinotecan induced apoptosis

[132]. Curcumin reduced the cancer stemness of mismatch repair deficient colon cancer cells and reversed their cisplatin resistance [136]. Another polyphenol hispidin induced apoptosis and reduced stemness in synergism with gemcitabine in pancreatic cancer cells [36]. An alkaloid, honokiol, exhibited suppression of stemness, invasion, migration, colony formation, and potentiation of cisplatin via downregulation of ALDH1/CD44/IL-6 /pSTAT pathway in OSCC cell lines [155]. Curcumin combined with carboplatin increased its sensitivity, reduced cell viability, retarded metastasis, and induced apoptosis in A549 lung cancer cells [133]. A prenylflavonoid xanthohumol inhibited growth, induced apoptosis, reduced cell invasion, and increased sensitization towards adriamycin and cisplatin in ALL cell lines and xenograft mouse model [53].The guanidine alkaloid cernumidine inhibited cancer cell division, invasion, and cell migration and induced apoptosis in cisplatin resistant bladder cancer cells T24 [121]. Neferine is a bis-benzylisoquinoline alkaloid inflicted chemosensitization towards oxaliplatin by inducing cytotoxicity, apoptosis and by reducing EMT induced migration & invasion [120]. Cryptotanshinone was observed to provide chemosensitivity towards paclitaxel in tongue squamous cell carcinoma [TSCC] by inhibition of EMT, cell migration, and apoptosis [159]. It also sensitized cisplatin in ovarian cancer by increasing DNA damage, reducing cell invasion and inducing apoptosis [109].

4.4.2.7 Downregulation of Multiple Signaling Pathways

Chemoresistance is a concerted action of multiple signaling networks. Phytochemicals by virtue of pleiotropic nature may affect multitude of signaling crosstalks involved with chemoresistance. EGCG, a principal green tea catechin sensitized resistant A549 lung cancer cells by downregulating nuclear factor erythroid 2-related factor 2 [Nrf2] in both Kelch-like ECH-associated protein 1 [Keap-1] dependent and Keap1-independent pathways, suppressing MRP-1, pERK1/2, poly [ADP-ribose] polymerase [PARP], and X-ray repair cross-complementing protein 1[XRCC1] and upregulating p53 [137]. Resveratrol reversed 5-FU non-responsiveness in colon cancer cells by modulating miR-34a/E2F3/NAD-dependent deacetylase sirtuin-1 [Sirt1] cascade and downregulating of PI3k/Akt, MAPK/ERK1/2 pathway [139]. Resveratrol imparted chemosentitization towards 5-FU by accumulation of ROS and inhibition Akt kinase pathway [140]. Fisetin sensitized sorafenib by suppression of PI3K/Akt/mammalian target of rapamycin [mTOR], mitogen-activated protein kinase[MEK], ERK, in melanoma [42] in vitro and in vivo models. Nobiletin, a hexamethoxyflavon, made adriamycin functionally effective in SCLC cell line by reduced expression of N-myc proto-oncogene protein, MRP1, Akt and glycogen synthase kinase 3 beta [GSK3β] and increased expression of caspase3 [143]. Baicalein suppressed EMT, downregulated PI3K/Akt/NF-κB pathway, and sensitized resistant lung cancer cells toward cisplatin [118]. Chlorogenic acid inhibited proliferation and provided chemosensitization towards 5-FU by downregulating ERK1/2 in human liver cancer cells [147]. Gingerol imparted increase in intracellular doxorubicin accumulation and cytotoxicity in the resistant human uterine sarcoma—MES-SA/Dx5 [173]. Instead of combination of anethole and curcumin [111] or quercetin and thymoquinone [95], prior addition of single compound before administration of cisplatin and /or oxaliplatin imparted better sensitization in resistant human ovarian cell lines.

4.4.3 Clinical Trials of Natural Compounds for Adjunct Therapy Along with Chemotherapy

According to NIH website "ClinicalTrials.gov" majority of clinical trials carried so far have explored the role of curcumin as an adjunct therapy for increasing efficacy of conventional chemotherapeutics. Gemcitabine in combination with curcumin and celecoxib was investigated against pancreatic cancer in phase III trial conducted in Israel [NCT00486460]. A phase II clinical conducted in France compared the response rate in HER2-negative patients breast cancer recurrence treated with docetaxel and curcumin vs docetaxel alone [NCT00852332]. A phase II clinical trial of US assessed the efficacy of capecitabine and radiation therapy in absence/ presence of curcumin in locally advanced rectal cancer [NCT00745134]. Another phase II clinical trial investigated the effect of curcumin and gemcitabine against advanced pancreatic cancer [NCT00192842]. A phase II clinical trial is going on with Avastin/FOLFIRI and curcumin against unresectable metastasis in colorectal cancer [NCT02439385]. A phase I clinical trial on curcumin is investigating whether curcumin administration along with 5-FU alters inflammatory and epigenetic biomarkers in patients with chemoresistant metastatic colorectal cancer [NCT02724202]. In Armenia a phase II clinical trial explored the efficacy of curcumin with paclitaxel against advanced breast cancer and metastatic breast cancer [NCT03072992]. Oral curcumin along with oxaliplatin was studied for safe and long term tolerated regimen in inoperable metastatic colorectal patients [NCT01490996]. A clinical trial is going on in France [NCT03959618] in order to determine the role of *Desmodium adscendens*, [source of triterpene saponins, alkaloids, flavonoids, polyphenols, and tryptamine derivatives] as a neoadjuvant / adjuvant along with IV chemotherapy in breast cancer patients. A clinical trial carried out at the USA [NCT01426620] evaluated blueberry powder [rich source of anthocyanidins] as an adjunct therapy with paclitaxel/docetaxel to treat NSCLC. Erlotinib was used with polyphenon E [EGCG] against NSCLC [NCT00707252] in phase I and II and against head and neck cancers [NCT01116336] in phase I clinical trials. Genistein was used with gemcitabine against stage IV breast cancer in a phase II trial [NCT00244933]. Erlotinib/gemcitabine along with genistein was investigated for efficacy against in-situ or secondarily metastasized pancreatic cancer [NCT00376948] [174].

4.4.4 Limitations of Phytochemicals as Chemosensitizers in Adjunct Therapy against Cancer

A wide range of natural phytochemicals being a part of regular diet are almost universally safe, easily available, and economic. But the main challenge remains in their poor bioavailability and short physiological half-life owing to their poor pharmacokinetic and pharmacodynamic indexes. Therefore, their persistent availability is only achievable if they are administered in an exceptionally high dose almost continuously. This will inevitably bring in high dose-related side effects that will ultimately limit the efficacy of the compounds. A meta-analysis reported 97 phytochemical-induced hepato-toxicity cases in Korea. Coumarins, furanocoumarins, EGCG, and piperine inhibited drug detoxification enzyme CYP3A4 and delayed the half-life of the chemotherapeutic drug [175].

The screening of chemo-sensitization potential of phytochemicals in cell culture conditions may alter drug efficacy which may provide a negative impact on the identification of clinically effective compounds. There are significant differences between the in vivo tumor microenvironment and cell culture conditions. In native tumor microenvironment there are limited molecular transport, fluidic shear stress, cell–stroma interaction, physical structure, and soft material properties. In conventional in-vitro cancer models there are unlimited molecular transport, static condition, no consideration on cell–stroma interaction, flat substrate, and non-physiological material properties [19].

Moreover, the drug–compound interaction should be favorable with no antagonistic effect. Presence of single nucleotide polymorphisms [SNPs] in the detoxifying genes may directly affect the functionality of the agent. Therefore even after successful preclinical studies compounds may confront failure in clinical trials. Some phytochemicals have been also found to be genotoxic or even carcinogenic, e.g. capsaicin, amygdalin, methyleugenol, coumarin, etc. Lastly there should be a constant cost-effective method to check for their purity druggability and personalized toxicity of the phytochemicals.

4.5 Latest Advancements of Plant Derived Chemosensitizers

MDR is one big challenge in clinical practice where the increase of effective drug dose accumulates hazardous systemic toxicity during cancer management. Therefore, research on MDR reversal should essentially focus on drug sensitization as well as minimum drug distribution in healthy tissues. After systemic administration small molecule chemosensitizers have difficulty to effectively accumulate in tumor tissues. A recent approach used a polyvinyl pyrrolidone-based solid dispersion of Zn [II]-curcumin exhibited gut microbiota influenced sensitization towards doxorubicin in hepatocellular carcinoma [HCC] in vitro and in vivo models [176].The drawbacks of phytochemicals have been primarily tackled by designing stable synthetic analogs [maybe nano-formulations] which are capable of sustained release. In that case effort has to be taken to not conjugate any toxic and non-biodegradable nano-particle with the phytochemical.

4.5.1 Derivatives of Phytochemicals

The deficiencies of phytochemicals have been tried to met with introduction of different synthetic derivatives which are more effective than their natural counterpart. A novel curcumin derivative [1,7-bis[3-methoxy-4-[prop-2-yn-1-yloxy]phenyl]hepta-1,6-diene-3,5-dione promoted apoptosis, inflicted G2/M cell cycle arrest and chemosensitized MDR chronic K562 cells towards doxorubicin better than curcumin [177]. Curcumin analog [2 J] along with paclitaxel /vincristine induced better efficacy in reversal of adriamycin resistance in K562/Adr cells and showed less toxicity towards human peripheral mononuclear cells and human and rat red blood cells [178]. An early study reported that a curcumin metabolite, tetrahydrocurcumin inhibited drug efflux activity mediated by P-gp, MRP-1, MXR in MCF7AdrVp3000 and MCF7FL1000 and HEK-293 cells [179]. Trans-resveratrol derivative—resveratrol triacetate caused S phase arrest, increased 5- FU-mediated inhibition of SW480, SW620, and HCT116, colon cancer cell proliferation [180] and stemofoline derivatives- OH-A1, NH-B6, NH-D6, resensitized adriamycin resistant K562/Adr cells to doxorubicin by supressing P-gp action without affecting its expression [181]. 5-Bromotetrandrine, a derivative of an alkaloid tetrandrine imparted better activity of doxorubicin, paclitaxel, taxotere, vincristine, and epirubicin by inhibiting P-gp governed MDR in human epidermoid cancer-KBv200 cells, KBv200xenografts but not in KB cells [182]. A coumarine derivative, 1-[9H-fluoren-9-yl] piperazin, with a coumarin core attached to a specific butynyl-amino chain exhibited improved antitumor activity and reversed MDR in resistant colon cancer cells LoVo/Dx [183]. Chalcone derivatives bearing cinnamaldehyde scaffold showed high potential of reversing cisplatin resistance by inducing G2/M cell cycle arrest followed by apoptosis in cisplatin resistant cell line A2780/cis [184]. A quinone-containing compound, 4-[2356-tetrafluoro-4-[4-hydroxyphenoxy] phenoxy]phenol [TFPP] and camptothecin in combination downregulated ERK pathway and aggravated ROS mediated apoptosis in A549 lung cancer cells [185]. Marine phytochemical, ningalin B derivatives, 23 with dimethoxy groups at rings A and B and tri-substitution at ring C with ortho-bromo, meta-methoxy, and para-trimethoxybenzyloxy groups was found effective in downregulating P-gp membrane-transport function and upregulating accumulation of chemotherapeutic drugs—paclitaxel, doxorubicin, vinblastine, and vincristine resistance in LCC6MDR breast cancer cells [186].

4.5.2 Multimode Nano-Drug Delivery Systems

The multimode nano-drug delivery systems can be used for conjoint-delivery of functional drugs and target genes to treat MDR. They are characterized by (a) potential of ratiometric loading for more than one agents, (b) tumor site specific targeting and improved uptake by membrane alteration, and (c) pegylation mediated hemodynamic stability which facilitates their longevity, half-life, and incorporation in specific cancer tissues [27]. Amphiphilic poly[curcumin-dithiodipropionic acid]-b-poly[ethylene glycol]-biotin copolymer efficiently codelivered curcumin and doxorubicin to breast cancer cells resistant to adriamycin [MCF-7/ADR] as well as to nude mice xenografted with MCF-7/ADR. The polymeric curcumin loaded nano-drug delivery system sensitized them towards doxorubicin by downregulating ATP, P-gp, Ki-67 and by inducing apoptosis [187]. Similarly, another study reported a combination of tumor cell type-specific targeted peptide, PEGylated polymeric prodrug, and curcumin loaded in a single nanocarrier elicited chemotherapeutic efficacy of doxorubicin against resistant cell lines, i.e. MCF-7/ADR [188]. Paclitaxel and curcumin delivery through cationic PEGylated niosomes augmented apoptosis in breast cancer cells MCF-7 relatively higher than the normal breast cells MCF-10A [189]. A nanocomposite containing magnetic [Fe_3O_4] and fluorescent contrast [FITC] working as a dual imaging probe, biotin helping the target specific delivery of nanoparticles, doxorubicin a chemotherapeutic drug, and quercetin as a naturally obtained chemosensitizer imparted greater apoptosis, suppressed tumor cell proliferation, and stimulated the therapeutic efficacy of doxorubicin against A549 cells [190]. PLGA polymeric nanoparticles loaded with rapamycin and piperine, inhibited P-gp activity which in turn caused sustained release and better absorption of rapamycin by MDA-MB-231 cells [191]. Nanoformulation of quercetin [phytosome] depleted Nrf2 downstream targets—MRP1 and NQO1, enhanced infiltration of doxorubicin into MCF-7 cells, augmented endocytosis and cytotoxicity [192].

4.6 Conclusion

Among the plethora of phytochemicals, flavonoids, terpenoids, and alkaloids primarily contributes towards chemosensitization and reversal of MDR. Majority of the natural compounds improvize sensitization of the chemotherapeutic drugs by [i] modulating drug efflux pumps; [ii] evoking cell cycle check-point arrest and apoptosis; and [iii] suppressing malignant stemness, EMT, invasion along with metastasis of cancer cells. The clinical trials have also proved the competence of the phytochemicals as adjunct therapy in cancer management to some extent. However, the issues of low bioavailability, optimum dosimetry, low pharmacokinetic indexes of these natural compounds are yet to be resolved. The advancement of structural chemistry has introduced many derivatives of the natural compounds for better efficacy. Nanoformulations of phytochemicals and multifunctional nano-drug delivery systems have facilitated targeted therapy with improved drug distribution and drug uptake by cancer cells. In future, more clinical trials are required which may translate the valuable pre-clinical data for effective cancer management.

Conflict of Interest The authors declare no conflict of interest.

References

1. IARC (2018) GLOBOCAN 2018: Latest global cancer data. Press release No 263. Int Agency Res Cancer
2. Ajithkumar TV, Hatcher HM (2011) Specialist training in oncology. Elsevier Inc., New York, pp 1–395
3. Liptak JM, Page E (2009) The principles of surgical oncology: diagnosis and staging. Cancer 31(August):1–13
4. Mehta SR, Suhag V, Semwal M, Sharma N (2010) Radiotherapy: basic concepts and recent advances. Med J Armed Forces India 66(2):158–162
5. Kerns SL, Ostrer H, Rosenstein BS (2014) Radiogenomics: using genetics to identify cancer patients at risk for development of adverse effects following radiotherapy. Cancer Discov 4 (2):155–165
6. Tobias J, Hochhauser D (2014) Cancer and its management. Wiley, New Jersey, 627 p
7. Arruebo M, Vilaboa N, Sáez-Gutierrez B, Lambea J, Tres A, Valladares M et al (2011) Assessment of the evolution of cancer treatment therapies. Cancers [Basel] 3(3):3279–3330
8. Caspi RR (2008) Immunotherapy of autoimmunity and cancer: the penalty for success. Nat Rev Immunol 8(12):970–976
9. Zugazagoitia J, Guedes C, Ponce S, Ferrer I, Molina-Pinelo S, Paz-Ares L (2016) Current challenges in Cancer treatment. Clin Ther 38(7):1551–1566
10. Bharti AC, Vishnoi K, Singh SM, Aggarwal BB (2018) Pathways linked to cancer chemoresistance and their targeting by nutraceuticals. In: Role of nutraceuticals in chemoresistance to cancer. Elsevier, London, pp 1–30
11. Gillet JP, Gottesman MM (2010) Mechanisms of multidrug resistance in cancer. Methods Mol Biol 596:47–76
12. Mondal J, Panigrahi AK (2014) Conventional chemotherapy: problems and scope for combined therapies with certain herbal products and dietary supplements. Austin J Mol Cell Biol 1 (1):10
13. Kyle AH, Huxham LA, Yeoman DM, Minchinton AI (2007) Limited tissue penetration of taxanes: a mechanism for resistance in solid tumors. Clin Cancer Res 13(9):2804–2810
14. Hamed AR, Abdel-Azim NS, Shams KA, Hammouda FM (2019) Targeting multidrug resistance in cancer by natural chemosensitizers. Bull Natl Res Cent 43(1):8
15. Marin JJ, Monte MJ, Blazquez AG, Macias RI, Serrano MA, Briz O (2014) The role of reduced intracellular concentrations of active drugs in the lack of response to anticancer chemotherapy. Acta Pharmacol Sin 35:1–10
16. Gillet J-P, Gottesman MM (2010) Mechanisms of multidrug resistance in cancer. In: Methods in molecular biology. Humana Press, Clifton, pp 47–96
17. Mansoori B, Mohammadi A, Davudian S, Shirjang S, Baradaran B (2017) The different mechanisms of cancer drug resistance: a brief review. Adv Pharm Bull 7(3):339–348
18. Valletti A, Marzano F, Pesole G, Sbisà E, Tullo A (2019) Targeting chemoresistant tumors: could TRIM proteins-p53 axis be a possible answer? Int J Mol Sci 20(7):1776
19. Jo Y, Choi N, Kim K, Koo HJ, Choi J, Kim HN (2018) Chemoresistance of cancer cells: requirements of tumor microenvironment-mimicking in vitro models in anti-cancer drug development. Theranostics 8(19):5259–5275
20. Vinod BS, Maliekal TT, Anto RJ (2013) Phytochemicals as chemosensitizers: from molecular mechanism to clinical significance. Antioxidants Redox Signal 18(11):1307–1348
21. Derdak Z, Mark NM, Beldi G, Robson SC, Wands JR, Baffy G (2008) The mitochondrial uncoupling protein-2 promotes chemoresistance in cancer cells. Cancer Res 68(8):2813–2819
22. Raghunand N, Gillies RJ (2000) pH and drug resistance in tumors. Drug Resist Updat 3 (1):39–47
23. Wachsberger PR, Landry J, Storck C, Davis K, O'Hara MD, Owen CS et al (1997) Mammalian cells adapted to growth at pH 6.7 have elevated HSP27 levels and are resistant to cisplatin. Int J Hyperth 13(3):251–259

24. Butler EB, Zhao Y, Muñoz-Pinedo C, Lu J, Tan M (2013) Stalling the engine of resistance: targeting cancer metabolism to overcome therapeutic resistance. Cancer Res 73(9):2709–2717
25. Guerra F, Arbini AA, Moro L (2017) Mitochondria and cancer chemoresistance. Biochim Biophys Acta Bioenerg 1858(8):686–699
26. Dewanjee S, Dua TK, Bhattacharjee N, Das A, Gangopadhyay M, Khanra R et al (2017) Natural products as alternative choices for P-glycoprotein [P-gp] inhibition. Molecules 22 (6):1–93
27. Khan IU, Khan RU, Asif H, Khalid SH, Asghar S et al (2017) Co-delivery strategies to overcome multidrug resistance in ovarian cancer. Int J Pharm 533:111–124
28. Kathawala RJ, Gupta P, Ashby CR, Chen ZS (2015) The modulation of ABC transporter-mediated multidrug resistance in cancer: a review of the past decade. Drug Resist Updat 18 (2015):1–17
29. Kumar A, Jaitak V (2019) Natural products as multidrug resistance modulators in cancer. Eur J Med Chem 176:268–291
30. Ye Q, Liu K, Shen Q, Li Q, Hao J, Han F et al (2019) Reversal of multidrug resistance in Cancer by multi-functional flavonoids. Front Oncol 9:487
31. Karthikeyan S, Hoti S (2015) Development of fourth generation ABC inhibitors from natural products: a novel approach to overcome cancer multidrug resistance. Anti Cancer Agents Med Chem 15(5):605–615
32. Guo Q, Cao H, Qi X, Li H, Ye P, Wang Z et al (2017) Research progress in reversal of tumor multi-drug resistance via natural products. Anti Cancer Agents Med Chem 17(11):1466–1476
33. de Oliveira Júnior RG, Christiane Adrielly AF, da Silva Almeida JRG, Grougnet R, Thiéry V, Picot L (2018) Sensitization of tumor cells to chemotherapy by natural products: a systematic review of preclinical data and molecular mechanisms. Fitoterapia 129(January):383–400
34. Dhima I, Zerikiotis S, Lekkas P, Simos YV, Gkiouli M, Vezyraki P et al (2019) Curcumin acts as a chemosensitizer for leiomyosarcoma cells in vitro but fails to mediate antioxidant enzyme activity in cisplatin-induced experimental nephrotoxicity in rats. Integr Cancer Ther 18:153473541987281
35. Summary C. 3/1/2020 Curcumin | IC21H20O6 - PubChem. 2020
36. Chandimali N, Huynh DL, Jin WY, Kwon T (2018) Combination effects of hispidin and gemcitabine via inhibition of stemness in pancreatic cancer stem cells. Anticancer Res 38 (7):3967–3975
37. Summary C. PubChem CID: 54685921 Structure: 2020
38. Hernandez-Valencia J, Garcia-Villa E, Arenas-Hernandez A, Garcia-Mena J, Diaz-Chavez J, Gariglio P (2018) Induction of p53 phosphorylation at serine 20 by resveratrol is required to activate p53 target genes, restoring apoptosis in MCF-7 cells resistant to cisplatin. Nutrients 10 (9):1148
39. Summary C. PubChem CID: 445154 Structure: 2020;[1]:1–67
40. Summary C. PubChem CID: 185617 Structure: 2020
41. Xie Z, Guo Z, Lei J, Yu J (2019) Scutellarin synergistically enhances cisplatin effect against ovarian cancer cells through enhancing the ability of cisplatin binding to DNA. Eur J Pharmacol:9–16
42. Pal HC, Baxter RD, Hunt KM, Agarwal J, Elmets CA, Athar M et al (2015) Fisetin, a phytochemical, potentiates sorafenib-induced apoptosis and abrogates tumor growth in athymic nude mice implanted with BRAF-mutated melanoma cells. Oncotarget 6 (29):28296–28311
43. Summary C. PubChem CID: 5281614 Structure: 2020;1–33
44. Summary C. PubChem CID: 5280805 Structure: 2020
45. Ganeshpurkar A, Saluja AK (2017) The pharmacological potential of Rutin. Saudi Pharm J 25:149–164
46. Kawabata K, Mukai R, Ishisaka A (2015) Quercetin and related polyphenols: new insights and implications for their bioactivity and bioavailability. Food Funct [Internet] 6(5):1399–1417. [cited 2020 Mar 1], Available from: http://www.ncbi.nlm.nih.gov/pubmed/25761771

47. Structures FS. 3/1/2020 Quercetin | C15H10O7 - PubChem. 2020
48. Liu H, Dong Y, Gao Y, Du Z, Wang Y, Cheng P et al (2016) The fascinating effects of baicalein on cancer: A review. Int J Mol Sci 17:1681
49. Summary C. PubChem CID: 5281605 Structure: 2020;
50. Agarwal C, Wadhwa R, Deep G, Biedermann D, Gažák R, Křen V et al (2013) Anti-Cancer efficacy of Silybin derivatives - a structure-activity relationship. PLoS One 8(3):e60074
51. Semwal DK, Semwal RB, Combrinck S, Viljoen A (2016) Myricetin: a dietary molecule with diverse biological activities. Nutrients 8:90
52. Louisa M, Soediro TM, Suyatna FD (2014) In vitro modulation of P-glycoprotein, MRP-1 and BCRP expression by mangiferin in doxorubicin-treated MCF-7 cells. Asian Pacific J Cancer Prev 15(4):1639–1642
53. Benelli R, Ven R, Ciarlo M, Carlone S, Barbieri O, Ferrari N (2012) The AKT/NF-κB inhibitor xanthohumol is a potent anti-lymphocytic leukemia drug overcoming chemoresistance and cell infiltration. Biochem Pharmacol 83(12):1634–1642
54. Summary C. PubChem CID: 639665 Structure: 2020;20
55. Nagle DG, Ferreira D, Zhou Y-D (2006) Epigallocatechin-3-gallate [EGCG]: chemical and biomedical perspectives. Phytochemistry 67(17):1849–1855
56. Summary C. [−]-Epigallocatechin gallate. 2020;2020
57. Santana-Gálvez J, Cisneros-Zevallos L, Jacobo-Velázquez DA (2017) Chlorogenic acid: recent advances on its dual role as a food additive and a nutraceutical against metabolic syndrome. Molecules 22:358
58. Makkar HPS, Siddhuraju P, Becker K (2007) Chlorogenic Acid:89–91
59. Li S, Yu H, Ho CT (2006) Nobiletin: efficient and large quantity isolation from orange peel extract. Biomed Chromatogr 20(1):133–138
60. Summary C. PubChem CID: 72344 Structure: 2020;1–25
61. Hernández-Aquino E, Muriel P (2018) Beneficial effects of naringenin in liver diseases: molecular mechanisms. World J Gastroenterol 24(16):1679–1707
62. Summary C. PubChem CID: 439246 Structure: 2020;16:1–35
63. Jiang J, Wang X, Cheng K, Zhao W, Hua Y, Xu C et al (2016) Psoralen reverses the P-glycoprotein-mediated multidrug resistance in human breast cancer MCF-7/ADR cells. Mol Med Rep 13(6):4745–4750
64. Summary C. PubChem CID: 6199 Structure: 2020;
65. Barthomeuf C, Grassi J, Demeule M, Fournier C, Boivin D, Béliveau R (2005) Inhibition of P-glycoprotein transport function and reversion of MDR1 multidrug resistance by cnidiadin. Cancer Chemother Pharmacol 56(2):173–181
66. Summary C. Cnidiadin. 2020;1–11
67. Barthomeuf C, Demeule M, Grassi J, Saidkhodjaev A, Beliveau R (2006) Conferone from Ferula schtschurowskiana enhances vinblastine cytotoxicity in MDCK-MDR1 cells by competitively inhibiting P-glycoprotein transport. Planta Med 72(7):634–639
68. Summary C. PubChem CID: 3108117 Structure: 2020;
69. Manogaran P, Beeraka NM, Huang CY, Vijaya PV (2019) Neferine and isoliensinine enhance 'intracellular uptake of cisplatin' and induce 'ROS-mediated apoptosis' in colorectal cancer cells – a comparative study. Food Chem Toxicol 132(May):110652
70. Summary C. PubChem CID: 159654 Structure: 2020;1–16
71. Summary C. PubChem CID: 5274591 Structure: 2020;1–17
72. Miranda MA, Mondal A, Sachdeva M, Cabral H, Neto Y, Khan I et al (2019) Chemosensitizing effect of Cernumidine extracted from Solanum cernuum on bladder Cancer cells in vitro. Chem Biodivers 16(10):e1900334
73. Summary C. Cernumidine. 2020;1–11
74. Song TH, Chen XX, Lee CKF, Sze SCW (2019) Feng Y bin, Yang ZJ, et al. Dendrobine targeting JNK stress signaling to sensitize chemotoxicity of cisplatin against non-small cell lung cancer cells in vitro and in vivo. Phytomedicine 53:18–27
75. Summary C. PubChem CID: 442523 Structure: 2020;2020:1–13

76. Summary C. PubChem CID: 72303 Structure: 2020;
77. Arora S, Singh S, Piazza GA, Contreras CM, Panyam J, Singh AP (2012) Honokiol: a novel natural agent for Cancer prevention and therapy. Curr Mol Med 12(10):1244–1252
78. Yodkeeree S, Wongsirisin P, Pompimon W, Limtrakul P (2013) Anti-invasion effect of crebanine and O-methylbulbocapnine from stephania venosa via down-regulated matrix metalloproteinases and urokinase plasminogen activator. Chem Pharm Bull 61 (11):1156–1165
79. Summary C. PubChem CID: 333544 Structure: 2020;2020[20]:1–15
80. Summary C. O-Methylbulbocapnine. 2020;1–12
81. Umsumarng S, Pitchakarn P, Yodkeeree S, Punfa W, Mapoung S, Ramli RA et al (2017) Modulation of P-glycoprotein by Stemona alkaloids in human multidrug resistance leukemic cells and structural relationships. Phytomedicine 34:182–190
82. Structures FS. 1 Structures 1.1. 2019;2020:1–25
83. Structures FS. 3/1/2020 Capsaicin | C18H27NO3 - PubChem. 2020
84. Antonio AS, LSM W, Veiga Junior VF (2018) The genus: capsicum: a phytochemical review of bioactive secondary metabolites. RSC Adv 8:25767–25784
85. PubChem. 11/4/2019 Piperine | C17H19NO3 - PubChem. 2019;1–38
86. Jeon HJ, Kim K, Kim YD, Lee SE (2019) Naturally occurring piper plant amides potential in agricultural and pharmaceutical industries: perspectives of piperine and piperlongumine. Appl Biol Chem 62:1–7
87. Summary C. PubChem CID: 637858 Structure: 2020
88. Summary C. PubChem CID: 73078 Structure: 2020
89. Liu T, Liu X, Li W (2016) Tetrandrine, a Chinese plant-derived alkaloid, is a potential candidate for cancer chemotherapy. Oncotarget 7(26):40800
90. Summary C. PubChem CID: 457825 Structure: 2020;1–17
91. Wang TX, Yang XH (2008) Reversal effect of isotetrandrine, an isoquinoline alkaloid extracted from caulis Mahoniae, on P-glycoprotein-mediated doxorubicin-resistance in human breast cancer [MCF-7/DOX] cells. Yaoxue Xuebao 43(5):461–466
92. Su J, Lai H, Chen J, Li L, Wong YS, Chen T et al (2013) Natural Borneol, a Monoterpenoid compound, potentiates Selenocystine-induced apoptosis in human hepatocellular carcinoma cells by enhancement of cellular uptake and activation of ROS-mediated DNA damage. PLoS One 8(5):e63502
93. Summary C. PubChem CID: 64685 Structure: 2020;1–51
94. Summary C. 3/1/2020 Thymoquinone | C10H12O2 - PubChem. 2020;1–34
95. Nessa MU, Beale P, Chan C, Yu JQ, Huq F (2011) Synergism from combinations of cisplatin and oxaliplatin with quercetin and thymoquinone in human ovarian tumour models. Anticancer Res 31(11):3789–3797
96. Summary C. PubChem CID: 9848024 Structure: 2020;1–17
97. Pfisterer PH, Rollinger JM, Schyschka L, Rudy A, Vollmar AM, Stuppner H (2010) Neoandrographolide from andrographis paniculata as a potential natural chemosensitizer. Planta Med 76(15):1698–1700
98. Li C, Guan X, Xue H, Wang P, Wang M, Gai X (2017) Reversal of P-glycoprotein-mediated multidrug resistance is induced by saikosaponin D in breast cancer MCF-7/adriamycin cells. Pathol Res Pract 213(7):848–853
99. Summary C. Saikosaponin d. Nat Compd. 2013;502–502
100. Teng YN, Sheu MJ, Hsieh YW, Wang RY, Chiang YC, Hung CC (2016) β-Carotene reverses multidrug resistant cancer cells by selectively modulating human P-glycoprotein function. Phytomedicine 23(3):316–323
101. Structures FS. 3/1/2020 beta-Carotene | C40H56 - PubChem. 2020;
102. He R, Shi X, Zhou M, Zhao Y, Pan S, Zhao C et al (2018) Alantolactone induces apoptosis and improves chemosensitivity of pancreatic cancer cells by impairment of autophagy-lysosome pathway via targeting TFEB. Toxicol Appl Pharmacol 356:159–171
103. Summary C. PubChem CID: 72724 Structure: 2020

104. Long S, Sousa E, Kijjoa A, Pinto MMM (2016) Marine natural products as models to circumvent multidrug resistance. Molecules 21(7):892
105. Wang L-H, Yang J-Y, Yang S-N, Li Y, Ping G-F, Hou Y et al (2014) Suppression of NF-κB signaling and P-glycoprotein function by Gambogic acid synergistically potentiates Adriamycin -induced apoptosis in lung Cancer. Curr Cancer Drug Targets 14 (1):91–103
106. Summary C. PubChem CID: 9852185 Structure: 2020;1–20
107. Račkauskas R, Zhou D, Uselis S, Strupas K, Herr I, Schemmer P (2017) Sulforaphane sensitizes human cholangiocarcinoma to cisplatin via the downregulation of anti-apoptotic proteins. Oncol Rep 37(6):3660–3666
108. Summary C. PubChem CID: 5350 Structure: 2020;98
109. Jiang G, Liu J, Ren B, Zhang L, Owusu L, Liu L et al (2017) Anti-tumor and chemosensitization effects of Cryptotanshinone extracted from Salvia miltiorrhiza Bge. On ovarian cancer cells in vitro. J Ethnopharmacol 205:33–40
110. Summary C. PubChem CID: 160254 Structure: 2020;1–23
111. Nessa MU, Beale P, Chan C, Yu JQ, Huq F (2012) Studies on combination of platinum drugs cisplatin and oxaliplatin with phytochemicals anethole and curcumin in ovarian tumour models. Anticancer Res 32(11):4843–4850
112. Summary C. PubChem CID: 637563 Structure: 2020;1–50
113. Summary C. PubChem CID: 6474107 Structure: 2020
114. Miyamoto K, Yamashita T, Tsukiyama T, Kitamura N, Minami N, Yamada M et al (2008) Reversible membrane Permeabilization of mammalian cells treated with Digitonin and its use for inducing nuclear reprogramming by *Xenopus* egg extracts. Cloning Stem Cells 10 (4):535–542
115. Wang L, Ma Q (2018) Clinical benefits and pharmacology of scutellarin: a comprehensive review. Pharmacol Ther 190:105–127
116. Jeng LB, Kumar Velmurugan B, Chen MC, Hsu HH, Ho TJ, Day CH et al (2018) Fisetin mediated apoptotic cell death in parental and Oxaliplatin/irinotecan resistant colorectal cancer cells in vitro and in vivo. J Cell Physiol 233(9):7134–7142
117. Khan F, Niaz K, Maqbool F, Hassan FI, Abdollahi M, Nagulapalli Venkata KC et al (2016) Molecular targets underlying the anticancer effects of quercetin: an update. Nutrients 8:529
118. Yu M, Qi B, Xiaoxiang W, Xu J, Liu X (2017) Baicalein increases cisplatin sensitivity of A549 lung adenocarcinoma cells via PI3K/Akt/NF-κB pathway. Biomed Pharmacother 90:677–685
119. Ma W, Feng S, Yao X, Yuan Z, Liu L, Xie Y (2015) Nobiletin enhances the efficacy of chemotherapeutic agents in ABCB1 overexpression cancer cells. Sci Rep 5:18789
120. Deng G, Zeng S, Ma J, Zhang Y, Qu Y, Han Y et al (2017) The anti-tumor activities of Neferine on cell invasion and oxaliplatin sensitivity regulated by EMT via snail signaling in hepatocellular carcinoma. Sci Rep 7(December 2016):1–14
121. Miranda MA, Mondal A, Sachdeva M, Cabral H, Neto YAAH, Khan I et al (2019) Chemosensitizing effect of Cernumidine extracted from *Solanum cernuum* on bladder Cancer cells *in Vitro*. Chem Bfiodivers [Internet] 16(10):e1900334. 12 [cited 2020 Jan 11]Available from: https://onlinelibrary.wiley.com/doi/abs/10.1002/cbdv.201900334
122. Gorgani L, Mohammadi M, Najafpour GD, Nikzad M (2017) Piperine—the bioactive compound of black pepper: from isolation to medicinal formulations. Compr Rev Food Sci Food Saf 16(1):124–140
123. Huang KJ, Kuo CH, Chen SH, Lin CY, Lee YR (2018) Honokiol inhibits and growth of oral squamous cell carcinoma through induction of apoptosis, cell cycle arrest and autophagy. J Cell Mol Med 22(3):1894–1908
124. Yodkeeree S, Pompimon W, Limtrakul P (2014) Crebanine, an aporphine alkaloid, sensitizes TNF-α-induced apoptosis and suppressed invasion of human lung adenocarcinoma cells A549 by blocking NF-κB-regulated gene products. Tumor Biol 35(9):8615–8624

125. Chanmahasathien W, Ampasavate C, Greger H, Limtrakul P (2011) Stemona alkaloids, from traditional Thai medicine, increase chemosensitivity via P-glycoprotein-mediated multidrug resistance. Phytomedicine 18(2–3):199–204
126. O'Neill J, Brock C, Olesen AE, Andresen T, Nilsson M, Dickenson AH (2012) Unravelling the mystery of capsaicin: a tool to understand and treat pain. Pharmacol Rev 64(4):939–971
127. Piska K, Gunia-Krzyżak A, Koczurkiewicz P, Wójcik-Pszczoła K, Pękala E (2018) Piperlongumine [piplartine] as a lead compound for anticancer agents – synthesis and properties of analogues: a mini-review. Eur J Med Chem 156:13–20
128. Wang T-X, Yang X-H (2008) Reversal effect of isotetrandrine, an isoquinoline alkaloid extracted from caulis Mahoniae, on P-glycoprotein-mediated doxorubicin-resistance in human breast cancer [MCF-7/DOX] cells. Yao Xue Xue Bao 43(5):461–466
129. Wang L, Ci X, Lv H, Wang X, Qin FX, Cheng G (2016) Isotetrandrine ameliorates tert-butyl hydroperoxide-induced oxidative stress through upregulation of heme oxygenase-1 expression. Exp Biol Med 241(14):1568–1576
130. Li DG, Wang ZR, Lu HM (2001) Pharmacology of tetrandrine and its therapeutic use in digestive diseases. World J Gastroenterol 7(5):627–629
131. Wen C, Fu L, Huang J, Dai Y, Wang B, Xu G et al (2019) Curcumin reverses doxorubicin resistance via inhibition the efflux function of ABCB4 in doxorubicin-resistant breast cancer cells. Mol Med Rep 19(6):5162–5168
132. Su P, Yang Y, Wang G, Chen X, Ju Y (2018) Curcumin attenuates resistance to irinotecan via induction of apoptosis of cancer stem cells in chemoresistant colon cancer cells. Int J Oncol 53 (3):1343–1353
133. Kang JH, Kang HS, Kim IK, Lee HY, Ha JH, Yeo CD et al (2015) Curcumin sensitizes human lung cancer cells to apoptosis and metastasis synergistically combined with carboplatin. Exp Biol Med 240(11):1416–1425
134. Shah K, Mirza S, Desai U, Jain N, Rawal R (2015) Synergism of Curcumin and Cytarabine in the down regulation of multi-drug resistance genes in acute myeloid leukemia. Anti Cancer Agents Med Chem 16(1):128–135
135. Chen P, Li J, Jiang HG, Lan T, Chen YC (2015) Curcumin reverses cisplatin resistance in cisplatin-resistant lung caner cells by inhibiting FA/BRCA pathway. Tumor Biol 36 (5):3591–3599
136. Shakibaei M, Buhrmann C, Kraehe P, Shayan P, Lueders C, Goel A (2014) Curcumin chemosensitizes 5-fluorouracil resistant MMR-deficient human colon cancer cells in high density cultures. PLoS One 9(1):1–12
137. Datta S, Sinha D (2019) EGCG maintained Nrf2-mediated redox homeostasis and minimized etoposide resistance in lung cancer cells. J Funct Foods 62(September):103553
138. Qian F, Wei D, Zhang Q, Yang S (2005) Modulation of P-glycoprotein function and reversal of multidrug resistance by [−]-epigallocatechin gallate in human cancer cells. Biomed Pharmacother 59(3):64–69
139. Kumazaki M, Noguchi S, Yasui Y, Iwasaki J, Shinohara H, Yamada N et al (2013) Anti-cancer effects of naturally occurring compounds through modulation of signal transduction and miRNA expression in human colon cancer cells. J Nutr Biochem 24(11):1849–1858
140. Santandreu FM, Valle A, Oliver J, Roca P (2011) Resveratrol potentiates the cytotoxic oxidative stress induced by chemotherapy in human colon cancer cells. Cell Physiol Biochem 28(2):219–228
141. de Oliveira JMP F, Pacheco AR, Coutinho L, Oliveira H, Pinho S, Almeida L et al (2018) Combination of etoposide and fisetin results in anti-cancer efficiency against osteosarcoma cell models. Arch Toxicol 92(3):1205–1214
142. Te Lin M, Lin CL, Lin TY, Cheng CW, Yang SF, Lin CL et al (2016) Synergistic effect of fisetin combined with sorafenib in human cervical cancer HeLa cells through activation of death receptor-5 mediated caspase-8/caspase-3 and the mitochondria-dependent apoptotic pathway. Tumor Biol 37(5):6987–6996

143. Moon JY, Hung LVM, Unno T, Cho SK (2018) Nobiletin enhances chemosensitivity to adriamycin through modulation of the Akt/GSK3β/β–catenin/ MYCN/MRP1 signaling pathway in A549 human non-small-cell lung cancer cells. Nutrients 10(12):1829
144. Iriti M, Kubina R, Cochis A, Sorrentino R, Varoni EM, Kabała-Dzik A et al (2017) Rutin, a Quercetin glycoside, restores Chemosensitivity in human breast Cancer cells. Phyther Res 31 (10):1529–1538
145. Molavi O, Narimani F, Asiaee F, Sharifi S, Tarhriz V, Shayanfar A et al (2017 Jan) Silibinin sensitizes chemo-resistant breast cancer cells to chemotherapy. Pharm Biol 55(1):729–739
146. Li J, Li B, Xu WW, Chan KW, Guan XY, Qin YR et al (2016) Role of AMPK signaling in mediating the anticancer effects of silibinin in esophageal squamous cell carcinoma. Expert Opin Ther Targets 20(1):7–18
147. Yan Y, Li J, Han J, Hou N, Song Y, Dong L (2015) Chlorogenic acid enhances the effects of 5-fluorouracil in human hepatocellular carcinoma cells through the inhibition of extracellular signal-regulated kinases. Anti-Cancer Drugs 26(5):540–546
148. Abaza MSI, Orabi KY, Al-Quattan E, Al-Attiyah RJ (2015) Growth inhibitory and chemosensitization effects of naringenin, a natural flavanone purified from Thymus vulgaris, on human breast and colorectal cancer. Cancer Cell Int 15(1):1–19
149. Huang H, Chen AY, Ye X, Li B, Rojanasakul Y, Rankin GO et al (2015) Myricetin inhibits proliferation of cisplatin-resistant cancer cells through a p53-dependent apoptotic pathway. Int J Oncol 47(4):1494–1502
150. Wang L, Feng J, Chen X, Guo W, Du Y, Wang Y et al (2014) Myricetin enhance chemosensitivity of 5-fluorouracil on esophageal carcinoma in vitro and in vivo. Cancer Cell Int 14(1):1–8
151. Neshati V, Matin MM, Bahrami AR, Iranshahi M, Rassouli FB, Saeinasab M (2012) Increasing the cisplatin cytotoxicity and cisplatin-induced DNA damage by conferone in 5637 cells. Nat Prod Res 26(18):1724–1727
152. Zhang P, Shi L, Zhang T, Hong L, He W, Cao P et al (2019 Dec) Piperlongumine potentiates the antitumor efficacy of oxaliplatin through ROS induction in gastric cancer cells. Cell Oncol 42(6):847–860
153. Xie Z, Wei Y, Xu J, Lei J, Yu J (2019) Alkaloids from Piper nigrum synergistically enhanced the effect of paclitaxel against paclitaxel-resistant cervical Cancer cells through the Downregulation of Mcl-1. J Agric Food Chem 67(18):5159–5168
154. Li S, Lei Y, Jia Y, Li N, Wink M, Ma Y (2011) Piperine, a piperidine alkaloid from Piper nigrum re-sensitizes P-gp, MRP1 and BCRP dependent multidrug resistant cancer cells. Phytomedicine 19(1):83–87
155. Te Chang M, Lee SP, Fang CY, Hsieh PL, Liao YW, Lu MY et al (2018) Chemosensitizing effect of honokiol in oral carcinoma stem cells via regulation of IL-6/Stat3 signaling. Environ Toxicol 33(11):1105–1112
156. Zhu X, Sui M, Fan W (2005) In vitro and in vivo characterizations of tetrandrine on the reversal of P-glycoprotein-mediated drug resistance to paclitaxel. Anticancer Res 25 (3 B):1953–1962
157. Huang XC, Sun YL, Salim AA, Chen ZS, Capon RJ (2013) Parguerenes: marine red alga bromoditerpenes as inhibitors of P-glycoprotein [ABCB1] in multidrug resistant human cancer cells. Biochem Pharmacol 85(9):1257–1268
158. Li D, Yang J-Y, Zhou C-J, Ma H-Z, Sun C-L, Feng R-J (2016) Sulforaphane enhances cisplatin sensitivity in human osteosarcoma cells through upregulation of p53-p21 pathway by enhancing G1 arrest. Int J Clin Exp Med 9:1621–1629
159. Wang Y, Lu HL, Liu YD, Yang LY, Jiang QK, Zhu XJ et al (2017) Cryptotanshinone sensitizes antitumor effect of paclitaxel on tongue squamous cell carcinoma growth by inhibiting the JAK/STAT3 signaling pathway. Biomed Pharmacother 95(August):1388–1396
160. Wang LH, Li Y, Yang SN, Wang FY, Hou Y, Cui W et al (2014) Gambogic acid synergistically potentiates cisplatin-induced apoptosis in non-small-cell lung cancer through suppressing NF-κB and MAPK/HO-1 signalling. Br J Cancer 110(2):341–352

161. Mon MT, Yodkeeree S, Punfa W, Pompimon W, Limtrakul P (2018) Alkaloids from Stephania venosa as chemo-sensitizers in SKOV3 ovarian cancer cells via Akt/NF-κB signaling. Chem Pharm Bull 66(2):162–169
162. Li H, Krstin S, Wang S, Wink M (2018) Capsaicin and piperine can overcome multidrug resistance in cancer cells to doxorubicin. Molecules 23(3):1–11
163. Li H, Krstin S, Wink M (2018) Modulation of multidrug resistant in cancer cells by EGCG, tannic acid and curcumin. Phytomedicine 50:213–222
164. Chen H, Landen CN, Li Y, Alvarez RD, Tollefsbol TO (2013) Epigallocatechin gallate and sulforaphane combination treatment induce apoptosis in paclitaxel-resistant ovarian cancer cells through hTERT and Bcl-2 down-regulation. Exp Cell Res 319(5):697–706
165. Alimbetov D, Askarova S, Umbayev B, Davis T, Kipling D (2018) Pharmacological targeting of cell cycle, apoptotic and cell adhesion signaling pathways implicated in chemoresistance of cancer cells. Int J Mol Sci 19(6):1690
166. Hientz K, Mohr A, Bhakta-Guha D, Efferth T (2017) The role of p53 in cancer drug resistance and targeted chemotherapy. Oncotarget 8(5):8921–8946
167. Morisaki T, Katano M (2005) Mitochondria-targeting therapeutic strategies for overcoming Chemoresistance and progression of Cancer. Curr Med Chem 10(23):2517–2521
168. Angelini A, Di Ilio C, Castellani ML, Conti P, Cuccurullo F (2010) Modulation of multidrug resistance p-glycoprotein activity by flavonoids and honokiol in human doxorubicin- resistant sarcoma cells [MES-SA/Dx-5]: implications for natural sedatives as chemosensitizing agents in cancer therapy. J Biol Regul Homeost Agents 24(2):197–205
169. Mohammad RM, Muqbil I, Lowe L, Yedjou C, Hsu H-Y, Lin L-T et al (2015) Broad targeting of resistance to apoptosis in cancer HHS public access. Semin Cancer Biol [Internet] 35:78–103. [cited 2020 Jan 10]. Available from: http://globocan.iarc.fr/
170. Liao YX, Yu HY, Lv JY, Cai YR, Liu F, He ZM et al (2019) Targeting autophagy is a promising therapeutic strategy to overcome chemoresistance and reduce metastasis in osteosarcoma. Int J Oncol 55(6):1213–1222
171. Pakravan K, Mahjoub MA, Jahangiri B, Babashah S (2015) Cancer stem cells: a quick walk through the concepts. In: Cancer stem cells: emerging concepts and future perspectives in translational oncology [internet]. Springer International Publishing, Cham, pp 3–11. [cited 2020 Jan 11]. Available from: http://link.springer.com/10.1007/978-3-319-21030-8_1
172. Nunes T, Hamdan D, Leboeuf C, El Bouchtaoui M, Gapihan G, Nguyen TT et al (2018) Targeting cancer stem cells to overcome chemoresistance. Int J Mol Sci 19(12):4036
173. Angelini A, Conti P, Ciofani G, Cuccurullo F, Di Ilio C (2013) Modulation of multidrug resistance P-glycoprotein activity by antiemetic compounds in human doxorubicin-resistant sarcoma cells [MES-SA/Dx-5]: implications on cancer therapy. J Biol Regul Homeost Agents 27(4):1029–1037
174. Shen M, Chan TH, Dou QP (2012) Targeting tumor Chemosensitization pathway with polyphenols for. Anti Cancer Agents Med Chem 12:891–901
175. Lin S, Chang C, Hsu C, Tsai M, Cheng H, Leong MK et al (2019) Natural compounds as potential adjuvants to cancer therapy: preclinical evidence. Br J Pharmacol 177(6):1409–1423
176. Wu R, Mei X, Ye Y, Xue T, Wang J, Sun W et al (2019) Zn[II]-curcumin solid dispersion impairs hepatocellular carcinoma growth and enhances chemotherapy by modulating gut microbiota-mediated zinc homeostasis. Pharmacol Res 150:104454
177. Lopes-Rodrigues V, Oliveira A, Correia-da-Silva M, Pinto M, Lima RT, Sousa E et al (2017) A novel curcumin derivative which inhibits P-glycoprotein, arrests cell cycle and induces apoptosis in multidrug resistance cells. Bioorganic Med Chem. 25(2):581–596
178. Mapoung S, Pitchakarn P, Yodkeeree S, Ovatlarnporn C, Sakorn N, Limtrakul P (2016) Chemosensitizing effects of synthetic curcumin analogs on human multi-drug resistance leukemic cells. Chem Biol Interact 244:140–148
179. Limtrakul P, Chearwae W, Shukla S, Phisalphong C, Ambudkar SV (2007) Modulation of function of three ABC drug transporters, P-glycoprotein [ABCB1], mitoxantrone resistance

protein [ABCG2] and multidrug resistance protein 1 [ABCC1] by tetrahydrocurcumin, a major metabolite of curcumin. Mol Cell Biochem 296(1–2):85–95
180. Colin D, Gimazane A, Lizard G, Izard JC, Solary E, Latruffe N et al (2009) Effects of resveratrol analogs on cell cycle progression, cell cycle associated proteins and 5fluoro-uracil sensitivity in human derived colon cancer cells. Int J Cancer 124(12):2780–2788
181. Umsumarng S, Pitchakarn P, Sastraruji K, Yodkeeree S, Ung AT, Pyne SG et al (2015) Reversal of human multi-drug resistance leukaemic cells by stemofoline derivatives via inhibition of p-glycoprotein function. Basic Clin Pharmacol Toxicol 116(5):390–397
182. Chen LM, Liang YJ, Zhang X, Su XD, Dai CL, Wang FP et al (2009) Reversal of P-gp-mediated multidrug resistance by bromotetrandrine in vivo is associated with enhanced accumulation of chemotherapeutical drug in tumor tissue. Anticancer Res 29(11):4597–4604
183. Bisi A, Cappadone C, Rampa A, Farruggia G, Sargenti A, Belluti F et al (2017) Coumarin derivatives as potential antitumor agents: growth inhibition, apoptosis induction and multidrug resistance reverting activity. Eur J Med Chem 127:577–585
184. Shin SY, Jung H, Ahn S, Hwang D, Yoon H, Hyun J et al (2014) Polyphenols bearing cinnamaldehyde scaffold showing cell growth inhibitory effects on the cisplatin-resistant A2780/Cis ovarian cancer cells. Bioorganic Med Chem 22(6):1809–1820
185. Chou HL, Fong Y, Wei CK, Tsai EM, Chen JYF, Chang WT et al (2017) A quinone-containing compound enhances camptothecin-induced apoptosis of lung cancer through modulating endogenous ROS and ERK signaling. Arch Immunol Ther Exp 65(3):241–252
186. Yang C, Wong ILK, Peng K, Liu Z, Wang P, Jiang T et al (2017) Extending the structure–activity relationship study of marine natural ningalin B analogues as P-glycoprotein inhibitors. Eur J Med Chem 125:795–806
187. Guo S, Lv L, Shen Y, Hu Z, He Q, Chen X (2016) A nanoparticulate pre-chemosensitizer for efficacious chemotherapy of multidrug resistant breast cancer. Sci Rep 6(January):1–11
188. Yang M, Yu L, Guo R, Dong A, Lin C, Zhang J (2018) A modular coassembly approach to all-in-one multifunctional nanoplatform for synergistic codelivery of doxorubicin and curcumin. Nanomaterials 8(3):1–18
189. Alemi A, Zavar Reza J, Haghiralsadat F, Zarei Jaliani H, Haghi Karamallah M, Hosseini SA et al (2018) Paclitaxel and curcumin coadministration in novel cationic PEGylated niosomal formulations exhibit enhanced synergistic antitumor efficacy. J Nanobiotechnol 16(1):28
190. Daglioglu C (2017) Enhancing tumor cell response to multidrug resistance with pH-sensitive quercetin and doxorubicin conjugated multifunctional nanoparticles. Colloids Surfaces B Biointerfaces 156:175–185
191. Katiyar SS, Muntimadugu E, Rafeeqi TA, Domb AJ, Khan W (2016) Co-delivery of rapamycin- and piperine-loaded polymeric nanoparticles for breast cancer treatment. Drug Deliv 23(7):2608–2616
192. Minaei A, Sabzichi M, Ramezani F, Hamishehkar H, Samadi N (2016) Co-delivery with nano-quercetin enhances doxorubicin-mediated cytotoxicity against MCF-7 cells. Mol Biol Rep 43 (2):99–105

Dietary Phytochemicals as Epigenetic Modulators in Cancer Prevention: Emerging Research Trends, Gaps, and Future Perspectives

5

Ritwija Bhattacharya, Shrinjana Dhar, Aniruddha Mukhopadhyay, and Pritha Bhattacharjee

Abstract

Studies over the last few decades have revealed that cancer is not only genetic, this is an epigenetic phenomenon as well. Present studies have elucidated the involvement of epigenetic regulations in cancer progression. Several epigenetic modifications involving histone, DNA, and non-coding RNA have been identified, of which methylation, histone modifications, and miRNA mediated regulations have been commonly reported. Several studies have reported a close association between diet rich in phytochemicals and reduced prevalence of cancer. It has been found that phytochemicals alter gene expression and signalling pathways. Nutriepigenetics is becoming a promising field and the interplay of bioactive molecules with epigenome plays a potential role in cancer chemoprevention. Most of the phytochemicals, including alkaloids, polyphenols, organosulfurs, terpenes, etc., have shown promising effects as epigenetic modulators in the prevention of cancer. In this endeavor, we have collated the studies on the epigenetic potential of phytochemicals available from our daily dietary sources. In addition to this, we have emphasized the importance of in silico studies in epitherapeutic drug designing and its applications in translational research.

Keywords

Cancer · Bioinformatics · Dietary sources · Epigenetics · Phytochemicals

R. Bhattacharya · S. Dhar · A. Mukhopadhyay · P. Bhattacharjee (✉)
Department of Environmental Science, Ballygunge Science College, University of Calcutta, Kolkata, India

M. Kumar et al. (eds.), *Pharmacotherapeutic Botanicals for Cancer Chemoprevention*, https://doi.org/10.1007/978-981-15-5999-0_5

5.1 Introduction

Over the last few years, the immense emphasis has been put on epigenetics and its impact on the regulation of numerous diseases. "*Epigenetics*" as described in its usual definition refers to alterations in gene expression and chromatin organization without any changes occurring in the DNA sequence which are heritable upon cell division [1]. This phenomenon of epigenetics made it a potential target for the regulation of different diseases. Epigenetics takes part in multiple physiological processes such as development, imprinting, chromosomal stability, regulation of gene transcription, and X-chromosome inactivation [2] Therefore, this phenomenon becomes a potential target for various disease regulations. The epigenetic machinery mainly involves DNA methylation, histone modifications, chromatin structure remodeling, and non-coding RNA interactions. Epigenetic patterns of cancer cells are significantly disrupted which results in multiple anomalies and imbalance in DNA methylation and histone modifications. The reversible nature of epigenetic mechanisms makes it feasible for a therapeutic target [3]. It is evident from various studies that environmental factors, diet, heavy metals exposure induce epigenetic changes. Interaction of nutrients with epigenome can play a significant role in cancer chemotherapy. In both developing and developed countries, cancer is one of the major causes of death. In spite of rigorous intervention, a huge number of patients are suffering from poor diagnosis. Hence, the demand for early detection and new anticancer agents with fewer side effects is sustained. Recently, several in vivo and in vitro studies have highlighted the chemotherapeutic potential of "*Phytochemicals*" present in our daily diet as they can inhibit cell proliferation and tumor growth. However, the scientific validation of preventing the risk of cancer remains an issue of debate.

5.2 Epigenetic Mechanisms in Cancer Regulation: Introduction to the Major Epigenetic Phenomena in Cancer Progression

5.2.1 DNA Methylation and Its Role in Cancer

DNA methylation can take part in embryonic development, inactivation of chromosome X, and silencing of genes. DNA methylation can be often explained as the addition of –CH3 molecule to the fifth carbon atom of the cytosine ring by the DNA-methyl transferase (DNMT) enzyme. DNA-methyl transferases include *DNMT1*, *DNMT2*, *DNMT3A*, *DNMT3B,* and *DNMT3L* which plays a pivotal role in the methylation process. The CpG islands in the promoter regions of some genes show the maximum amount of methylation. Methylation blocks other transcription factors to bind to the promoter region resulting in gene silencing. There are mainly two patterns of DNA methylation as studied so far, hypomethylation and hypermethylation [4, 5].

5.2.1.1 Hypermethylation of DNA

Hypermethylation is the most frequently considered epigenetic mechanism associated with cancerous outcome [6, 7]. Hypermethylation of CpG island in the promoter region results in silencing of various tumor suppressor genes having important functions in the regulation of cell cycle (*p16INK4a*, *p14ARF*, *p15INK4b*, *RB1*), DNA repair (*hMLH1*, *BRCA1*, *MGMT*, *WRN*), apoptosis (*DAPK*, *TMS1*, *SFRP1*), etc. [8]. Some DNMT-inhibitor drugs have been identified for cancer treatment. Gene silencing may vary depending upon different types of cancer. In some cases, like colon cancer, hypermethylation can be used as one of the potential biomarkers [8, 9]. DNA methylation of various genes in different types of cancer can be utilized as a diagnostic tool for early prediction. Banerjee et al., 2009 in his article, have highlighted the methylation markers specific to different cancers [10] (Table 5.1).

5.2.1.2 Hypomethylation of DNA

Hypomethylation in the promoter region of a particular gene will turn on gene expression. DNA methylation has been found to be involved in various diseases like cancers, lupus, muscular dystrophy, and other lifestyle and inherited diseases [4, 5]. Compared with the normal cells, cancer cells are prone to global hypomethylation.

Studies have shown that DNA hypomethylation can induce chromosomal instability, loss of imprinting, and reactivation of transposable elements in cancer [40]. Loss of imprinting of *IGF2* (the insulin-like growth factor gene) has been observed among the patients with hereditary Beckwith-Wiedemann syndrome, showing an elevated risk of cancer development. Loss of methylation has turned on repetitive DNA sequences and introns [41] which causes open chromatins in those particular regions resulting in DNA breaks and mitotic recombination of

Table 5.1 DNA methylation markers of different cancers

S. No.	Types of cancer	Methylation markers	References
1	Breast cancer	*APC, BRCA1, BCSG1, CDH1, RASSF1a, RUNX3CXCL12, HIC-1, PROX1, RARβ, TMS1, BCSG1, Cyclin D2, GSTP1*	[11, 12]
2	Brain cancer	*MSN, POU3F4, S100A10, S100A6, HTATIP2, CDH1, LXN, HTR2C, COL1A2*	[13]
3	Cervical cancer	*SOX1, RASSF1A, ONECUT1NKX6-1, WT1, DNMT3L, HS3ST2, CDH1, SPARC, TFP12, PAX1, LMX1A, RRAD, SFRP1, SFRP2, SFRP3, MINT31, PTEN*	[14–18]
4	Colon cancer	*PGP95, LRRC3B, HACE1, BAGE, hMLH1, NGFR, BMP3, EYA2, MAL, TMEFF2, ALX4, APC, DAPK, MGMT, LINE1*	[19–27]
5	Lung cancer	*CDH13, RASSF1, p15, MGMT RARβ, DAPK, p16,*	[28–31]
6	Ovarian cancer	*OPCML, FANCF, IGFBP-3, GSTP1, ER-α, MLH1, ANGPTL-2, hMLH1*	[32–38]
7	Prostate cancer	*GSTP1, RASF1A, CDH1, MDR1*	[39]

homologous repetitive sequences and increases the risk of cancer. Hypomethylation in cancer cells, causes reactivation of endoparasitic DNA, *L1* (long interspersed nuclear elements), and *Alu* (recombinogenic sequence) repeat which increases chromosomal instability by the translocation of these unmethylated transposons to other genomic regions [42].

5.2.2 Histone Modifications and Its Role in Cancer

Histone modification is another significant tool in epigenetic mechanisms. Modification of histones may occur by several mechanisms such as acetylation, methylation, SUMOylation, ADP-ribosylation, ubiquitylation, and phosphorylation. Histone modifications influence the recruitment of several transcriptional regulatory binding proteins, thereby significantly regulate gene expression pattern.

When H3 is trimethylated at lysine 4 (*H3K4me3*) the promoters remain activated and again inactivated upon methylation of H3 at Lys27 (*H3K27me3).* Distinct aberrant histone modifications have been reported in cancer cells such as losses in histone methylation and acetylation in H4 at the acetylated Lys16 and trimethylated Lys20 residues have been reported [43]. It was found that the deletion of *EZH2*, a *H3K27* methyltransferase, increased the frequency of spontaneous T-cell leukemia [44]. In basal carcinoma, histone acetylation (*H3K18ac* and *H4K12ac*) and histone methylation (*H3K4me2, H4K20me3, H4R3me2*) were reported [45]. Patients with low levels of H3K4 acetylation and patients with high levels of H3K27me3 showed association with the progression of oral squamous cell carcinoma [46]. Alterations in the histone-modifying enzymes such as histone deacetylases (HDACs), histone acetyltransferases (HATs), and histone methyltransferases (HMTs) have been commonly reported in cancer [47]. HDAC is a major therapeutic target for cancer prevention due to its ubiquitous expression in malignant cancer cells.

However, several studies have reported that HAT in association with HDAC can regulate the levels of histone acetylation. In leukemia, chromosomal translocation of HAT and HAT related genes (*MOZ, p300, MORF, CBP*) causes abnormality in the formation of fusion proteins [48]. The aberrant proteins alter global histone acetylation. Alterations of HMT activity cause tumor suppressor gene silencing which thereby results in different types of cancer. HMTs are over expressed in breast and prostate cancer and show a significant role in leukemic cancer progression. Several histone demethylase (HDM) have also been identified such as Lysine *demethylase 1* and *Jumonji C* domain proteins. HMTs work in association with lysine-specific demethylase and regulate global histone methylation [49]. Targeting HDMs for drug therapy can be a promising strategy in the future.

5.2.3 microRNA and Its Role in Cancer Epigenetics

miRNAs are small non-coding RNAs that can alter post-transcriptional gene expression and play a crucial role in the advancement and prevention of diseases.

Alterations in miRNA can be studied to understand the progression as well as the diagnosis of cancer. Both in vivo and in vitro studies displayed an association of miRNA with tumorigenesis. miRNA regulation of genes that are responsible for cell proliferation, apoptosis, and transcriptional regulation can lead to carcinogenesis. miRNA have roles in regulating tumor suppressors and oncogenes as well [50, 51].

5.2.3.1 Down-Regulation of miRNA

Many tumor suppressor genes are deregulated in cancer. miRNA 15 and 16 which repress *BCl2* (anti-apoptotic gene) are downregulated in leukemia and miRNA let-7 that represses an oncogene is downregulated in lung cancer [52]. miRNA of let-7 family was also found to be downregulated in breast cancer which was inversely correlated to *ERα* expression [53]. It was reported that the expression of miR-127, a tumor suppressor miRNA, was silenced by DNA methylation in cancer [54].

5.2.3.2 Upregulation of miRNA

Few studies revealed that oncogenic miRNAs such as miR-17, iR-92 are upregulated. The changes in DNA methylation status and chromatin remodeling among cancer and normal cell line play an imperative role underlying miRNA expression. In a microarray profiling of DNMT1- and DNMT3B-knockout HCT116 colon cancer cell line, 18 out of 320 human miRNA were found to be upregulated. There are several factors such as chromosomal aberrations, transcription factors that bring changes in miRNA expression which again depends on tissue specificity [51]. The crosstalk between the epigenetic machinery plays a very significant role in the progression of cancer (Fig. 5.1). Reactivation of miRNA in several cancer cells was reported with the treatment of DNMT and HDAC inhibitor drugs.

5.3 Epigenetic Modulators in Cancer Prevention

The immense prospective for epigenetic therapies lies within the certain fact that, contrasting to, genetic anomalies, epigenetic modifications are reversible. Various agents focusing on the epigenetic pathways are under progress toward the inside of clinical trials in late 70s, the study on the possibility to reverse epigenetic changes began with the finding of the agents that undo DNA methylation [55]. Recently, "Nutri-epigenetics," focusing on the influence of the nutritional factors on epigenome has come forward as a novel field of interest. Targeting abnormal epigenetic modifications has attained importance in cancer chemoprevention research. The perception of "Epi drugs" is at the forefront of drug which have been approved and some are undergoing the clinical trial. The universal demand for harmless, cost-effective, and readily accessible therapeutics has a renewed concern in plant-based drugs leading towards chronic use [56]. Through several attempts to recognize and build up anticancer agents that can effectively kill cancer cells without any damage to normal cells, naturally existing phytochemicals in food and plant sources have been identified [57].

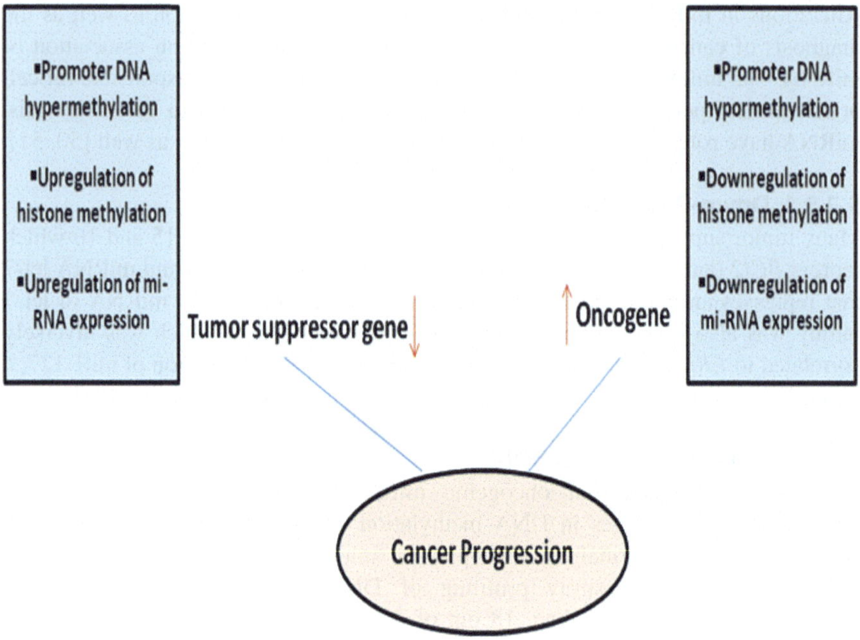

Fig. 5.1 Interplay of epigenetic machinery in the progression of cancer

According to the long established research studies plentiful bioactives from medicinal plant sources have been reported to have anticarcinogenic effects. Several in vitro and in vivo experiments pointed out the anti-proliferative, antimetastatic, antiangiogenic activities of some phytochemical compounds in the prevention of carcinogenesis. Nevertheless, a very few number of studies have been conducted in cancerous patients which shows inadequate evidence to prove their clinical efficiency.

5.3.1 Classification of Phytochemical and their Significance

Large groups of chemical compounds that are present in the fruits, vegetables, whole grain, seeds, nuts, plant-based beverages, fungi, herbs, and spices are often responsible for color, flavor, and aroma [58]. Based on chemical structure and characteristics dietary phytochemicals are classified into five major groups, viz. alkaloids, polyphenols, terpenes, organosulfur compounds, and other phytochemicals (carotenoid, saponins) as demonstrated in Fig. 5.2.

5.3.1.1 Alkaloids

Alkaloids are large groups of various chemical molecules having alkalies like properties and a nitrogen atom in a heterocyclic ring structure. Most of the natural products consist of around 18% alkaloids. The structure of alkaloids is often

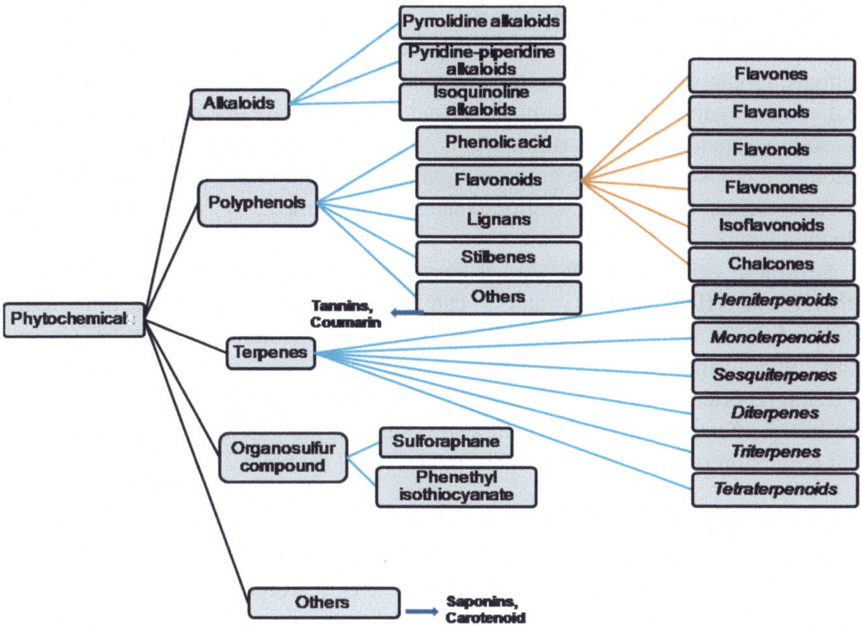

Fig. 5.2 Classification of phytochemicals

mimicked in the synthesis of chemical drugs as they have anticancer properties. They are found in more than 150 plant families, majorly found in higher dicotyledon plant families (*Apocynaceae, Papaveraceae, Papilionaceae, Ranunculaceae, Rubiaceae, Rutaceae, and Solanaceae*) and not commonly found in lower plants and fungi [59]. Depending on the heterocyclic ring patterns, they are categorized into different classes as follows: Pyrrolidine alkaloids (tetrahydropyrrole ring), Pyridine-piperidine alkaloids (pyridine ring joined to a piperidine ring), Isoquinoline alkaloids (heterocyclic isoquinoline ring). Alkaloids are widely known for their various pharmacological properties that include indole for anti-hypertensive effect, quinidine, spareien for antiarrhythmic effect, quinine for antimalarial effect and dimeric indoles, vincristine, vinblastine for anticancer effects.

Some of these components such as caffeine and morphine show stimulant properties [60]. Recently, alkaloids have gained popularity due to increased interest for herbal therapeutics. Cellular apoptosis is promoted by alkaloids via DNA damage pathway that seems to be a potential target for cancer therapy [61]. Also, these compounds effectively showed oncogene suppression by regulating the key signaling pathways of the cell cycle, cell proliferation, and metastasis which makes them the key target biomolecules for cancer therapy [60, 62]. Apart from this, alkaloids can also prevent cancer through epigenetic modulation such as inhibition of DNMT, HDAC activity, dysregulation of cancer related miRNA [63–65]. However, specific molecular mechanisms of dietary alkaloids inducing cancer chemoprevention are still under research.

5.3.1.2 Organosulfur Compounds

Organosulfurs are the sulfur containing compounds found in both plants and animals. Organosulfurs such as llicin, ajoene, and isothiocyanates show antimicrobial effects against both gram-positive and gram-negative bacteria. Sulforaphane (1-isothiocyanato-4-(methyl-sulfinyl)) is a widely studied biomolecule which belongs to the isothiocyanate group of organosulfur compounds. After the ingestion of cruciferous vegetables, sulforaphanes are formed by the hydrolysis of glucoraphanin.

Several in vitro and in vivo studies have reported that sulforaphane can suppress DNMT and HDAC activity and it can also regulate a cluster of microRNA (Table 5.2). In some cruciferous vegetables, Phenyl isothiocyanate (PEITC) and some other forms of isothiocyanates are found.

It has reported that *PEITC* can demethylate and activate *GSTP1. PEITC* shows an inhibitory effect against HDAC and also contributes to promoter hypomethylation and regulation of miR-194, miR-423-5p, etc. [127] (Table 5.2).

5.3.1.3 Terpenes

Terpenes are a class of hydrocarbons made of five carbon atoms that are attached to eight hydrogen atoms [CH2 = C(CH3)-CH=CH2]. Terpenes are available from both plant and animal sources. The oxygenated derivatives of these compounds are called terpenoids. Terpenes, mainly terpenoids, are present in around 27% of all natural compounds. Terpenoids are categorized into six major subgroups depending on the number of isoprene groups. They are hemiterpenoids (single isoprene unit), monoterpenoids (two isoprene units), sesquiterpenes (three isoprene units), diterpenes (have four *isoprene* units), and triterpenes (six isoprene units), tetraterpenoids (eight isoprene units). Medicinal values of terpenoids were reported. Perilla alcohol showed anti-carcinogenic properties, glycyrrhizin have anti-microbial activity, sesquiterpenoid has been used as an anti-malarial drug [60]. It was reported that terpenoids have DNMT-inhibitory effects and regulate miRNA in human and animal cell lines [127]. For example, Capsaicin, an active component of chili peppers, can enhance hMOF activity, HDAC, HAT activity and suppress LINE-1 retrotransposition in both human cell line and mouse model study [69–71].

5.3.1.4 Polyphenols

The most important phytochemicals that are present in vegetables, fruits, and beverages are the polyphenols. Polyphenols consist of large multiple phenol units. The physical, chemical, and biological properties of polyphenols vary depending on the number and structures of these phenol units. Polyphenols are categorized into phenolic acid, flavonoids, lignans, stilbenes, and others like tannins based on chemical composition. Flavonoid is again classified into flavones, flavanols, flavonols, flavonones, isoflavonoids, and chalcones. Fruits such as grapes, apples, cherries, pears, and beverages (mainly tea) are a good source of polyphenols. Polyphenols play a very significant role as epigenetic modulators. Studies reported that polyphenols were able to alter methylation, histone modifications, and miRNA regulation.

Table 5.2 Source of dietary phytochemicals and their role in cancer prevention: understanding the underlying epigenetic mechanisms

Compound name	Chemical structure	Group	Sources	Effects	Model system	References
Apigenin	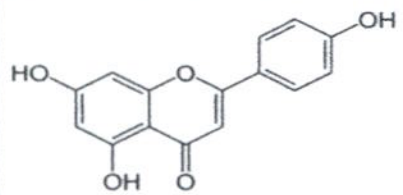	Polyphenol	Fruits and vegetables; especially parsley, celery, celeriac, chamomile tea	*Nrf2* promoter hypermethylation ↓, DNMT ↓, HDAC activity ↓, *p21/waf1* expression ↑, histone H3 acetylation ↑	Mouse cell line, human cell line, mice	[66–68]
Capsaicin	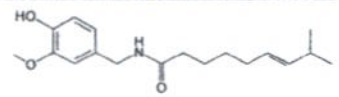	Terpenes	Pepper (Capsicum frutescens), including varieties like cayenne, green or red chili, spur or tabasco peppers, etc.	hMOF activity ↑, HDAC ↑, LINE-1 retrotransposition ↓, HAT ↑	Human cell line, mice	[69–71]
Curcumin	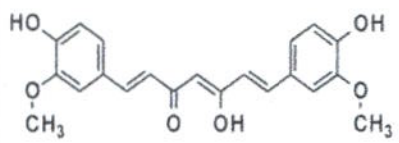 Enol form 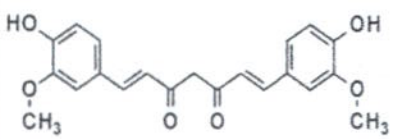 Keto form	Polyphenol	Turmeric	HAT ×, HDAC1 ↓, HDAC3 ↓, HDAC2 ×, histone H3 and H4 acetylation ↓, DNMT ×, miRNA-22 ↑, miRNA-199a* ×, miR-21 promoter activity ↓, miR-29B ↑, miR-145 ↑, miR-9 ↑, miR-125a-5p ↓, miR-146a ↑, miR-181b ↑	Human cell line, mice	[72–75]
EGCG	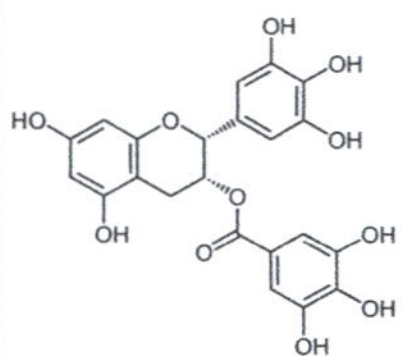	Polyphenol	Tea, apple skin, plums, onions, hazelnuts, pecans, and carob powder	HDAC1 ×, DNMT3B ×	Human cell line, Mice	[76–78]
Genistein	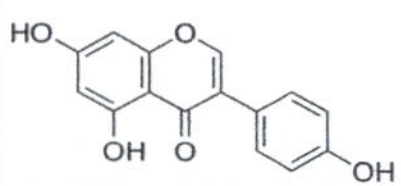	Polyphenol	Lupin, fava beans, soybeans, kudzu, and psoralea	HDAC1 ↓, DNMT ↓, Oncogenic miRNA ↓, HAT ↑	Human cell line	[79–81]

(continued)

Table 5.2 (continued)

Compound name	Chemical structure	Group	Sources	Effects	Model system	References
Kaempferol	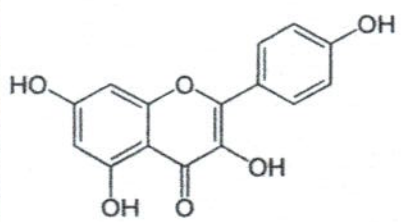	Polyphenol	Apples, grapes, tomatoes, green tea, potatoes, onions, broccoli, Brussels sprouts, squash, cucumbers, lettuce, green beans, peaches, blackberries, raspberries, and spinach.	HDAC ×, DNMT3B ×, miR-21 ↓, miR-101 ↑, miR-146a ↓, miR-27 ↑, miR-340 ↑, miR-26a-5p ↑	Human cell line, mice	[82–86]
Glabridin	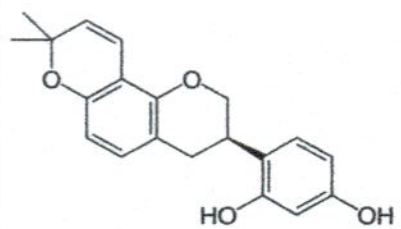	Polyphenol	Leguminous plants	miR-148a ↑, demethylation of miR-200c promoter ↑, DNMT1 ↓, DNMT3A ↓	Human cell line	[71, 87]
Lycopene	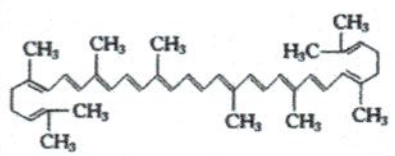	Flavonoid	Tomatoes, red carrots, watermelons, melons, papayas, asparagus and parsley.	Promoter demethylation ↑, global methylation of LINE-1 ×	Human cell line	[88]
Quercetin	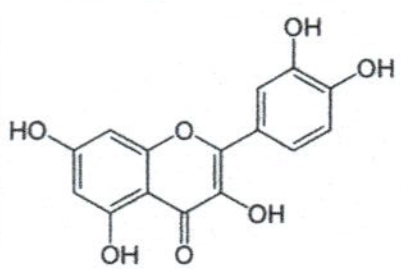	Polyphenol	Onions, Grapes, Berries, Cherries, Broccoli, And Citrus Fruits	HDAC ×, DNMT ×	Human cell line	[89]
Phenethyl isothiocyanate	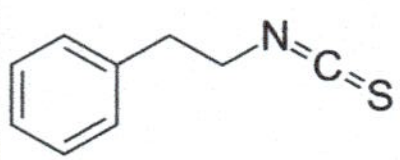	Organo-sulfur compound	Cruciferous vegetables such as broccoli, Brussels sprouts, and cabbages	Promoter methylation ↓, miR-194 ↑, miR-423-5p ↓, HDAC ×	Microarray, human cell line	[90–93]
Resveratrol	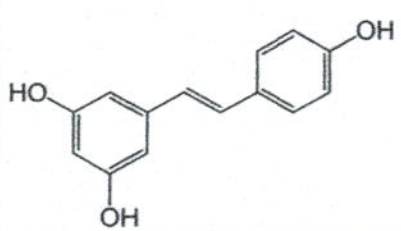	Polyphenol	Skin of grapes, blueberries, raspberries, mulberries, and peanuts.	KDAC ↓, KAT2A/3B ↑, DNMT1 ↓, miRNA-200 ↑, miR-122-5p ×, miR-124 ↑, miR-22-3p ×	Human Cell line, mouse model	[92, 94–98]

Rosmarinic acid	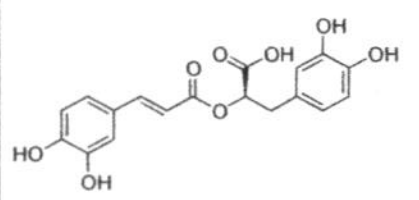	Polyphenol	*Ocimum basilicum* (basil), *Ocimum tenuiflorum* (holy basil), *Melissa officinalis* (lemon balm), *Rosmarinus officinalis* (rosemary), *Origanum majorana* (marjoram), *Salvia officinalis* (sage), thyme and peppermint.	DNMT ×, miR-155 ×, miR-6785-5p ↓, miR-642a-3p ↓, HDAC ↓	Human cell line, mice model	[99–101]
Silibinin	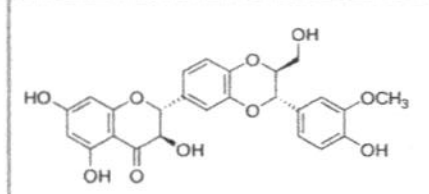	Flavono-lignan	Milk thistle, *Silybum marianum*	H3K27me3 ↑, DNMT ↑, HDAC ↓, miR-21 ↓, miR-15a ↓, miR-141 ↓, miR-200c ↑, miR-155 ↓, miR-7-1-3p ↑, miR-125b ↓, miR-182 ↓, miR-494 ↑	Human cell line, mice model	[102–106]
Sulforaphane	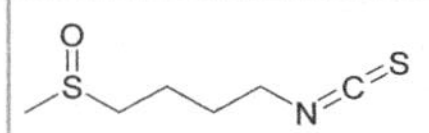	Organo-sulfur compound	Cruciferous vegetables	H4R3me2s ↓, Promoter methylation ↓, DNMT ↓, HDAC ↓, miR-9-3 ↓, miR-23b ↓, miR-92b ↓, miR-381 ↓, miR-382 ↓, miR-616-5p ↓, miR200c ↓, miR-9 ↓, miR-326 ↓, miR-21 ↓, miR-214 ↑, miRNA-124-3p ↑	Human cell line, mice model	[25, 28, 70, 107–116]
Catechins	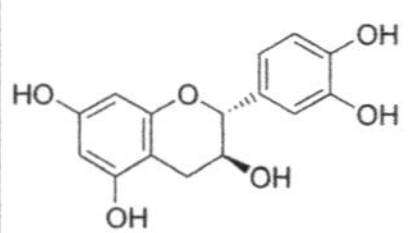	Polyphenol	Tea, coffee, pome fruit	HDAC ↓, DNMT1 ×, miR-182 ↑, miR-let7-a ↑	Human cell line	[117–122]
Theaflavins	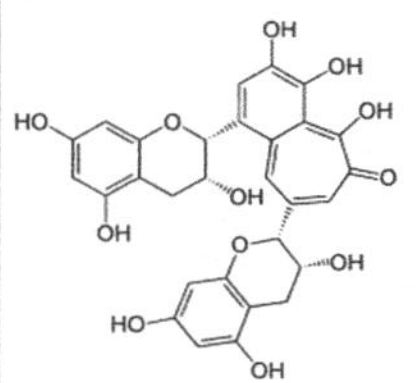	Polyphenol	Tea leaves	DNMT3a ↓,miRNA-128-3p ↑	Human cell line, mouse model	[123, 124]

(continued)

Table 5.2 (continued)

Compound name	Chemical structure	Group	Sources	Effects	Model system	References
Anthocyanins	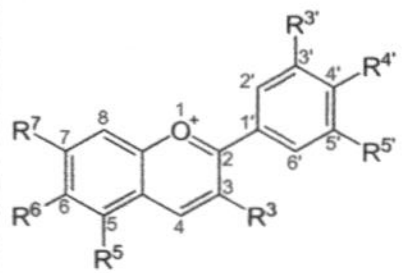	Polyphenol	Blueberry, cranberry, and bilberry, black raspberry, red raspberry, blackberry; blackcurrant, cherry, eggplant peel, black rice, ube, okinawan sweet potato, concord grape, muscadine grape, red cabbage, violet petals	DNMT ×, Histone acetylation ↑, HDAC ×	Human cell line	[29, 125, 126]

[128]. For example, curcumin, the major component of turmeric, inhibits HAT, DNMT activity and downregulates H3, H4 acetylation [72, 73]. A cluster of oncogenic miRNAs is being regulated by dietary polyphenols (Table 5.2).

5.4 Bioinformatics Platform for Drug Designing and Modifications

Discovery and development of drugs is an intricate, time consuming and risk taking practice. It requires hi-tech proficiency, funding, and human resources to develop a new drug.

Prior to the launching of a new drug in the market, a stringent observance to testing and manufacturing standards is required.

Two branches that brought an optimistic effect on drug designing procedure, cutting down the cost and risk, are Bioinformatics and Pharmacogenomics.

The information on medicinal plants acquired over the years and research outcome from contemporary methodologies is scattered and mostly embedded unstructured within the literature. This has become a major challenge in creating a common platform that requires a consolidated and integrated access to ethnobotanical knowledge. It is imperative to develop a technique that will help to extract, store, and present data in a constructive format. Several databases of ethnobotany, pharmacological uses, bioactive metabolites, genomic or transcript-based information, molecular targets of active ingredients, etc. have been developed to enhance data mining techniques. This can elucidate targeting bioprospecting and screening strategies for the application of plant medicines. Information on medicinal plants can also be extracted from the realms of databases of other categories such as taxonomy, chemical and molecular records (Fig. 5.3). To identify putative genes and pathways related to the production of secondary metabolites, transcript-level data can be applied. Analysis of transcriptome data of bioactive metabolites will help in the identification of transcription factors, putative genes, and response elements of the metabolite synthesis pathway. The application of "Omic" approaches can help to extract molecular information on the target molecules of medicinal plants [129].

Bioinformatics is an interdisciplinary science that includes genomics, proteomics, transcriptomics, molecular phylogenetics, etc. The major roles of bioinformatics are (1) Connecting between diseases and genetic and epigenetic modifications, (2) Identification of drugs that can restore cell function or abort the malfunctioning cells (e.g., malignant cell), (3) design and predict the action of drugs on target molecules, (4) Drug resistance and assessment of environmental hazards. The techniques involving assembling of protein, RNA structures, simulation with small molecules and metabolites, protein-ligand docking paved the way for more realistic screening of drugs. The application of dietary components for cancer prevention has some limitations. Most of the bioactive molecules have the potential for cancer treatment. Few are only effective at considerably high doses which are not achievable through diet. Currently, some of the studies focus on the combinatorial effects of phytochemical compounds. Humans do not consume one food item, so the necessity of

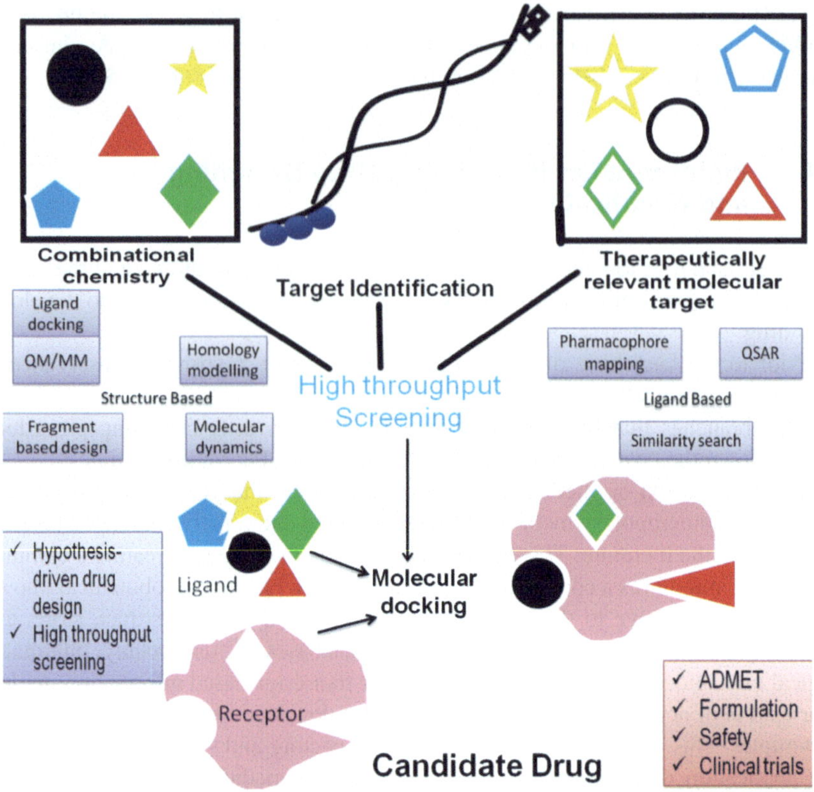

Fig. 5.3 Steps of drug discovery processes and the role of bioinformatics

combinatorial studies will help to understand the interaction between multiple epigenome-modifying bioactive molecules. These compounds may show synergistic or even antagonistic effects with each other. This area of epigenetics remains unexplored and evokes interest for further research [130].

5.5 Future Prospects of Epigenetics in Translational Research

Detection of epigenetic changes is a very promising tool not only for understanding cancer biology but also in therapeutic applications. DNA hypermethylation is a very distinct marker for the diagnosis of cancer. Hypermethylation of genes is specific to cancer types, for example, prostate cancer patients exhibited de novo methylation of *GSTP1* in most of the cases [131, 132]. Detection of hypermethylation of *BRCA1* could help in the diagnosis of the biopsy specimens of breast cancer patients [133]. Lung, colorectal, and brain cancer have been linked to the hypermethylation of *DAPK*, *p16INK4a*, and *EMP3* [40]. DNA hypermethylation from breast ductal fluids and from exfoliated cells for cervical cancer patients could be identified as a conventional cytological analysis in the patients [134, 135]. DNA methylation profiling by microarray analysis will help in the early detection of cancer [136]. The chemotherapeutic response of hypermethylation of *MGMT* and other DNA repair pathway genes with different *DNMT*-inhibitors have also been studied [137]. Therapeutics such as DNMT-inhibitors or HDAC-inhibitors has been studied in cell line and animal model systems. In spite of countless promises and the verified accomplishment in vitro and pre-clinical studies, there has been modest to no improvement in the switch of phytochemicals to the clinic as the first-line treatment. Phytochemicals are easily available in natural food sources and they have fewer side effects. However, there are some limitations as well. The extraction of particular bioactive molecules may not be cost-effective. Moreover, mechanisms are studied in cell line or model systems; therefore, optimization of dosing under human physiological circumstances are difficult to predict. There are some practical and ethical restrictions on the application of bioactive compounds involving human subjects without sufficient evidence that makes the research area confined to in vitro studies. Thus, it is imperative to create a pre-clinical model system that can mimic systemic exposure to phytochemicals with metabolomic and pharmacokinetic consequences.

5.6 Conclusion

Epigenetic regulations of gene expressions are maintained by the crosstalk between DNA methylation and histone modification that helps to maintain the cellular functions. The beneficial role of phytochemicals in the prevention of cancer has been reported in several experiments based on cell lines and animal models. Advancement of in silico epigenetic modeling system could further mimic and predict the plausible impact of these phytochemicals on the human system and help to fulfill the gap between in vitro and translational research.

Conflict of Interest The authors declare no conflict of interest.

References

1. Roberti A, Valdes AF, Torrecillas R et al (2019) Epigenetics in cancer therapy and nanomedicine. Clin Epigenetics 11(1):81
2. Fardi M, Solali S, Hagh MF (2018) Epigenetic mechanisms as a new approach in cancer treatment: an updated review. Genes Dis 5(4):304–311
3. Herceg Z, Vaissière T (2011) Epigenetic mechanisms and cancer: an interface between the environment and the genome. Epigenetics 6(7):804–819
4. Sharma S, Kelly TK, Jones PA (2010) Epigenetics in cancer. Carcinogenesis 31(1):27–36
5. Shenker N, Flanagan JM (2012) Intragenic DNA methylation: implications of this epigenetic mechanism for cancer research. Br J Cancer 106(2):248–253
6. Jones PA, Baylin SB (2007) The epigenomics of cancer. Cell 128:683–692
7. Laird PW (2005) Cancer epigenetics. Hum Mol Genet 14:R65–R76
8. Estécio MR, Yan PS, Ibrahi AE et al (2007) High-throughput methylation profiling by MCA coupled to CpG island microarray. Genome Res 17(10):1529–1536

9. Esteller M (2007) Cancer epigenomics: DNA methylomes and histone-modification maps. Nat Rev Genet 8(4):286
10. Banerjee HN, Verma M (2009) Epigenetic mechanisms in cancer. Biomark Med 3(4):397–410
11. Rodriguez BA, Cheng AS, Yan PS et al (2008) Epigenetic repression of the estrogen-regulated Homeobox B13 gene in breast cancer. Carcinogenesis 29(7):1459–1465
12. Gupta A, Godwin AK, Vanderveer L et al (2003) Hypomethylation of the synuclein γ gene CpG island promotes its aberrant expression in breast carcinoma and ovarian carcinoma. Cancer Res 63(3):664–673
13. Anderton JA, Lindsey JC, Lusher ME et al (2008) Global analysis of the medulloblastoma epigenome identifies disease-subgroup-specific inactivation of COL1A2. Neuro-oncology 10 (6):981–994
14. Overmeer RM, Henken FE, Snijders PJ et al (2008) Association between dense CADM1 promoter methylation and reduced protein expression in high-grade CIN and cervical SCC. J Pathol 215(4):388–397
15. Gokul G, Gautami B, Malathi S et al (2007) DNA methylation profile at the DNMT3L promoter: a potential biomarker for cervical cancer. Epigenetics 2(2):80–85
16. Shivapurkar N, Sherman ME, Stastny V et al (2007) Evaluation of candidate methylation markers to detect cervical neoplasia. Gynecol Oncol 107(3):549–553
17. Chung MT, Sytwu HK, Yan MD et al (2009) Promoter methylation of SFRPs gene family in cervical cancer. Gynecol Oncol 112(2):301–306
18. Yang XJ (2004) The diverse superfamily of lysine acetyltransferases and their roles in leukemia and other diseases. Nucleic Acids Res 32(3):959–976
19. Tian XQ, Zhang Y, Sun D, Zhao S et al (2009) Epigenetic silencing of LRRC3B in colorectal cancer. Scand J Gastroenterol 44(1):79–84
20. Hibi K, Sakata M, Sakuraba K et al (2008) Aberrant methylation of the HACE1 gene is frequently detected in advanced colorectal cancer. Anticancer Res 28(3A):1581–1584
21. Grunau C, Brun ME, Rivals I et al (2008) BAGE hypomethylation, a new epigenetic biomarker for colon cancer detection. Cancer Epidemiol Prev Biomark 17(6):1374–1379
22. Guan H, Ji M, Hou P et al (2008) Hypermethylation of the DNA mismatch repair gene hMLH1 and its association with lymph node metastasis and T1799A BRAF mutation in patients with papillary thyroid cancer. Cancer 113(2):247–255
23. Lind GE, Ahlquist T, Kolberg M et al (2008) Hypermethylated MAL gene–a silent marker of early colon tumorigenesis. J Transl Med 6(1):13
24. Lofton-Day C, Model F, DeVos T et al (2008) DNA methylation biomarkers for blood-based colorectal cancer screening. Clin Chem 54(2):414–423
25. Zhou JW, Wang M, Sun NX et al (2019) Sulforaphane induced epigenetic regulation of Nrf2 expression by DNA methyltransferase in human Caco 2 cells. Oncol Lett 18(3):2639–2647
26. Arnold CN, Goel A, Niedzwiecki D et al (2004) APC promoter hypermethylation contributes to the loss of APC expression in colorectal cancers with allelic loss on 5q1. Cancer Biol Ther 3 (10):960–964
27. Ogino S, Nosho K, Kirkner GJ et al (2008) A cohort study of tumoral LINE-1 hypomethylation and prognosis in colon cancer. J Natl Cancer Inst 100(23):1734–1738
28. Liu CM, Peng CY, Liao YW et al (2017) Sulforaphane targets cancer stemness and tumor initiating properties in oral squamous cell carcinomas via miR-200c induction. J Formos Med Assoc 116(1):41–48
29. Subramaniam D, Thombre R, Dhar A et al (2014) DNA methyltransferases: a novel target for prevention and therapy. Front Oncol 4:80
30. Suzuki M, Shigematsu H, Iizasa T (2006) Exclusive mutation in epidermal growth factor receptor gene, HER-2, and KRAS, and synchronous methylation of nonsmall cell lung cancer. Cancer 106(10):2200–2207
31. Toyooka S, Tokumo M, Shigematsu H et al (2006) Mutational and epigenetic evidence for independent pathways for lung adenocarcinomas arising in smokers and never smokers. Cancer Res 66(3):1371–1375

32. Choi YL, Kang SY, Choi JS (2006) Aberrant hypermethylation of RASSF1A promoter in ovarian borderline tumors and carcinomas. Virchows Arch 448(3):331–336
33. Fiegl H, Windbichler G, Mueller-Holzner E et al (2008) HOXA11 DNA methylation—a novel prognostic biomarker in ovarian cancer. Int J Cancer 123(3):725–729
34. Su HY, Lai HC, Lin YW et al (2009) An epigenetic marker panel for screening and prognostic prediction of ovarian cancer. Int J Cancer 124(2):387–393
35. Cui Y, Ying Y, Van Hasselt A et al (2008) OPCML is a broad tumor suppressor for multiple carcinomas and lymphomas with frequently epigenetic inactivation. PLoS One 3(8):e2990
36. Sellar GC, Watt KP, Rabiasz GJ et al (2003) OPCML at 11q25 is epigenetically inactivated and has tumor-suppressor function in epithelial ovarian cancer. Nat Genet 34(3):337
37. Gifford G, Paul J, Vasey PA et al (2004) The acquisition of hMLH1 methylation in plasma DNA after chemotherapy predicts poor survival for ovarian cancer patients. Clin Cancer Res 10(13):4420–4426
38. Pattamadilok J, Huapai N, Rattanatanyong P et al (2008) LINE-1 hypomethylation level as a potential prognostic factor for epithelial ovarian cancer. Int J Gynecol Cancer 18(4):711–717
39. Verma M, Manne U (2006) Genetic and epigenetic biomarkers in cancer diagnosis and identifying high risk populations. Crit Rev Oncol Hematol 60(1):9–18
40. Esteller M (2008) Epigenetics in cancer. N Engl J Med 358(11):1148–1159
41. Bjornsson HT, Brown LJ, Fallin MD et al (2007) Epigenetic specificity of loss of imprinting of the IGF2 gene in Wilms tumors. J Natl Cancer Inst 99(16):1270–1273
42. Howard G, Eiges R, Gaudet F et al (2008) Activation and transposition of endogenous retroviral elements in hypomethylation induced tumors in mice. Oncogene 27(3):404
43. Fraga MF, Esteller M (2005) Towards the human cancer epigenome: a first draft of histone modifications. Cell Cycle 23(10):1377–1381
44. Simon C, Chagraoui J, Krosl J, Gendron P, Wilhelm B, Lemieux S, Boucher G, Chagnon P, Drouin S, Lambert R, Rondeau C (2012) A key role for EZH2 and associated genes in mouse and human adult T-cell acute leukemia. Genes Dev 1(7):651–656
45. Elsheikh SE, Green AR, Rakha EA, Powe DG, Ahmed RA, Collins HM, Soria D, Garibaldi JM, Paish CE, Ammar AA, Grainge MJ (2009) Global histone modifications in breast cancer correlate with tumor phenotypes, prognostic factors, and patient outcome. Cancer Res 1 (9):3802–3809
46. Chen H, Li J, Jiao L, Petersen RB, Li J, Peng A, Zheng L, Huang K (2014) Apelin inhibits the development of diabetic nephropathy by regulating histone acetylation in Akita mouse. J Physiol 592:505–521
47. Bolden JE, Peart MJ, Johnstone RW (2006) Anticancer activities of histone deacetylase inhibitors. Nat Rev Drug Discov 5(9):769–784
48. Yang XJ, Seto EH (2007) HATs and HDACs: from structure, function and regulation to novel strategies for therapy and prevention. Oncogene 26(37):5310
49. Schulte JH, Lim S, Schramm A, Friedrichs N, Koster J, Versteeg R, Ora I, Pajtler K, Klein-Hitpass L, Kuhfittig-Kulle S, Metzger E (2009) Lysine-specific demethylase 1 is strongly expressed in poorly differentiated neuroblastoma: implications for therapy. Cancer Res 69 (5):2065–2071
50. Calin GA, Croce CM (2006) MicroRNA signatures in human cancers. Nat Rev Cancer 6 (11):857
51. Lu J, Getz G, Miska EA et al (2005) MicroRNA expression profiles classify human cancers. Nature 435(7043):834
52. Lujambio A, Ropero S, Ballestar E et al (2007) Genetic unmasking of an epigenetically silenced microRNA in human cancer cells. Cancer Res 67(4):1424–1429
53. Zhao Y, Deng C, Wang J et al (2011) Let-7 family miRNAs regulate estrogen receptor alpha signaling in estrogen receptor positive breast cancer. Breast Cancer Res Treat 127(1):69–80
54. Saito Y, Suzuki H, Tsugawa H, Nakagawa I, Matsuzaki J, Kanai Y, Hibi T (2009) Chromatin remodeling at Alu repeats by epigenetic treatment activates silenced microRNA-512-5p with downregulation of Mcl-1 in human gastric cancer cells. Oncogene 28(30):2738–2744

55. Ahuja N, Sharma AR, Baylin SB (2016) Epigenetic therapeutics: a new weapon in the war against cancer. Annu Rev Med 67(1):73–89
56. Jayasinghe CD, Udalamaththa A, Imbulana IB, Suetake I (2015) Dietary phytochemicals as epi-drugs: role in modulating the epigenetic mechanisms of human diseases. Int J Curr Pharm Rev Res 7(1):50–58
57. Thakur VS, Deb G, Babcook MA, Gupta S (2014) Plant phytochemicals as epigenetic modulators: role in cancer chemoprevention. AAPS J 16(1):151–163
58. Barbieri R, Coppo E, Marchese A, Daglia M, Sobarzo-Sánchez E, Nabavi SF, Nabavi SM (2017) Phytochemicals for human disease: an update on plant-derived compounds antibacterial activity. Microbiol Res 196:44–68
59. Mohan K, Jeyachandran R, Deepa R (2012) Alkaloids as anticancer agents. Ann Phytomed 1 (1):46–53
60. Koche D, Shirsat R, Kawale M (2016) An overerview of major classes of phytochemicals: their types and role in disease prevention. Hislopia J 9:1–11
61. Habli Z, Toumieh G, Fatfat M et al (2017) Emerging cytotoxic alkaloids in the battle against cancer: overview of molecular mechanisms. Molecules 22(2):250
62. Lu JJ, Bao JL, Chen XP et al (2012) Alkaloids isolated from natural herbs as the anticancer agents. Evid Based Complement Alternat Med 2012:485042
63. Liu D, Meng X, Wu D et al (2019) A natural isoquinoline alkaloid with antitumor activity: studies of the biological activities of berberine. Front Pharmacol 10:9–9
64. Singh AK, Bishayee A, Pandey AK (2018) Targeting histone deacetylases with natural and synthetic agents: an emerging anticancer strategy. Nutrients 10(6):731
65. Lee HS (2002) Rat lens aldose reductase inhibitory activities of Coptis japonica root-derived isoquinoline alkaloids. J Agric Food Chem 50(24):7013–7016
66. Paredes-Gonzalez X, Fuentes F, Su ZY et al (2014) Apigenin reactivates Nrf2 anti-oxidative stress signaling in mouse skin epidermal JB6 P+ cells through epigenetics modifications. AAPS J 16(4):727–735
67. Pandey M, Kaur P, Shukla S et al (2012) Plant flavone apigenin inhibits HDAC and remodels chromatin to induce growth arrest and apoptosis in human prostate cancer cells: in vitro and in vivo study. Mol Carcinog 51(12):952–962
68. Tseng TH, Chien MH, Lin WL et al (2017) Inhibition of MDA-MB-231 breast cancer cell proliferation and tumor growth by apigenin through induction of G2/M arrest and histone H3 acetylation-mediated p21WAF1/CIP1 expression. Environ Toxicol 32(2):434–444
69. Kim YM, Hwang JT, Kwak DW et al (2007) Involvement of AMPK signaling cascade in capsaicin-induced apoptosis of HT-29 colon cancer cells. Ann N Y Acad Sci 1095(1):496–503
70. Wang DX, Zou YJ, Zhuang XB et al (2017) Sulforaphane suppresses EMT and metastasis in human lung cancer through miR-616-5p-mediated GSK3β/β-catenin signaling pathways. Acta Pharmacol Sin 38(2):241
71. Pop S, Enciu AM, Tarcomnicu I et al (2019) Phytochemicals in cancer prevention: modulating epigenetic alterations of DNA methylation. Phytochem Rev 18(4):1005–1024
72. Boyanapalli SS, Kong ANT (2015) "Curcumin, the king of spices": epigenetic regulatory mechanisms in the prevention of cancer, neurological, and inflammatory diseases. Curr Pharmacol Rep 1(2):129–139
73. Hassan FU, Rehman MSU, Khan MS et al (2019) Curcumin as an alternative epigenetic modulator: mechanism of action and potential effects. Front Genet 10:514
74. Reuter S, Gupta SC, Park B et al (2011) Epigenetic changes induced by curcumin and other natural compounds. Genes Nutr 6(2):93–108
75. Teiten MH, Dicato M, Diederich M (2013) Curcumin as a regulator of epigenetic events. Mol Nutr Food Res 57(9):1619–1629
76. Khan MA, Hussain A, Sundaram MK et al (2015) (−)-Epigallocatechin-3-gallate reverses the expression of various tumor-suppressor genes by inhibiting DNA methyltransferases and histone deacetylases in human cervical cancer cells. Oncol Rep 33(4):1976–1984

77. Liu L, Zhao W, Liu J et al (2018) Epigallocatechin-3 gallate prevents pressure overload-induced heart failure by up-regulating SERCA2a via histone acetylation modification in mice. PLoS One 13(10):e0205123
78. Negri A, Naponelli V, Rizzi F et al (2018) Molecular targets of epigallocatechin—Gallate (EGCG): a special focus on signal transduction and cancer. Nutrients 10(12):1936
79. Sundaram MK, Ansari MZ, Al Mutery A et al (2018) Genistein induces alterations of epigenetic modulatory signatures in human cervical cancer cells. Anti Cancer Agents Med Chem 18(3):412–421
80. Hardy TM, Tollefsbol TO (2011) Epigenetic diet: impact on the epigenome and cancer. Epigenomics 3(4):503–518
81. Majid S, Kikuno N, Nelles J et al (2008) Genistein induces the p21WAF1/CIP1 and p16INK4a tumor suppressor genes in prostate cancer cells by epigenetic mechanisms involving active chromatin modification. Cancer Res 68(8):2736–2744
82. Qiu W, Lin J, Zhu Y et al (2017) Kaempferol modulates DNA methylation and downregulates DNMT3B in bladder cancer. Cell Physiol Biochem 41(4):1325–1335
83. Zhu G, Liu X, Li H et al (2018) Kaempferol inhibits proliferation, migration, and invasion of liver cancer HepG2 cells by down-regulation of microRNA-21. Int J Immunopathol Pharmacol 32:2058738418814341
84. Gan CC, Ni TW, Yu Y et al (2017) Flavonoid derivative (Fla-CN) inhibited adipocyte differentiation via activating AMPK and up-regulating microRNA-27 in 3T3-L1 cells. Curr Pharmacol Rep 797:45–52
85. Han X, Liu CF, Gao N et al (2018) Kaempferol suppresses proliferation but increases apoptosis and autophagy by up-regulating microRNA-340 in human lung cancer cells. Biomed Pharmacother 108:809–816
86. Zhong X, Zhang L, Li Y et al (2018) Kaempferol alleviates ox-LDL-induced apoptosis by up-regulation of miR-26a-5p via inhibiting TLR4/NF-κB pathway in human endothelial cells. Biomed Pharmacother 108:1783–1789
87. Jiang F, Li Y, Mu J et al (2016) Glabridin inhibits cancer stem cell-like properties of human breast cancer cells: an epigenetic regulation of miR-148a/SMAd2 signaling. Mol Carcinog 55 (5):929–940
88. Fu LJ, Ding YB, Wu LX et al (2014) The effects of lycopene on the methylation of the GSTP1 promoter and global methylation in prostatic cancer cell lines PC3 and LNCaP. Int J Endocrinol 2014:620165
89. Alvarez M, Maso V, Torello CO et al (2018) The polyphenol quercetin induces cell death in leukemia by targeting epigenetic regulators of pro-apoptotic genes. Clin Epigenetics 10(1):139
90. Fuentes F, Paredes-Gonzalez X, Kong ANT (2015) Dietary glucosinolates sulforaphane, phenethyl isothiocyanate, indole-3-carbinol/3, 3′-diindolylmethane: Antioxidative stress/ inflammation, Nrf2, epigenetics/epigenomics and in vivo cancer chemopreventive efficacy. Curr Pharmacol Rep 1(3):179–196
91. Zhang W, Jiang H, Chen Y et al (2019) Resveratrol chemosensitizes adriamycin-resistant breast cancer cells by modulating miR-122-5p. J Cell Biochem 120(9):16283–16292
92. Feng MH, Li JW, Sun HT et al (2019) Sulforaphane inhibits the activation of hepatic stellate cell by miRNA-423-5p targeting suppressor of fused. Hum Cell 32(4):403–410
93. Gupta R, Bhatt LK, Momin M (2019) Potent antitumor activity of Laccaic acid and Phenethylisothiocyanate combination in colorectal cancer via dual inhibition of DNA methyltransferase-1 and Histone deacetylase-1. Toxicol Appl Pharm 377:114631
94. Chatterjee B, Ghosh K, Kanade SR (2019) Resveratrol modulates epigenetic regulators of promoter histone methylation and acetylation that restores BRCA1, p53, p21CIP1 in human breast cancer cell lines. Biofactors 45(5):818–829
95. Izquierdo-Torres E, Hernández-Oliveras A, Meneses-Morales I, Rodríguez G, Fuentes-Jang YG, Hwang KA, Choi KC (2018) Rosmarinic acid, a component of rosemary tea, induced the cell cycle arrest and apoptosis through modulation of HDAC2 expression in prostate cancer cell lines. Nutrients 10(11):1784

96. Fudhaili A, Yoon N, Kang S et al (2019) Resveratrol epigenetically regulates the expression of zinc finger protein 36 in non-small cell lung cancer cell lines. Oncol Rep 41(2):1377–1386
97. Gandy KAO, Zhang J, Nagarkatti P et al (2019) Resveratrol (3, 5, 4′-trihydroxy-trans-stilbene) attenuates a mouse model of multiple sclerosis by altering the miR-124/sphingosine kinase 1 axis in encephalitogenic T cells in the brain. J Neuroimmune Pharmacol 14(3):462–477
98. Wen W, Chen X, Huang Z (2019) Resveratrol regulates muscle fiber type conversion via miR-22-3p and AMPK/SIRT1/PGC-1α pathway. J Nutr Biochem 77:108297
99. Paluszczak J, Krajka-Kuźniak V, Baer-Dubowska W (2010) The effect of dietary polyphenols on the epigenetic regulation of gene expression in MCF7 breast cancer cells. Toxicol Lett 192 (2):119–125
100. Yu C, Chen DQ, Liu HX et al (2019) Rosmarinic acid reduces the resistance of gastric carcinoma cells to 5-fluorouracil by downregulating FOXO4-targeting miR-6785-5p. Biomed Pharmacother 109:2327–2334
101. Jang YG, Hwang KA, Choi KC (2018) Rosmarinic acid, a component of rosemary tea, induced the cell cycle arrest and apoptosis through modulation of HDAC2 expression in prostate cancer cell lines. Nutrients 10(11):1784
102. Anestopoulos I, Sfakianos A, Franco R et al (2017) A novel role of silibinin as a putative epigenetic modulator in human prostate carcinoma. Molecules 22(1):62
103. YazdiRouholamini SE, Moghassemi S, Maharat Z et al (2018) Effect of silibinin-loaded nano-niosomal coated with trimethyl chitosan on miRNAs expression in 2D and 3D models of T47D breast cancer cell line. Artif Cells Nanomed Biotechnol 46(3):524–535
104. Zadeh MM, Motamed N, Ranji N et al (2016) Silibinin-induced apoptosis and downregulation of microRNA-21 and microRNA-155 in MCF-7 human breast cancer cells. J Breast Cancer 19 (1):45–52
105. Hossainzadeh S, Ranji N, NaderiSohi A et al (2019) Silibinin encapsulation in polymersome: a promising anticancer nanoparticle for inducing apoptosis and decreasing the expression level of miR-125b/miR-182 in human breast cancer cells. J Cell Physiol 234(12):22285–22298
106. Chang YC, Jan CI, Peng CY et al (2015) Activation of microRNA-494-targeting Bmi1 and ADAM10 by silibinin ablates cancer stemness and predicts favourable prognostic value in head and neck squamous cell carcinomas. Oncotarget 6(27):24002
107. Saha K, Fisher ML, Adhikary G et al (2017) Sulforaphane suppresses PRMT5/MEP50 function in epidermal squamous cell carcinoma leading to reduced tumor formation. Carcinogenesis 38(8):827–836
108. Wong CP, Hsu A, Buchanan A et al (2014) Effects of sulforaphane and 3, 3′-diindolylmethane on genome-wide promoter methylation in normal prostate epithelial cells and prostate cancer cells. PLoS One 9(1):e86787
109. Royston K, Udayakumar N, Lewis K et al (2017) A novel combination of withaferin A and sulforaphane inhibits epigenetic machinery, cellular viability and induces apoptosis of breast cancer cells. Int J Mol Sci 18(5):1092
110. Gao L, Cheng D, Yang J et al (2018) Sulforaphane epigenetically demethylates the CpG sites of the miR-9-3 promoter and reactivates miR-9-3 expression in human lung cancer A549 cells. J Nutr Biochem 56:109–115
111. Jiang LL, Zhou SJ, Zhang XM et al (2016) Sulforaphane suppresses in vitro and in vivo lung tumorigenesis through downregulation of HDAC activity. Biomed Pharmacother 78:74–80
112. Lewinska A, Adamczyk-Grochala J, Deregowska A et al (2017) Sulforaphane-induced cell cycle arrest and senescence are accompanied by DNA hypomethylation and changes in microRNA profile in breast cancer cells. Theranostics 7(14):3461
113. Kiani S, Akhavan-Niaki H, Fattahi S et al (2018) Purified sulforaphane from broccoli (Brassica oleracea var. italica) leads to alterations of CDX1 and CDX2 expression and changes in miR-9 and miR-326 levels in human gastric cancer cells. Gene 678:115–123
114. Lan F, Pan Q, Yu H et al (2015) Sulforaphane enhances temozolomide-induced apoptosis because of down-regulation of miR-21 via Wnt/β-catenin signaling in glioblastoma. J Neurochem 134(5):811–818

115. Li QQ, Xie YK, Wu Y et al (2017) Sulforaphane inhibits cancer stem-like cell properties and cisplatin resistance through miR-214-mediated downregulation of c-MYC in non-small cell lung cancer. Oncotarget 8(7):12067
116. Li X, Zhao Z, Li M et al (2018) Sulforaphane promotes apoptosis, and inhibits proliferation and self-renewal of nasopharyngeal cancer cells by targeting STAT signal through miRNA-124-3p. Biomed Pharmacother 103:473–481
117. Oya Y, Mondal A, Rawangkan A et al (2017) Down-regulation of histone deacetylase 4,− 5 and− 6 as a mechanism of synergistic enhancement of apoptosis in human lung cancer cells treated with the combination of a synthetic retinoid, Am80 and green tea catechin. J Nutr Biochem 42:7–16
118. Cordero-Herrera I, Chen X, Ramos S et al (2017) (−)-Epicatechin attenuates high-glucose-induced inflammation by epigenetic modulation in human monocytes. Eur J Nutr 56 (3):1369–1373
119. Yiannakopoulou EC (2015) Targeting DNA methylation with green tea catechins. Pharmacology 95(3–4):111–116
120. Lee WJ, Shim JY, Zhu BT (2005) Mechanisms for the inhibition of DNA methyltransferases by tea catechins and bioflavonoids. Mol Pharmacol 68(4):1018–1030
121. Gheysarzadeh A, Yazdanparast R (2015) STAT5 reactivation by catechin modulates H 2 O 2-induced apoptosis through miR-182/FOXO1 pathway in SK-N-MC cells. Cell Biochem Biophys 71(2):649–656
122. Appari M, Babu KR, Kaczorowski A et al (2014) Sulforaphane, quercetin and catechins complement each other in elimination of advanced pancreatic cancer by miR-let-7 induction and K-ras inhibition. Int J Oncol 45(4):1391–1400
123. Li R, Li X, Wu H et al (2019) Theaflavin attenuates cerebral ischemia/reperfusion injury by abolishing miRNA 128 3p mediated Nrf2 inhibition and reducing oxidative stress. Mol Med Rep 20(6):4893–4904
124. Rajavelu A, Tulyasheva Z, Jaiswal R et al (2011) The inhibition of the mammalian DNA methyltransferase 3a (Dnmt3a) by dietary black tea and coffee polyphenols. BMC Biochem 12 (1):16
125. Wang LS, Kuo CT, Cho SJ et al (2013) Black raspberry-derived anthocyanins demethylate tumor suppressor genes through the inhibition of DNMT1 and DNMT3B in colon cancer cells. Nutr Cancer 65(1):118–125
126. Zhan W, Liao X, Xie RJ et al (2017) The effects of blueberry anthocyanins on histone acetylation in rat liver fibrosis. Oncotarget 8(57):96761
127. Guo Y, Su ZY, Kong ANT (2015) Current perspectives on epigenetic modifications by dietary chemopreventive and herbal phytochemicals. Curr Pharmacol Rep 1(4):245–257
128. Rajendran P, Alzahrani AM, Rengarajan T et al (2019) Polyphenols and Cancer. Front Anti-Cancer Drug Discov 10(10):62
129. Sharma V, Sarkar IN (2012) Bioinformatics opportunities for identification and study of medicinal plants. Brief Bioinform 14(2):238–250
130. Lewis KA, Tollefsbol TO (2017) The influence of an epigenetics diet on the cancer epigenome. Epigenomics 9(9):1153–1155
131. Lee WH, Morton RA, Epstein JI et al (1994) Cytidine methylation of regulatory sequences near the pi-class glutathione S-transferase gene accompanies human prostatic carcinogenesis. Proc Natl Acad Sci 91(24):11733–11737
132. Esteller M, Corn PG, Urena JM et al (1998) Inactivation of glutathione S-transferase P1 gene by promoter hypermethylation in human neoplasia. Cancer Res 58(20):4515–4518
133. Esteller M, Fraga MF, Guo M et al (2001) DNA methylation patterns in hereditary human cancers mimic sporadic tumorigenesis. Hum Mol Genet 10(26):3001–3007
134. Fackler MJ, Malone K, Zhang Z et al (2006) Quantitative multiplex methylation-specific PCR analysis doubles detection of tumor cells in breast ductal fluid. Clin Cancer Res 12 (11):3306–3310

135. Feng Q, Balasubramanian A, Hawes SE et al (2005) Detection of hypermethylated genes in women with and without cervical neoplasia. J Natl Cancer Inst 97(4):273–282
136. Maier S, Nimmrich I, Koenig T et al (2007) DNA-methylation of the homeodomain transcription factor PITX2 reliably predicts risk of distant disease recurrence in tamoxifen-treated, node-negative breast cancer patients–technical and clinical validation in a multi-Centre setting in collaboration with the European Organisation for Research and Treatment of Cancer (EORTC) PathoBiology group. Eur J Cancer 43(11):1679–1686
137. Esteller M, Hamilton SR, Burger PC et al (1999) Inactivation of the DNA repair gene O6-methylguanine- DNA methyltransferase by promoter hypermethylation is a common event in primary human neoplasia. Cancer Res 59:793–797

Modulation of Cancer Cell Metabolism and Microenvironment by Phytochemicals

6

Amarjeet Shrama, Priyanshi, and Chandi C. Mandal

Abstract

The quiddity of neoplastic disease is represented by chronic and often times uncontrolled cell multiplication. It includes deregulated proliferative capability of the tumorigenic cells, along with the modulation in metabolic process for cellular proliferation and growth. Discovery of this metabolic alteration as one of the hallmarks of cancer cells has provided the new way to target tumor cell metabolism for therapeutic purpose. Tumors are not only insular masses of proliferating cancer cells; they also have another dimension of intricacy: they recruit ostensibly normal cells that help in acquiring hallmarks characteristics by forming the microenvironment near by the tumor cells which is called as tumor microenvironment (TME). Beside of these existing diversity of tumors reported in the identical or distinct locations and organs in the body, similar components and properties have been revealed in the tumor microenvironment of epithelial originated cancer. Various modern therapeutics have been developed targeting cancer cell metabolism and modulating tumor microenvironment. In recent years, several phytochemicals are being identified as the possible therapeutic compounds with their anticancer properties. Phytochemicals are the crucial ingredients of traditional medicine, and are also widely used to synthesize the novel therapeutics to treat numerous diseases. Recent investigations have been made by using distinct cancer models to highlight the role of these phytochemicals as the modulator of cancer cell metabolism and how these phytochemicals are altering the tumor microenvironment. Here, we have selected some phytochemicals based on their distinct anti-tumorigenic nature and discussed their effects in modulating the TME microenvironment and cancer

Amarjeet Shrama and Priyanshi contributed equally with all other contributors.

A. Shrama · Priyanshi · C. C. Mandal (✉)
Department of Biochemistry, Central University of Rajasthan, Ajmer, Rajasthan, India

M. Kumar et al. (eds.), *Pharmacotherapeutic Botanicals for Cancer Chemoprevention*, https://doi.org/10.1007/978-981-15-5999-0_6

cell metabolism. The prominent targets of the phytochemicals, i.e. the site of action where they modulate the cancer cell metabolism and tumor microenvironment are also being highlighted. With the integrated understanding of effects of phytochemicals on tumor cells, advancements can be made in better cancer treatment.

Keywords

Cancer cell metabolism · Tumor microenvironment · Phytochemicals

Abbreviations

α	Ketoglutarate
ACS	Acetyl-CoA synthetase
ACS	Acyl-CoA synthetases
ALT	Alanine aminotransferase
ANGPT	Angiopoietin
ATP	Adenosine triphosphate
EMT	Epithelial to mesenchymal transition
FAS	Fatty acid synthase
FBAA	Fructose bisphosphate aldolase
GDH	Glutamate dehydrogenase
HIF 1α	Hypoxia inducible factor
HK	Hexokinase
NADH	Nicotine amino dehydrogenase
nSMase	Neutral sphingomyelinase
PFK	Phosphofructokinase
PFKB	6-phosphofructo-2-kinase/fructose-2,6,bisphophatases
PG	Phosphoglycerate
ROS	Reactive oxygen species
SAM	S-adenosyl methionine
SCDs	Stearoyl-CoA desaturase
SPT	Serine palmitoyal transferase
TCA cycle	Tricarboxylic acid cycle
TME	Tumor microenvironment
VEGF	Vascular endothelial growth factor
VEGFR	Vascular endothelial growth factor receptor
WHO	World Health Organization

6.1 Introduction

In today's world, according to WHO, cancer is the second leading cause of the death throughout the globe, which means that cancer is a big health issue. This disease is responsible for a roughly calculated 9.6 million deaths in 2018 that means 1 from 6 deaths are occurring due to cancer only. Many factors are responsible for the cancer, including Tobacco consumption, low fruits and vegetables intake, and less exercise. Research regarding to cancer treatment areas has been progressively increased throughout the decade. But there is no such improvement has been seen in the cancer survival rate. Resistant against the molecular targeting therapies has been a big concern [1], which forced the scientist to look for a better treatment approach like targeting the cancer cells with certain effective molecules which are naturally present on the earth, and have certain properties to treat the cancer. These molecules are known as phytochemicals. Beside the cancer preventive effect of some phytochemicals, directly obtained from the dietary sources, some phytochemicals can target the various metabolic pathways in tumor cells along with a significant targeting of distinct cancer cell signaling, while showing zero or least side effects in healthy active cells [2]. The modulation of metabolism and TME in cancer cells with respect to the normal cell is now become a good therapeutic target. Many phytochemicals have been reported as the inhibitors of metabolic enzymes, widely used in the cancer cells for deregulated growth. Thus, phytochemicals are good therapeutic molecules to treat the cancer. According to statistical analysis, form around 175 approved cancer therapeutic molecules, till the year 2014, 85 from them were only the phytochemicals, and their derivatives [2]. Good effects have been observed in the laboratories, but not achieved on the same rate in clinical trials due to the change in certain physiological conditions between in vivo and ex vivo experiments. Here in this chapter, we are discussing precisely about the distinct targets, activity, and mechanism of various phytochemicals which have been reported as a good anticancer compounds, in terms of tumor microenvironment and deregulated tumor cell metabolism by highlighting the, metabolic enzyme inhibition, cellular cross talk inhibition in TME, and ROS formation because of the phytochemicals and how it is affecting the cancer cell survival. Firstly, a discussion is there about cancer cell metabolism and how it is modulated from the normal cell, then there is a discussion about the TME and the relation between TME and cancer cell metabolism. A brief introduction of phytochemicals and followed by individual discussion of the phytochemicals and their targets have been addressed (Figs. 6.1 and 6.2). Phytochemicals are selected of three different classes i.e., polyphenols, terpenoids, and alkaloids. Finally, the chapter is concluding with the future prospects of the use of phytochemicals as cancer therapeutics.

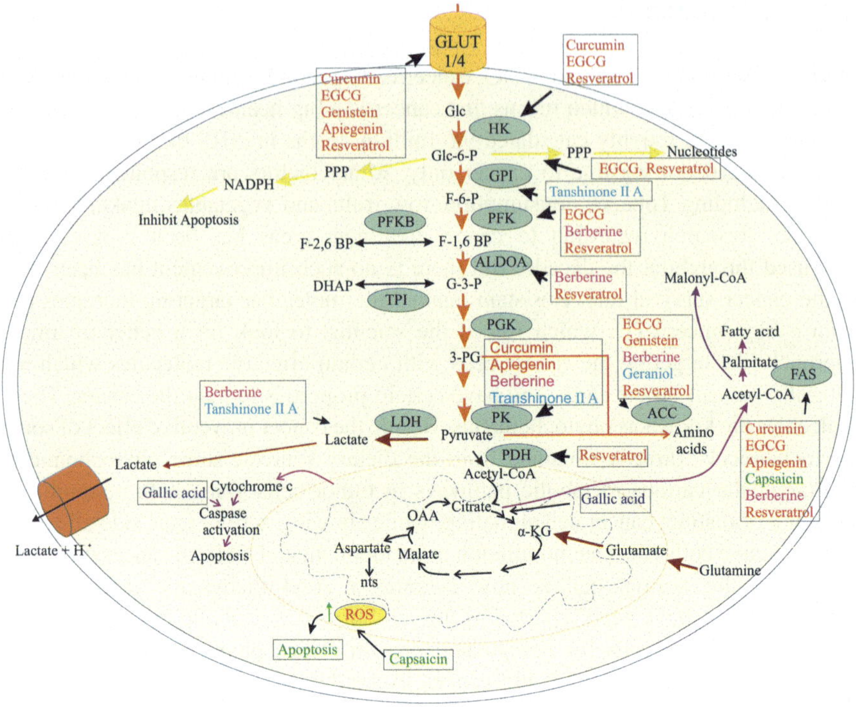

Fig. 6.1 Schematic view of cancer cell metabolism and targets of different phytochemicals. Red and blue phenolic compounds, Pink and green alkaloid compounds, Light blue isoprenoids. *Glc* glucose, *HK* hexokinase, *Glc-6-P* glucose-6-phosphate, *GPI* glucose-6-phosphate isomerase, *F-6-P* fructose-6-phosphate, *PFK* phosphofructokinase, *F-1,6 BP* fructose-1,6-bisphosphate, *ALDOA* aldolase, G-3-P-glyceraldehyde-3-phosphate, *DHAP* dihydroxy acetone phosphate, *TPI* triose phosphate isomerase, *3-PG* 3 phosphoglycerate, *PGK* phosphoglycerate kinase, *PK* pyruvate kinase, *LDH* lactate dehydrogenase, *PDH* pyruvate dehydrogenase, *ACC* Acetyl-CoA carboxylase, *FAS* fatty acyl synthase, *ROS* reactive oxygen species, *PPP* pentose phosphate pathway

6.2 Cancer Cell Metabolism

Cancer cells need abundant energy and macromolecules for their growth and to assist in their proliferation this energy is obtained by cells through ATP hydrolysis. To cope up with the increasing demand for energy, they have to alter their metabolism. So, modification in cancer cell metabolism is of the essence for tumor initiation and progression [3]. Reprogramming of metabolism is considered as a hallmark of cancer cell. Tumor cells alter their several metabolic pathways along with their interactions with TME in response to the need of enhance uptake of various nutrients including glucose and glutamine [4]. Various alterations in cancer cell metabolism are associated with cancer linked oncogenic signaling pathways [5]. These changes

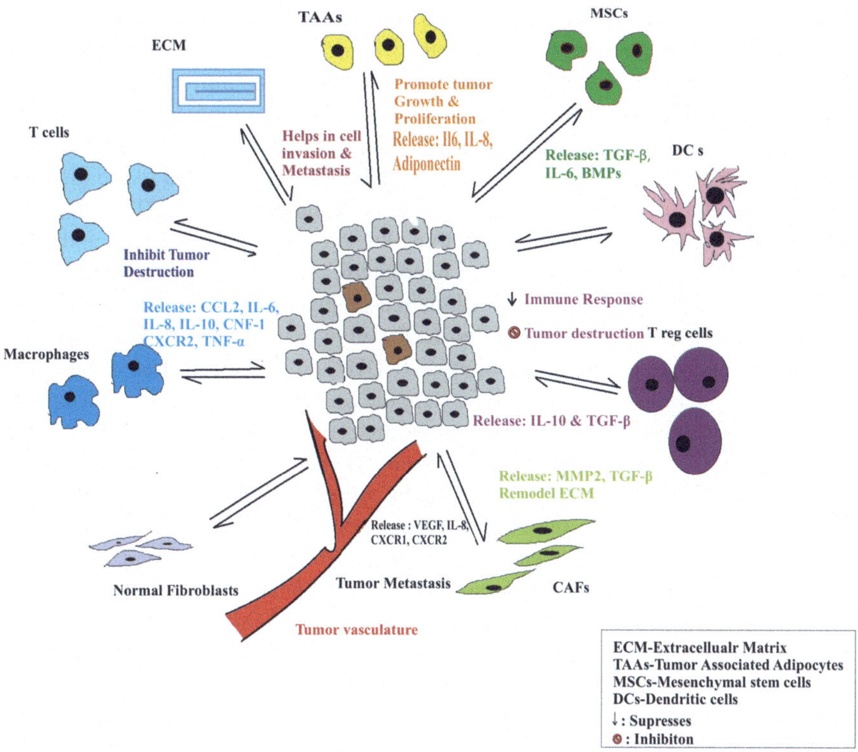

Fig. 6.2 Influence of various phytochemicals on tumor microenvironment. Blue line: targets of different phytochemicals. *IL* interleukins, *TGF-β* transforming growth factor beta, *BMP* bone morphogenetic protein, and *MMP* matrix metalloproteinases

in cancer cells are thought to be the edge point of oncogenic activation, beside this these changes provide them assessment in tumor development. Hence, much novel therapeutics can be made targeting altered metabolic pathways in cancer cells (Figs. 6.1 and 6.2).

6.2.1 Glycolysis and Warburg Effect

Glucose exists as major constituent of nutrients present in the blood, and it is the fundamental substrate for cells metabolism in mammals. In aerobic conditions when oxygen is present, normal cells attain energy by oxidative phosphorylation of glucose derived pyruvate which enters in mitochondria's TCA cycle before undergoing oxidative phosphorylation in mitochondria. But as observed by Otto Warburg in some proliferative cells mainly in tumor cells, there is more glycolytic flux and they conversely metabolize the glucose derived pyruvate into lactate even when oxygen is present [6–8]. This effect is acknowledged as "Warburg effect." The preferred shift to aerobic glycolysis from oxidative phosphorylation is the most inspected altered metabolism of cancer cells. Distinct classes of hexose transporters, like glucose transporters (GLUTs) and transporters for sodium-mediated glucose transportation (SGLTs), are reported to be responsible for glucose transport across the cellular membrane. GLUTs all have identical transmembrane anatomy, but there are many isoforms of them which vary in substrate recognition and tissue distribution [9, 10]. Tumor cells have aggravated glucose intake and GLUT 1 is mostly overexpressed in cancer cells [11]. The upregulated expression of GLUT1 and other enzymes involved in glycolysis such as hexokinase (HK) 1-2 and phosphofructokinase is mediated by KRAS. Along with increased glycolytic flux, cancer cells also have elevated level of enzymes related to the glycolytic pathway. The important elevated enzymes are lactate dehydrogenase (LDH) and pyruvate kinase isoform M2 (PKM2 [12, 13]. In very first glycolytic step, the formation of glucose -6-phosphate via glucose as a substrate, which is first committed (irreversible) step and is mediated by hexokinase, can be done by its two isoforms which are hexokinase 1 and 2. Out of the two isoforms, hexokinase 2 is mostly highly upregulated in malignant cells while hexokinase1 is abundantly present in normal cells [14]. Cancer cells have high proliferation rates due to which they encounter oxygen tension and have high expression of hypoxia inducible factor (HIF)-1α. After conversion of glucose to glucose-6-phosphate another checkpoint of glycolysis is the step related to formation of fructose-1,6 bisphosphate from fructose-6 phosphate, mediated by the enzyme phosphofructokinase-1(PFK-1). The expression of PFK-1 can be regulated by many mechanisms. In malignant cells its activity has been found to be increased because of the activity of oncogenes. Beside of this, HIF-1α is also reported as the responsible factor for the same [15]. The conversion of fructose to fructose 1,6- bisphosphate by PFK-1 is negatively regulated by ATP and it is shown that Akt kinase might activate PFK-1 by phosphorylating and releasing the inhibition of PFK-1 by ATP[16]. The most effective allosteric regulator of PFK-1 is fructose 2,6-bisphosphate which is produced and degraded by a class of bi-functional PFKB enzyme. In human cancers PFKB is often overexpressed which not only upregulates the expression of fructose 2,6-bisphosphate, but also plays an crucial part in proliferation of tumor cells through the regulation of cell cycle [17]. After glycolysis, pyruvate formed is preferentially converted into lactate by the enzyme lactate dehydrogenase which is also overexpressed in cancer cells, in this reaction NAD^+ is regenerated which helps in maintaining the glycolytic flux [18]. There are numerous benefits of "Warburg effect" to the cancer cells. Reduction in mitochondrial oxidation and increase in aerobic glycolysis lead to the accumulation of precursors which are required for the synthesis of major cellular products such as amino acids, nucleotides, and lipids. This altered metabolism helps in fueling the anabolic pathways like serine biosynthesis pathway along with the phosphogluconate pathway (hexose monophosphate shunt) [19]. Reduction mitochondrial respiration also leads to decrease in ROS production which is advantageous for rapidly proliferating cells. Exacerbate glycolytic flux and lactate production in cancer cells is advantageous for prompt energy production, accumulation of biomass, and redox maintenance [14].

6.2.2 Altered Glutamine Metabolism

In spite of the glucose, an important substrate known as glutamine is recorded as a very crucial substrate for cancer cell metabolism and growth. To be used in the form of anaplerotic precursor, glutamine first converts to glutamate and further it converts into α-ketoglutarate. Various enzymes are there to carry out these reactions, including glutamine dehydrogenase (GDH), and numerous transaminases [20]. Because of the higher ALT [20] and GDH [21] activity, metabolism of glutamine is found to be elicited in cancer cells, it results in large glutamine-derived α-KG formation, and further this α-KG is used up in the tricarboxylic acid (TCA) cycle. Intermediates and bi-products of the TCA cycle are precursors of different anabolic pathways, result in the synthesis of lipid, nucleotide, and other alpha nitrogen containing acids [21]. Therefore, the metabolism of glutamine in cancer cells is a relatively speedy and preferable source of carbon to fulfill the demand of various biosynthetic pathways. Along with these, glutamine metabolism also promotes the accumulation of lactate via malate formation and aggravates glycolysis along with NADPH generation via glutaminolysis. Most of the α-nitrogen from glutamine degradation is secreted from the cancer cells as ammonia and alanine which means that secretion of amino groups from Glutamine is a required phenomenon of the use of glutamine as an anapleoritc precursor and to generate NADPH more preferably than maintaining the amino acid pool inside the cancer cell [20].

6.2.3 Lipid Metabolism Alterations

To proliferate rapidly cancer cells need a great amount of lipid and steroids so that they can fulfill the demand of different phospholipids for the formation of lipid bilayer and signaling molecules derived from the lipids [22]. The expression of the enzymes related to the lipid synthesis as well as the transcription factors controlling the lipid homeostasis is deregulated in the cancer cells [23]. The expression of these enzymes and transcription factors is increased in tumor cells. Beside the deregulated metabolism of the lipids in cancer cells, the lipid related factors are also very crucial for the accretion of important lipid messengers like prostaglandin and lysophosphatidic acid [23].

The enzyme ATP citrate lyase (ACL) is responsible for the conversion of citrate into acetyl-CoA (a lipogenic precursor) and reported to be highly expressed in various human cancers, the expression of this enzyme is reported under the positive regulation of PI3K/Akt pathway [24]. Another enzyme called acetyl-CoA carboxylase (ACC) is also overexpressed in many human cancers [25], this enzyme is responsible for the formation of malonyl-CoA from acetyl-CoA, which is considered as rate limiting step in fatty acid production pathway. One more enzyme, FAS is also described as a key regulatory lipogenic enzyme, it is a multicomplex enzyme which is required for the successive condensation of malonyl-CoA and acetyl-CoA as a substrate to form palmitate [26]. FAS expression is upregulated in most of the cancer cells [23] and it is being described as a potential biomarker to diagnose cancer for therapeutic mean [27–29]. Due to the ubiquitous upregulation of the FAS in most of the cancer cells, several inhibitors of FAS are being developed to test their efficacy on the preclinical models [30, 31]. Apart to these core key regulatory enzymes, one transcription factor termed as sterol regulatory element binding transcription protein 1 (SREBP1) is also upregulated in the cancer cells [22, 32], and the target gene of this transcription factor promotes the tumorigenesis [33]. Some other enzymes promoting lipogenesis which is known to be involved in metabolic deregulation, are also upregulated in cancer cells; it includes enzymes related to fatty acid synthesis like ACS & SCDs [34]. Beside these, enzymes like choline kinase [for choline synthesis and 3-hydroxy-3methylglutaryl-CoA reductase (HMGCR) (for cholesterol synthesis)] are also highly expressed and there role in tumor development is also identified.

6.3 Influence of Tumor Microenvironment on Cancer

In earlier times there was a reductionist view of cancer according to which tumor was considered as an assemblage of relatively homogenous cancer cells. According to the reductionist view by knowing about the autonomous properties of a cancer cell we can understand almost all the tumor biology. By the advancement in techniques it is now being found out that tumors have multiple complexities. With the help of different types of cells present in tumor it creates a microenvironment which often promotes its growth and invasive properties during tumorigenesis. To understand about the tumor biology, it is important to know about the function of each cell type which are present within it [3]. The tumor microenvironment includes different types of cells and factors present in the locality of tumor which help in tumor invasion and metastasis. The different types of cells which are present in tumor microenvironment are cancer stem cells, immune cells, endothelial cells, cancer associated fibroblast, pericytes, cancer associated macrophages, cancer associated adipocytes, etc. Various cytokines/growth factors/metabolites performing different signaling are available in TME. Different cells and cytokines along with their effect on tumor are being discussed below.

6.4 Cancer Stem Cells (CSCs)

CSCs have almost similar transcriptional profiles as the normal tissue stem cell populations and are present in tumor along with cancer cells which carry oncogenic and tumor suppressor mutations. They have the capability of seeding new tumors after injecting them in the preclinical models like mice. These cells are the tumor initiating cells and display different surface markers. There are two theories of cancer stem cells origin one is that they may arise from normal tissue stem cell by ongoing oncogenic transformation, other one is that they may arise from partially differentiated progenitor cells that may be subjected to oncogenic transformation acquiring more stem like characteristics. After formation of primary tumors, cancer stem cell can do two functions: they can self-renew themselves maintaining their population or can differentiate into more derivatives. Cancer stem cells support EMT (epithelial to mesenchymal transition) by initiating tumor at some sites other than their site of origin. These cells can also be the reason of recurrence of tumor even after chemotherapy because they are more resistant to the chemotherapy and may also persist for many years after therapy and may suddenly flare up and generate tumor. The growth of tumor is supported by these different kinds of cells within the tumor that have different functions and help in overall growth of tumor. In this way, tumors are not relied on the environment to provide different cells for their growth rather they take the support from stromal cells by forming different type of cells by modulation [3].

6.5 Endothelial Cells and Pericytes

The vasculature present in the tumor is formed by the endothelial cells. During the formation of tumor as there is increased growth rate and there is basal lamina between them and connective tissue as a result of which blood supply is reduced which causes hypoxia, due to which angiogenic switch occurs in the quiescent endothelial cells and they enter a biological program to form new blood vessels. Along with providing nutrients to the growing tumor endothelial cells also afftects infiltration of immune cells and composition of stroma. Endothelial cells associated with tumor express various cell surface receptors like VEGFR2 and various others cytokine receptors such as ACKR1 (DARC), ACKR3 (CXCR7), CXC4, and CCR2. The induction of angiogenesis for the growing tumor is done by the recruitment of endothelial cells is response to the released chemokines and these endothelial cells overexpress there receptors and turn on feedback loops for forming new blood vessels. Migration, recruitment, and proliferation of endothelial cells are also controlled by chemokines such as CCL2 and other CXC chemokine [35]. By targeting distinguishing cell surface receptors on tumor endothelial cells various drugs can be formulated. For example, there are many drugs targeting VEGFR2. The change in tumor associated tumor cells as compared to the normal endothelial cells can be done by the transcriptome analysis. [3]. Pericytes are specialized mesenchymal cell type. They are found wrapping around the endothelial tubing of blood vessels. Pericytes help in maintaining the integrity and function of vasculature. They are recruited at the newly forming vessels in response to platelet-derived growth factor B, and their interaction with the endothelial cells occurs with ANGPT-TIE2 system [36]. Generally, pericytes are embedded in basement membrane and secrete growth factors for endothelial cells and other ECM molecules which promote their survival and hamper their proliferation [36, 37]. Further pericytes recruitment is promoted by NG2 proteoglycan and neural cell adhesion molecule 1 (NCAM 1) secretion from the attached pericytes. The interaction of pericytes and endothelial cells is disrupted by the increased secretion of VEGFA which promotes the expression of ANGPT2 and ANGPT2 competes with ANGPT 1 for binding with TIE-2. In this way, the detachment of pericytes from endothelial cells further promotes neo-angiogenesis and unevens blood flow within the tumor, which further potentiates tumor development [36, 38, 39].

6.6 Immune Cells

Along with many subpopulations of stromal cells various immune cells are also there in tumor microenvironment, which performs both the functions of promoting the growth of tumor and antagonizing the growth of tumor, they do these functions in varying proportion. There are many tumor promoting leukocytes which promote oncogenesis. Various macrophages subtypes, T and B lymphocytes, mast cells and neutrophils are present in tumor, which modulate oncogenesis [40]. Various inflammatory cells secrete different signaling molecules including angiogenic growth factor VEGF, tumor growth factor EGF, and other pro-angiogenic factors: cytokines, FGF2, and chemokines. Other enzymes which help in invasion by degrading matrix are also secreted, which are matrix metalloproteinases, cysteine cathepsin proteases, and heparanase [41]. Immune cells associated with tumor not only promote angiogenesis and tumor progression, but also provide a path to escapes form destruction by immune cells. Main tumor promoting cells in TME are myeloid-derived suppressor cells (MDSCs) and regulatory T cells (Tregs). Tregs express Foxp3 surface marker along with CD4+ marker. It helps in maintaining peripheral tolerance. Foxp3 surface marker is important for specification of its lineage and suppressive signals regulation [42]. In normal tissues, Tregs can control the over-reactive immune cells and can suppress them in auto-immune diseases while in TME they prevent the cancer cells from cytotoxic T cells [43]. MDSCs are considered to have a crosstalk with the tumor which is important for its development [44]. It can promote many properties of tumor cells like angiogenesis by the production of MMP 9, vascular endothelial growth factor (VEGF) and prokinectin2, migration of cancer cells and metastasis [45]. They can also inhibit the function of T cells by the production of immunosuppressive cytokines (TGF-β, IL-10), Arginase and inducible nitric oxide synthase [46, 47]. There are mainly three strategies by which immune cells can be targeted for therapeutic purposes. First is the stimulation of anti-tumor response by administrating cytokines for non-specific immune stimulation such as GM-CSFs [48], Interleukins (IL-2, IL-12, IL-15 and IL-21) and interferon as an adjuvants while treating cancer [49]. Second is adoptive cell transfer strategy in which the T cell is modified to target the tumor cells by either deriving it from unmodified tumor-infiltrating T/N cells or the receptors can be modified genetically to recognize antigens on tumor cells such a modified T cell receptor and chimeric antigen receptor [50, 51]. Third is to modulate immune checkpoints. This can be done by interfering with the inhibitory checkpoints and suppressing them or by promoting the stimulatory checkpoints using antibodies. One example of blocking the inhibitory checkpoint is the blockade of CTLA-4 signaling by utilizing CTLA-4 antibody [51] and the blockade of PD-1 signaling by utilizing PD-1/PD-L1 antibody [52].

6.7 Cancer Associated Fibroblasts

There are two categories of fibroblast which are present in tumor and support its epithelial tissues by creating structural foundations. One is the normal fibroblast cells and other is myofibroblasts which have distinct α-smooth muscle actin expression (SAM). Myofibroblast are also present in some other tissues like: liver, lung, and kidney where they help in tissue repair but they cause problems in chronic inflammation and pathological fibrosis. In tumor cells, fibroblast cells enhance cancer cell growth, angiogenesis, metastasis, and invasion by secreting different extracellular matrix components [3].

6.8 Cancer Associated Adipocytes

Adipocytes stimulate the creation of a microenvironment hospitable for tumor progression by the secretion of several factors [53]. Increase in factors such as the chemokines CCL5 and CCL2 and IL-6 and TNF-α increases proliferation and invasion of tumor cells [54]. The characteristics of adipocyte tissue in obesity are the same as that in CAAs of tumorous tissue, in turn suggesting they play a part in tumor development in people with obesity. They have thus been emerging as targets to treat cancer. Some signals that the adipocytes receive from their neighbor tumor cells are those that activate metastasis in distant organs. Blocking of adipocyte-derived progression of cancer can open new pathways to treat the disease in obese patients.

6.9 Influence of Tumor Microenvironment on Cancer Cell Metabolism

Tumor cells show deregulated metabolism, which is a hallmark of cancer. This deregulation is basically coupled with the adaptation of cancer cells, against change in the environment where these cells are growing, what type of nutrients have been provided to the tumor cells along with the quantity of nutrients. TME is the main regulator for alteration of metabolism occurring in cancer cells. TME provides certain cytokines, regulatory molecules so that cancer cells can adopt to the difficult environmental conditions, CAFs (cancer associated fibroblasts) and ECM (extracellular matrix) are known to be the main regulatory elements for the changes related to metabolism in cancer cells [55]. Various metabolites like pyruvate, lactate, aspartate, and glutamine are secreted by the CAFs. And these metabolites are further used up by the cancer cells to achieve cellular growth and proliferation. Further the alteration in metabolism caused by CAFs leads to the increased metabolism of glutamine by upregulation of TCA cycle. Beside of this, the components of ECM like laminin, fibronectin, are reported as the proteins to be responsible for integrin mediated control of nutrients signaling. But despite this there is a lot more to be found soon, increase in growth of the tumor, and very less blood supply leads the formation of nutrients-deprived TME. This leads cancer cells to adopt various preventive strategies to achieve the required metabolic substrate for energy. Metastatic breast cancer cells represent increase in proliferation and migration capabilities when cells grow in the presence of obese adipocytes. The adipocyte secrets some fatty acid in the TEM, that transfer to the invasive breast cancer cells. Furthermore, it leads to the induction of ATGL-mediated lipid digestion and fatty acid oxidation in the cellular power house [56]. Cancer cells can use ECM lipids in starvation and hypoxic conditions. Lipid is required by the cancer cells to produce the cell membrane and to proliferate more.

6.10 Phytochemicals Affecting Cancer Cell Metabolism and Tumor Microenvironment

In concomitant with finding the alteration in cancer cell metabolism and changes in tumor microenvironment, the next logical objective is to find out the therapeutic and preventive purposes of this information. The specific alterations can be discern as the probable molecular targets of novel anticancer compounds [23]. Most suitable anticancer compounds can be found in the natural products and out of the natural products the class which has been profusely studied for its biological effects is phytochemicals. Various phytochemicals have entered in clinical trials for providing the therapeutic compounds for cancer [57]. Phytochemicals are classified into four major classes: alkaloids, phenolics, terpenoids, and tannins according to their chemical properties. Here the anti-cancerous properties of some of the phytochemicals are being discussed along with the molecular targets that how they are targeting the modulated cancer metabolism and tumor microenvironment. The following phytochemicals are mentioned for their effect on the cancer cell metabolism: curcumin, Gallic acid, EGCG, genistein, apigenin (Phenolic compounds), geraniol, oleanolic acid (Isoprenoids), piperine, berberine, capsaicin (Alkaloid), and the following phytochemicals are discussed for their effect on the tumor microenvironment and cancer cell metabolism (Figs. 6.1 and 6.2).

6.10.1 Curcumin

Curcumin (diferuloylmethane) belong to the class of phenolic compounds of phytochemicals and has been widely studied for its medicinal properties. Curcumin is obtained from the rhizome of plant *Curcuma longa* (Zingiberaceae). Curcumin has been shown to have multiple effects acting as an agent that can modify many distinct pathways of signaling which are occurring inside the cell and which further influence cellular growth, apoptotic response and inflammentry response. Due to its properties like anti-oxidant, anti-inflammatory it can be used for treating some diseases [58]. There are various studies which have shown the anti-cancerous properties for curcumin in many cell lines and animals. In cancer cells PKM2 expression is altered which is targeted by curcumin. Curcumin decreases the glycolytic flux by inhibiting

PKM2 expression due to which other metabolic pathways such as Hexose Phosphate Shunt for which metabolic precursors are provided by glycolysis are also inhibited. Curcumin is also reported to be an inhibitor of higher growth capability of cancer cells because of its modulatory properties thus results in the synthesis of products like nucleic acids, amino acids, and lipids by the intermediates of glycolysis. Curcumin is also known to reverse the inflammatory effect of TNF-α. TNF-α increases the glycolytic flux by increasing GLUT1 expression and curcumin reverses this effect by decreasing of GLUT1 expression at both transcriptional and translational levels in cells which have been stimulated by a cytokine, TNF-α [59]. In some cancer cell lines, curcumin have been observed to decrease the glucose transportation into the cells from outside environment related to tumor, lactate production and ATP production is also shown decreased in these cells. Expression of HK-2 is also decreased by curcumin along with its separation from mitochondria through activation of Akt signaling which leads to apoptosis mediated by cellular power house. Other than that curcumin can decrease the expression of FAS lipogenic enzyme in many tumor cells which deprive cancer cells of energy [60]. Apart from targeting altered metabolism curcumin also targets other pathways to inhibits the cancer development, such as induction of apoptosis, stabilization of mTOR, Wnt, Notch, inhibition of PI3K signaling, activation of AMPK, inhibition of cell cycle, downregulation of tumor promoting genes, inhibition of NF-κB activity, regulation decreasing the metastatic properties of cancer cells, inhibition of angiogenesis, regulation of miRNA, DNA damage and repair [61]. When we talk with respect to the tumor microenvironment, curcumin delivered as a conjugate with PEG and Trp2 peptide vaccine results in downregulation of IL-6 production and reduced expression of factors responsible for suppression of immune system in melanoma bearing mouse model [62]. Beside this curcumin in reported as a modulator of tumor cell microenvironment by showing an inhibitory on Hh signaling in pancreatic tumor [63] (Figs. 6.1 and 6.2).

6.10.2 Gallic Acid

There are many sources from which Gallic acid can be obtained such as gallnuts, sumac with hazel (*Hamamelis virginiana)*, clove (*Syzygium aromaticum)*, tealeaves, oak bark, sundew, and some other plants. It is present in many edibles also such as Blackberry, common walnut, hot chocolate, vinegar, wine and it belongs to the class of phenolic compounds. Gallic acid shows anti-tumorigenic activity through various mechanisms [64]. Gallic acid promotes apoptosis via mitochondrial pathway by increasing the expression of cleaved forms of caspase-9, caspase-3 PARP-1 and genes BAX and BAD which are required for apoptosis, along with the downregulation of genes Bcl-2 and Bcl-XL which are responsible for inhibition of apoptosis. Gallic acid is also responsible for the overexpression of distinct enzymes related to glycolysis which includes glucokinase, α-enolase, and aldolase. There are two reasons through which interactions between apoptosis and glycolysis can be explained, first is that BAD which is pro-apoptotic and glucokinase involved in glycolysis resides in mitochondrial complex, second reason is that while intensifying glycolysis Gallic acid is responsible for the production of reactive metabolite, i.e. methylglyoxal (MG), leading to apoptosis through mitochondrial intrinsic pathway [65] (Figs. 6.1 and 6.2).

6.10.3 Epigallocatechin Gallate (EGCG)

This phytochemical also belong to the class of phenolic compounds and it is obtained from green tea [23]. The anti-tumorigenic activity of EGCG has been reported in various cell lines highlighting the multiple targets of EGCG. In colon cancer cells (HT-29), it reduces the mRNA levels of GLUT1, and (VEGF) while foster the activation of AMPK. Phosphorylation of ACC is also elevated by EGCG which leads to decreased fatty acid synthesis [66]. When talk in context with amino acid metabolism, EGCG increases glutamine consumption and decreases the embellishment of Glutamate [67]. Various anabolic pathways which are upregulated in cancer cells to meet the increasing demand of energy due to high proliferation rates are also targeted by EGCG such as enzymes of hexose phosphate shunt, TKT, and glycolytic enzyme G6PD. Glycolysis can also be inhibited by depleting the expression of HK2 which is significantly regulated by PI3K/Akt signaling [68]. EGCG also decreases the anti-apoptotic effects of HK2 by decreasing its translocation to the mitochondrial outer membrane. Other enzymes which are downregulated by EGCG are PFK that can result in inhibition of glycolysis [22]. Other enzyme of glycolysis which is inhibited by EGCG is phosphoglycerate mutase 1(PGAM1), which converts 3-PG into 2-PG playing a crucial role in regulation of PPP flux [69]. Other than glycolysis EGCG is also known to inhibit the activity of FAS decreasing the lipid synthesis in cancer cells [70]. In some cancer cell lines such as pancreatic adenocarcinoma cells EGCG also decreases Acetyl-CoA, leading to the reduced synthesis of palmitate which is mainly required for cellular membranes [71]. With respect to TME, EGCG can also reduce the IL-6 level in the TME of breast cancer by affecting tumor associated macrophages (TAMs) [72]. EGCG has shown its suppressive effect on the myofibroblast differentiation in relation with prostate cancer [73] (Figs. 6.1 and 6.2).

6.10.4 Genistein

Genistein is a phytochemical which belongs to the family of flavonoids and is present in soybean and some forage plants. The main anti-tumorigenic effect of genistein is in the inhibition of cell viability and upregulated cellular proliferation. Genistein decreases GLUT1 expression in cancer cell lines, thus decreasing glycolytic flux in cancer cells while elevates the expression of pro-apoptotic genes which are p53 and p21. In many of the chemo-resistant cells COX 2 expression is increased which is reduced by genistein by doing the phosphorylation of AMPK. The possible signal which is present at upstream of AMPK is ROS hence genistein also increases ROS production which can induces death of cancer cells [74] (Figs. 6.1 and 6.2).

6.10.5 Apigenin

Apigenin also belongs to the family of flavonoids and is mostly found in fruits and vegetables. It possesses various properties such as anti-oxidative, anti-inflammatory, and anti-cancerous property also. Apigenin targets glycolytic flux by decreasing the uptake of glucose by downregulating the GLUT1 mRNA and protein levels mediated by PI3K/Akt pathway [75]. Decrease in glycolytic flux helps in decreasing ATP production which can lead to inhibition of growth, cause apoptosis and intensify the chemotherapeutic drugs sensitivity [76, 77]. In some cancer cell lines, apigenin can induce GLUT4 expression while downregulate GLUT1 expression; while in other cell line it downregulates expression of both GLU1 and GLUT4 [78]. Apigenin is also thought to be an allosteric inhibitor of PKM2. Inhibition of PKM2 results in decreased proliferation due to reduced glucose consumption, lactate production, and ATP generation [79]. Other than that, apigenin also targets FAS downregulating its expression [80]. Downregulation of FAS results in lipid synthesis inhibition which can cause membrane deformity and inhibition of various signaling pathways in cancer cells (Figs. 6.1 and 6.2).

6.10.6 Capsaicin

This compound belongs to the alkaloid class of secondary metabolites, which is obtained from the capsicum fruits, capsaicin is considered as the main pungent component of this fruit. Chili peppers of the genus capsicum are being used in the form of spices throughout the globe. Biologists have found that this plant has some remarkable clinical properties, thus contains some therapeutic significance as well [81]. In recent experiments, it is reported that this plant contains anticancer effects against various human cancers, therefor it has become a good point of attraction for many cancer researchers to look for some therapeutic role of this plant for the treatment of cancer. In the experimental analysis of the pancreatic tumor cells, a significant increase in the apoptosis was recorded in capsaicin treated cells [82], it was correlated with the amount of ROS produced and impairment of the mitochondrial functions. In further studies it was also observed that capsaicin was having some inhibitory role in the ROS defense mechanism of cancer cells which was responsible for the increased apoptosis [82]. Beside the role in mitochondrial dysfunction and apoptosis in the cancer cells, capsaicin also has a role in the downregulation of FAS protein level thus affects the lipogenesis [83] along with the decrease in de novo fatty acid synthesis (Figs. 6.1 and 6.2).

6.10.7 Berberine

Berberine, an alkaloid found in plants, is being reported as an inhibitor of cell-proliferative capabilities and is also shown as a good inducer of apoptosis in a very large number of distinct cancer cell lines [84]. Proteomics study have revealed that a number of glycolytic enzymes, like TPI, FBAA, and enolase-α were downregulated [85] when the cells were treated with this phytochemical and thus represents the anti-tumor effect of this natural extract. This was attributed to the compound's effect on pro apoptotic proteins and indicated a response wherein carbohydrates were passed on from glycolysis to some other distinct metabolic pathways that generated some molecules in the form of reducing power moieties which are highly required for prevention from oxidative stress which is being produced by ROS [85]. Experimental studies on breast cancer cells showed a changed level of several enzymes required for glycolysis, further validating the suppression of glycolysis via berberine. Berberine is also reported as a phytochemical responsible for increased ATP production, in tandem with over activation of OXPHOS and in direct opposition to what has been observed in mature lung cancer cells [86]. In breast tumor, it suppressed fatty acid synthesis by increasing p-ACC levels and decreasing p-ACL levels (Figs. 6.1 and 6.2).

6.10.8 Tanshinone IIA

Talking about the natural extracts with clinical significance, Tanshinone IIA is an important phytochemical due to its well-known growth suppressor activity. Tanshinone IIA is lipophilic compound which isolated from Salvia miltiorrhiza Bunge roots, it targets the tumor cell glucose metabolism and thus decreases the growth of cancer cell [87] by downregulating the enzyme glucose-6-phosphate isomerases (GPI), and further reduces the glucose metabolism and thus decreases the production of pyruvate. The repressive effect of Tanshinone IIA is found to be related with the overexpressed p53 which in turn downregulates the Akt signaling. Beside the role in glycolysis, other pathway called gluconeogenesis is also reported to be deregulated which is coupled with the overexpression of PCK2 gene, along with reduce expression of LDHB and MDH1 genes [88].

6.10.9 Geraniol

Geraniol belongs to the isoprenoid class of phytochemicals, it is a monoterpenoid, various plants produce the geraniol as the components of essential oils, which produce aroma, and these plants are widely used for the medical purpose in Ayurveda and natural therapies (1 online link). Geraniol is reported as an anticancer phytochemical due to its significant preventive role towards numerous cancers [89], although researchers are focusing to find out the proper reason of anticancer and growth suppressive activity of Geraniol, but very less is known about how this compound affects the cellular metabolism of cancer cells in HepG2 HCC cells, this compound is reported as a decreasing agent of fatty acid metabolism [90, 91]. Geraniol also shows its inhibitory effect on HMGCR and thus downregulates the mevalonate pathway which further results in reduction in the cell proliferation and prevention from apoptosis [91]. Apart from this, geraniol also helps in overexpression of p-ACC in some cancer cells via AMPK pathway activation [92] (Figs. 6.1 and 6.2).

6.10.10 Resveratrol

Resveratrol is very well-known phytochemical which is highly known for its modulatory action on lipid metabolism of the cancer cell metabolism, it is a phenolic compound, which has been reported as AMPK activator, which further results in the inhibitory phosphorylation of the ACC [62, 93]. Beside this, activated AMPK also inhibits the GSK3 [93]. Further, resveratrol also shows an inhibitory effect on FAS at both transcriptional and translational level and thus gives an anti-proliferative effect through the FAS Inhibition [94, 95]. Other studies also show that resveratrol increases the production of arachidonic acid. Furthermore, resveratrol is also responsible for the increase in ceramide level, which is a core component of the sphingolipids responsible for the cell proliferation and apoptosis by working as secondary messengers [96–98] this increase ceramide leads to its accumulation in the cells due to upregulation of two different enzymes, SPT & (nSMase) responsible for the ceramide biosynthesis. This increased accumulation of ceramide occurs due to the upregulated sphingomyelin degradation [98]. Beside its inhibitory role in the fatty acid synthesis, resveratrol is a well-known modulator of glucose metabolism where it targets many core glycolytic enzymes including, Hexokinase (HK), phosphoglycerokinase (PGK) enolase, and pyruvate kinase. Resveratrol can also inhibit the IL-18 production in melanoma tumor microenvironment and thus helps in the reduction of metastasis of cancer cells [99] (Figs. 6.1 and 6.2).

6.10.11 Plumbagin

Plumbagin is a plant secondary metabolite which forms yellow color dye, derived from naphthoquinone. The name plumbagin was given for the plant genus plumbago from which it was originally obtained. Other plants like Drosera and nepenthes are also known as a good source of it. Plumbagin is a propitious phytochemical for showing a clinically significant and useful effect on to the bone metastasis of breast cancer [100] via targeting the cancer microenvironment [101]. The main target of this compound is RANK and TRAF6, therefor inhibiting the MAPK and NF-κB Signaling [101]. Plumbagin is reported as an inhibitor of c-MYB activity therefor it can be a good therapeutic agent to modulate the tumor-stromal cross-talk present in the pancreatic tumor signaling [102] (Figs. 6.1 and 6.2).

6.11 Conclusion and Future Prospects

How phytochemicals reduce different properties required for the cancer development and progression is well defined. Analog derived from said compounds have been experimentally tested in various pre-clinical research models. These researches hint at the fact that many of the effects of these phytochemicals is due to their regulation of TME and metabolism of cancer cell. The targeting of the tumor cell metabolism is not yet completely explored, here in our chapter, we have mainly focused on the deregulation of metabolism which takes place in the early stage of cancer along with the role of TME in cancer growth, proliferation, and metastasis and how these alteration have been counteracted by using certain phytochemicals to achieve better chemoprevention. Despite all these knowledge's about phytochemicals and their role in cancer therapeutic by acting on altered metabolism and deregulated TME, effective agents need to be discovered which show better anticancer properties without having any toxic effects on the healthy cells. Although the TME is important in cancer progression, studies involving its regulation and participating components can get quite challenging. This is primarily due to the unavailability of assays that mimic the subtleties of the TME. For fast and efficient screening of new naturally extracted compounds, a 3D microfluidic device that has the capability to test multiple compounds in tandem for their anti-tumor potential has been reported to fulfill the above said assay needs [103]. A better representation of the TME can be obtained by co-culturing of endothelial/fibroblast/macrophages cells and cancer cells, with their 3D morphology, in this device. Advancement in experimental power has increased the hope for use of phytochemicals for treatment of cancer in the upcoming time. Although these phytochemicals have beneficial properties regarded to human health, but still some difficulties are present, which are affecting the use of these compound for therapeutic purpose these obstacles include bioavailability of the natural extracts and there absorption by the cellular system, stability and metabolism of these compounds by human body. Thus we conclude that additional research must needs to be performed for the use of these natural products for a better therapeutic purpose.

Conflict of Interest The author confirms that there is no conflict of interest in this book chapter.

References

1. Krepler C, Xiao M, Sproesser K et al (2016) Personalized preclinical trials in BRAF inhibitor-resistant patient-derived xenograft models identify second-line combination therapies. Clin Cancer Res 22(7):1592–1602
2. Newman DJ, Cragg GM (2016) Natural products as sources of new drugs from 1981 to 2014. J Nat Prod 79(3):629–661
3. Hanahan D, Weinberg RA (2011) Hallmarks of cancer: the next generation. Cell 144 (5):646–674
4. Pavlova NN, Thompson CB (2016) The emerging hallmarks of cancer metabolism. Cell Metab 23(1):27–47

5. Luengo A, Gui DY, Vander Heiden MG (2017) Targeting metabolism for cancer therapy. Cell Chem Biol 24(9):1161–1180
6. Cohen R, Neuzillet C, Tijeras-Raballand A et al (2015) Targeting cancer cell metabolism in pancreatic adenocarcinoma. Oncotarget 6(19):16832–16847
7. Warburg O, Wind F, Negelein E (1927) The metabolism of tumors in the body. J Gen Physiol 8(6):519–530
8. Warburg O (1956) On the origin of cancer cells. Science 123(3191):309–314
9. Adekola K, Rosen ST, Shanmugam M (2012) Glucose transporters in cancer metabolism. Curr Opin Oncol 24(6):650–654
10. Mueckler M, Thorens B (2013) The SLC2 (GLUT) family of membrane transporters. Mol Asp Med 34(2-3):121–138
11. Wang J, Ye C, Chen C et al (2017) Glucose transporter GLUT1 expression and clinical outcome in solid tumors: a systematic review and meta-analysis. Oncotarget 8 (10):16875–16886
12. Christofk HR, Vander Heiden MG, Harris MH et al (2008) The M2 splice isoform of pyruvate kinase is important for cancer metabolism and tumour growth. Nature 452(7184):230–233
13. Diaz-Ruiz R, Rigoulet M, Devin A (2011) The Warburg and Crabtree effects: on the origin of cancer cell energy metabolism and of yeast glucose repression. Biochim Biophys Acta 1807 (6):568–576
14. Vander Heiden MG, Cantley LC, Thompson CB (2009) Understanding the Warburg effect: the metabolic requirements of cell proliferation. Science 324(5930):1029–1033
15. Yalcin A, Telang S, Clem B et al (2009) Regulation of glucose metabolism by 6-phosphofructo-2-kinase/fructose-2,6-bisphosphatases in cancer. Exp Mol Pathol 86 (3):174–179
16. Scatena R, Bottoni P, Pontoglio A et al (2008) Glycolytic enzyme inhibitors in cancer treatment. Expert Opin Investig Drugs 17(10):1533–1545
17. Lincet H, Icard P (2015) How do glycolytic enzymes favour cancer cell proliferation by nonmetabolic functions? Oncogene 34(29):3751–3759
18. Miao P, Sheng S, Sun X et al (2013) Lactate dehydrogenase A in cancer: a promising target for diagnosis and therapy. IUBMB Life 65(11):904–910
19. Patra KC, Hay N (2014) The pentose phosphate pathway and cancer. Trends Biochem Sci 39 (8):347–354
20. DeBerardinis RJ, Mancuso A, Daikhin E et al (2007) Beyond aerobic glycolysis: transformed cells can engage in glutamine metabolism that exceeds the requirement for protein and nucleotide synthesis. Proc Natl Acad Sci U S A 104(49):19345–19350
21. Liu G, Zhu J, Yu M et al (2015) Glutamate dehydrogenase is a novel prognostic marker and predicts metastases in colorectal cancer patients. J Transl Med 13:144–144
22. Guerra AR, DuarteMF DIF (2018) Targeting tumor metabolism with plant-derived natural products: emerging trends in cancer therapy. J Agric Food Chem 66(41):10663–10685
23. Cerella C, Radogna F, Dicato M et al (2013) Natural compounds as regulators of the cancer cell metabolism. Int J Cell Biol 2013:639401
24. Chypre M, Zaidi N, Smans K (2012) ATP-citrate lyase: a mini-review. Biochem Biophys Res Commun 422(1):1–4
25. Currie E, Schulze A, Zechner R et al (2013) Cellular fatty acid metabolism and cancer. Cell Metab 18(2):153–161
26. Flavin R, Peluso S, Nguyen PL et al (2010) Fatty acid synthase as a potential therapeutic target in cancer. Future Oncol 6(4):551–562
27. Bandyopadhyay S, Pai SK, Watabe M et al (2005) FAS expression inversely correlates with PTEN level in prostate cancer and a PI 3-kinase inhibitor synergizes with FAS siRNA to induce apoptosis. Oncogene 24(34):5389–5395
28. Walter K, Hong SM, Nyhan S et al (2009) Serum fatty acid synthase as a marker of pancreatic neoplasia. Cancer Epidemiol Biomark Prev 18(9):2380

29. Kuhajda FP (2006) Fatty acid synthase and cancer: new application of an old pathway. Cancer Res 66(12):5977
30. Liu Q, Luo Q, Halim A et al (2017) Targeting lipid metabolism of cancer cells: a promising therapeutic strategy for cancer. Cancer Lett 401:39–45
31. Martinez-Outschoorn UE, Peiris-Pagés M, Pestell RG et al (2017) Cancer metabolism: a therapeutic perspective. Nat Rev Clin Oncol 14(1):11–31
32. Li Z, Zhang H (2016) Reprogramming of glucose, fatty acid and amino acid metabolism for cancer progression. Cell Mol Life Sci 73(2):377–392
33. Shao W, Espenshade PJ (2012) Expanding roles for SREBP in metabolism. Cell Metab 16 (4):414–419
34. Igal RA (2016) Stearoyl CoA desaturase-1: New insights into a central regulator of cancer metabolism. Biochim Biophys Acta 1861(12):1865–1880
35. Salazar N, Zabel BA (2019) Support of tumor endothelial cells by chemokine receptors. Front Immunol 10:147–147
36. De Palma M, Biziato D, Petrova TV (2017) Microenvironmental regulation of tumour angiogenesis. Nat Rev Cancer 17(8):457–474
37. Viallard C, Larrivee B (2017) Tumor angiogenesis and vascular normalization: alternative therapeutic targets. Angiogenesis 20(4):409–426
38. Klein D (2018) The tumor vascular endothelium as decision maker in cancer therapy. Front Oncol 8:367
39. Harrell CR, Simovic Markovic B, Fellabaum C et al (2018) Molecular mechanisms underlying therapeutic potential of pericytes. J Biomed Sci 25(1):21
40. Coffelt SB, Lewis CE, Naldini L et al (2010) Elusive identities and overlapping phenotypes of proangiogenic myeloid cells in tumors. Am J Pathol 176(4):1564–1576
41. Qian BZ, Pollard JW (2010) Macrophage diversity enhances tumor progression and metastasis. Cell 141(1):39–51
42. Samstein RM, Arvey A, Josefowicz SZ et al (2012) Foxp3 exploits a pre-existent enhancer landscape for regulatory T cell lineage specification. Cell 151(1):153–166
43. Wolf D, Sopper S, Pircher A et al (2015) Treg(s) in cancer: friends or foe? J Cell Physiol 230 (11):2598–2605
44. Talmadge JE, Gabrilovich DI (2013) History of myeloid-derived suppressor cells. Nat Rev Cancer 13(10):739–752
45. Zhou J, Nefedova Y, Lei A et al (2018) Neutrophils and PMN-MDSC: their biological role and interaction with stromal cells. Semin Immunol 35:19–28
46. Gabrilovich DI, Ostrand-Rosenberg S, Bronte VS (2012) Coordinated regulation of myeloid cells by tumours. Nat Rev Immunol 12(4):253–268
47. Gabrilovich DI (2017) Myeloid-derived suppressor cells. Cancer Immunol Res 5(1):3–8
48. Yan WL, Shen KY, Tien CY et al (2017) Recent progress in GM-CSF-based cancer immunotherapy. Immunotherapy 9(4):347–360
49. Floros T, Tarhini AA (2015) Anticancer cytokines: biology and clinical effects of interferon-alpha2, interleukin (IL)-2, IL-15, IL-21, and IL-12. Semin Oncol 42(4):539–548
50. Kalos M, June CH (2013) Adoptive T cell transfer for cancer immunotherapy in the era of synthetic biology. Immunity 39(1):49–60
51. Fang F, Xiao W, Tian Z (2017) NK cell-based immunotherapy for cancer. Semin Immunol 31:37–54
52. Topalian SL, Hodi FS, Brahmer JR et al (2012) Safety and activity of anti–PD-L1 antibody in patients with advanced cancer. N Engl J Med 366(26):2455–2465
53. Rybinska I, Agresti R, Trapani A et al (2020) Adipocytes in breast cancer, the thick and the thin. Cell 9(3):E560
54. Wu Q, Li B, Li Z et al (2019) Cancer-associated adipocytes: key players in breast cancer progression. J Hematol Oncol 12(1):95
55. Nazemi M, Rainero E (2020) Cross-talk between the tumor microenvironment, extracellular matrix, and cell metabolism in cancer. Front Oncol 10:239

56. Balaban S, Shearer RF, Lee LS et al (2017) Adipocyte lipolysis links obesity to breast cancer growth: adipocyte-derived fatty acids drive breast cancer cell proliferation and migration. Cancer Metab 5:1–1
57. Zubair H, Azim S, Ahmad A et al (2017) Cancer chemoprevention by phytochemicals: nature's healing touch. Molecules 22(3):395
58. Noorafshan A, Ashkani-Esfahani S (2013) A review of therapeutic effects of curcumin. Curr Pharm Des 19(11):2032–2046
59. Vaughan RA, Garcia-Smith R, Dorsey J et al (2013) Tumor necrosis factor alpha induces Warburg-like metabolism and is reversed by anti-inflammatory curcumin in breast epithelial cells. Int J Cancer 133(10):2504–2510
60. Guerra AR, Duarte MF, Duarte IF (2018) Targeting tumor metabolism with plant-derived natural products: emerging trends in cancer therapy. J Agric Food Chem 66(41):10663–10685
61. Siddiqui FA, Prakasam G, Chattopadhyay S et al (2018) Curcumin decreases Warburg effect in cancer cells by down-regulating pyruvate kinase M2 via mTOR-HIF1α inhibition. Sci Rep 8 (1):8323
62. Hou X, Xu S, Maitland-Toolan KA et al (2008) SIRT1 regulates hepatocyte lipid metabolism through activating AMP-activated protein kinase. J Biol Chem 283(29):20015–20026
63. Cao L, Xiao X, Lei J et al (2016) Curcumin inhibits hypoxia-induced epithelialmesenchymal transition in pancreatic cancer cells via suppression of the hedgehog signaling pathway. Oncol Rep 35(6):3728–3734
64. Subramanian AP, John AA, Vellayappan MV et al (2015) Gallic acid: prospects and molecular mechanisms of its anticancer activity. RSC Adv 5(45):35608–35621
65. Liu C, Lin JJ, Yang ZY et al (2014) Proteomic study reveals a co-occurrence of gallic acid-induced apoptosis and glycolysis in B16F10 melanoma cells. J Agric Food Chem 62 (48):11672–11680
66. Hwang JT, Ha J, Park IJ et al (2007) Apoptotic effect of EGCG in HT-29 colon cancer cells via AMPK signal pathway. Cancer Lett 247(1):115–121
67. Sánchez-Tena S, Alcarraz-Vizán G, Marín S et al (2013) Epicatechin gallate impairs colon cancer cell metabolic productivity. J Agric Food Chem 61(18):4310–4317
68. Gao F, Li M, Liu WB et al (2015) Epigallocatechin gallate inhibits human tongue carcinoma cells via HK2mediated glycolysis. Oncol Rep 33(3):1533–1539
69. Li X, Tang S, Wang QQ et al (2017) Identification of epigallocatechin-3- gallate as an inhibitor of phosphoglycerate mutase 1. Front Pharmacol 8:325–325
70. Brusselmans K, De Schrijver E, Heyns W et al (2003) Epigallocatechin-3-gallate is a potent natural inhibitor of fatty acid synthase in intact cells and selectively induces apoptosis in prostate cancer cells. Int J Cancer 106(6):856–862
71. Lu QY, Zhang L, Yee JK et al (2015) Metabolic consequences of LDHA inhibition by epigallocatechin gallate and oxamate in MIA PaCa-2 pancreatic cancer cells. Metabolomics 11(1):71–80
72. Jang JY, Lee JK, Jeon YK et al (2013) Exosome derived from epigallocatechin gallate treated breast cancer cells suppresses tumor growth by inhibiting tumor-associated macrophage infiltration and M2 polarization. BMC Cancer 13:421
73. Slezakova S, Ruda-Kucerova J (2017) Anticancer activity of artemisinin and its derivatives. Anticancer Res 37(11):5995–6003
74. Hwang JT, Ha J, Park OJ (2005) Combination of 5-fluorouracil and genistein induces apoptosis synergistically in chemo-resistant cancer cells through the modulation of AMPK and COX-2 signaling pathways. Biochem Biophys Res Commun 332(2):433–440
75. Melstrom LG, Salabat MR, Ding XZ et al (2008) Apigenin inhibits the GLUT-1 glucose transporter and the phosphoinositide 3-kinase/Akt pathway in human pancreatic cancer cells. Pancreas 37(4):426–431
76. FANG J, BAO YY, ZHOU SH et al (2015) Apigenin inhibits the proliferation of adenoid cystic carcinoma via suppression of glucose transporter-1. Mol Med Rep 12(5):6461–6466

77. Xu YY, Wu TT, Zhou SH et al (2014) Apigenin suppresses GLUT-1 and p-AKT expression to enhance the chemosensitivity to cisplatin of laryngeal carcinoma Hep-2 cells: an in vitro study. Int J Clin Exp Pathol 7(7):3938–3947
78. Gonzalez-Menendez P, Hevia D, Rodriguez-Garciae A et al (2014) Regulation of GLUT transporters by flavonoids in androgen-sensitive and -insensitive prostate cancer cells. Endocrinology 155(9):3238–3250
79. Shan S, Shi J, Yang P et al (2017) Apigenin restrains colon cancer cell proliferation via targeted blocking of pyruvate kinase M2-dependent glycolysis. J Agric Food Chem 65 (37):8136–8144
80. Menendez JA, Vazquez-Martin A, Oliveras-Ferraros C et al (2008) Analyzing effects of extra-virgin olive oil polyphenols on breast cancer-associated fatty acid synthase protein expression using reverse-phase protein microarrays. Int J Mol Med 22(4):433–439
81. Clark R, Lee SH (2016) Anticancer properties of capsaicin against human cancer. Anticancer Res 36(3):837–843
82. Pramanik KC, Boreddy SR, Srivastava SK (2011) Role of mitochondrial electron transport chain complexes in capsaicin mediated oxidative stress leading to apoptosis in pancreatic cancer cells. PLoS One 6(5):e20151
83. Impheng H, Pongcharoen S, Richert L et al (2014) The selective target of capsaicin on FASN expression and de novo fatty acid synthesis mediated through ROS generation triggers apoptosis in HepG2 cells. PLoS One 9(9):e107842
84. Ortiz LM, Lombardi P, Tillhon M et al (2014) Berberine, an epiphany against cancer. Molecules 19(8):12349–12367
85. Chou HC, Lu YC, Cheng CS et al (2012) Proteomic and redox-proteomic analysis of berberine-induced cytotoxicity in breast cancer cells. J Proteome 75(11):3158–3176
86. Fan LX, Liu CM, Gao AH et al (2013) Berberine combined with 2-deoxy-d-glucose synergistically enhances cancer cell proliferation inhibition via energy depletion and unfolded protein response disruption. Biochim Biophys Acta 1830(11):5175–5183
87. Lin LL, Hsia CR, Hsu CL et al (2015) Integrating transcriptomics and proteomics to show that tanshinone IIA suppresses cell growth by blocking glucose metabolism in gastric cancer cells. BMC Genomics 16:41
88. Zhang HS, Zhang FJ, Li H et al (2016) Tanshinone A inhibits human esophageal cancer cell growth through miR-122-mediated PKM2 down-regulation. Arch Biochem Biophys 598:50–56
89. Cho M, So I, Chun JN et al (2016) The antitumor effects of geraniol: modulation of cancer hallmark pathways (Review). Int J Oncol 48(5):1772–1782
90. Polo MP, de Bravo MG (2006) Effect of geraniol on fatty-acid and mevalonate metabolism in the human hepatoma cell line Hep G2. Biochem Cell Biol 84(1):102–111
91. Crespo R, Montero Villegas S, Abba MC et al (2013) Transcriptional and posttranscriptional inhibition of HMGCR and PC biosynthesis by geraniol in 2 Hep-G2 cell proliferation linked pathways. Biochem Cell Biol 91(3):131–139
92. Kim SH, Park EJ, Lee CR et al (2012) Geraniol induces cooperative interaction of apoptosis and autophagy to elicit cell death in PC-3 prostate cancer cells. Int J Oncol 40(5):1683–1690
93. Shin SM, Cho IJ, Kim SG (2009) Resveratrol protects mitochondria against oxidative stress through AMP-activated protein kinase-mediated glycogen synthase kinase-3beta inhibition downstream of poly(ADP-ribose)polymerase-LKB1 pathway. Mol Pharmacol 76(4):884–895
94. Pandey PR, Okuda H, Watabe et al (2011) Resveratrol suppresses growth of cancer stem-like cells by inhibiting fatty acid synthase. Breast Cancer Res Treat 130(2):387–398
95. Khan A, Aljarbou AN, Aldebasi YH et al (2014) Resveratrol suppresses the proliferation of breast cancer cells by inhibiting fatty acid synthase signaling pathway. Cancer Epidemiol 38 (6):765–772
96. Saunier E, Antonio S, Regazzetti A et al (2017) Resveratrol reverses the Warburg effect by targeting the pyruvate dehydrogenase complex in colon cancer cells. Sci Rep 7(1):6945

97. Scarlatti F, Sala G, Somenzi G et al (2003) Resveratrol induces growth inhibition and apoptosis in metastatic breast cancer cells via de novo ceramide signaling. FASEB J 17 (15):339–341
98. Mizutani N, Omori Y, Kawamoto Y et al (2016) Resveratrol-induced transcriptional up-regulation of ASMase (SMPD1) of human leukemia and cancer cells. Biochem Biophys Res Commun 470(4):851–856
99. Salado C, Olaso E, Gallot N et al (2011) Resveratrol prevents inflammation-dependent hepatic melanoma metastasis by inhibiting the secretion and effects of interleukin-18. J Transl Med 9:59
100. Yan W, Wang TY, Fan QM et al (2014) Plumbagin attenuates cancer cell growth and osteoclast formation in the bone microenvironment of mice. Acta Pharmacol Sin 35 (1):124–134
101. Li Z, Xiao J, Wu X et al (2012) Plumbagin inhibits breast tumor bone metastasis and osteolysis by modulating the tumor-bone microenvironment. Curr Mol Med 12(8):967–981
102. Azim S, Zubair H, Srivastava SK et al (2016) Deep sequencing and in silico analyses identify MYB-regulated gene networks and signaling pathways in pancreatic cancer. Sci Rep 6:28446
103. Niu Y, Bai J, Kamm RD et al (2014) Validating antimetastatic effects of natural products in an engineered microfluidic platform mimicking tumor microenvironment. Mol Pharm 11 (7):2022–2029

Role of Nutrigenetics and Nutrigenomics in Cancer Chemoprevention

7

Indranil Chattopadhyay

Abstract

Cancer is a multi-factorial complex genetic disorder that is characterized by multiple genetic alterations. Epigenetic and genetic alterations that drive expression of oncogenes are associated with malignant transformation of healthy cells. Nutritional deficiencies played a significant role in the progression and development of the tumor. Fruits, vegetables, cereals, nuts, green tea, and red wine showed an association with cancer chemoprevention and chemotherapy. Dietary nutrients showed significant inhibition of the growth of the tumor in both in vitro and in vivo model. Dietary nutrients are involved in the repair of DNA damage caused by environmental pollutants, free radicals, radiation, and infectious agents. Dietary chemicals alter gene expression patterns that drive the progression of cancer. The dietary natural compounds are involved in the inhibition of cell proliferation, invasion, angiogenesis, and metastasis of cancer as well as inhibit stem cell proliferation. Genetic polymorphisms influence the alteration of the response of the host to dietary components. Therefore, it is critical to assess the role of natural dietary compounds as chemotherapeutic agents in cancer treatment. In this chapter, the role of dietary natural compounds, macronutrients, and micronutrients in cancer chemoprevention in the context of nutrigenomics and nutrigenetics is focused. Nutrigenetics defines the effect of nutrition on the genetic constituents of cancer patients, whereas nutrigenomics represents the effect of dietary nutrition on the genomics and transcriptomic level of cancer patients. Nutrigenomics and nutrigenetic studies are required to understand the mode of action of dietary natural compounds on cancer genomes and their application in chemotherapy. Research in nutritional genetics and genomics

I. Chattopadhyay (✉)
Department of Life Sciences, Central University of Tamil Nadu, Thiruvarur, Tamil Nadu, India
e-mail: indranil@cutn.ac.in

M. Kumar et al. (eds.), *Pharmacotherapeutic Botanicals for Cancer Chemoprevention*, https://doi.org/10.1007/978-981-15-5999-0_7

may help to address how bioactive food components regulate genetic and genomic events in cancer.

Keywords

Cancer · Nutrigenomics · Nutrigenetic · Dietary Natural Compounds · Cancer Chemoprevention

7.1 Introduction

The uncontrolled growth of the cell through multiple genetic alterations causes cancer [1]. Multiple genetic and epigenetic factors are involved in the progression of tumors [2]. Several epidemiological studies reported that dietary natural compounds present in fruit and vegetables played a significant role in cancer prevention [3] particularly in the early stage of tumorigenesis [1]. Nutrigenetics is defined as the influence of nutrition on the gene level of the host [4], while nutrigenomics represents the molecular perception of a nutrition role at the genomic level of the individual genotype. Nutrigenomics is utilized in the personalized nutrition of cancer patients [5]. Fruit and vegetables, cereals, nuts, black and green tea, and red wine showed a correlation with cancer chemoprevention [6]. Carotenoids, phenolic compounds, or phytosterols derivatives are used in cancer chemoprevention [7]. Certain nutrients which are involved in tumor inhibitory mechanism such as inducing apoptosis and inhibition of angiogenesis are considered as a personalized medicine for cancer chemoprevention [8]. Natural nutrients inhibit carcinogenesis at the genetic, epigenetic, and transcriptomic level which enhance efficacy and reduce side effects of chemotherapeutic drugs commonly used in cancer treatment [9]. Dietary natural compounds affect 30–40% of cancers and influence the predictive tumor biomarkers [1]. This chapter aims to understand and highlight the effect of natural phytochemicals, micronutrients, and macronutrients on a genetic, epigenetic, and transcriptomic level in vitro and in vivo model of cancer cell and its role in chemoprevention (Fig. 7.1).

7.2 Definition of Nutrigenomics and Nutrigenetics

Nutrigenomics and Nutrigenetics are the studies of the interaction between the dietary compounds and genomic or genetic constituents of the host. The term nutrigenomics was first coined by Pelegrin in 2001 [10]. Nutrigenomics is used to determine the genomic alterations at DNA and RNA levels in the host cells in response to dietary natural compounds by using high throughput genomic technology and systems biology tools. Even though nutrigenomics showed similarity with pharmacogenomics, the main difference between the nutrigenomics and pharmacogenomics is that natural bioactive compounds are involved in nutrigenomics and synthetic chemicals are involved in pharmacogenomics

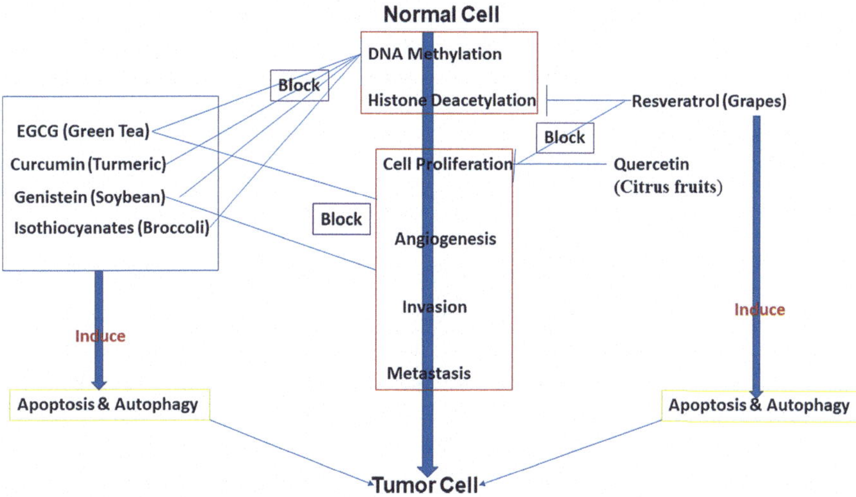

Fig. 7.1 Schematic representation of the effect of dietary natural compounds on epigenetic alterations and cellular processes that drive the transformation of a normal cell to tumor cell

[5]. Nutrigenetics is related to genetic and epigenetic alterations of host DNA in response to natural dietary components [11]. Gene expression microarrays and qRT-PCR evaluate the effect of natural compounds on the genomics level. The effects of natural compounds on epigenetic alterations such as DNA methylation, histone modification, nucleosome organization, and expression of methyl-binding proteins are evaluated by microarray, pyrosequencing, next-generation sequencing, and qRT-PCR-array panel [12]. The objectives of nutrigenomics are listed below:

(a) To identify the transcription factors and signaling pathways influenced by dietary nutrients at the cellular level
(b) To recognize cancer biomarkers controlled by dietary nutrients for early diagnosis

7.3 Role of Phytochemicals in Epigenetic Alterations of Cancer Prevention

Phytochemicals, polyphenols, terpenoids, and thiols are the sources of color, taste, and aroma of fruit and vegetables [13]. Flavones (kaempferol and quercetin), flavonols (apigenin and luteolin), flavanones (naringenin and hesperidin), isoflavone (daidzein and genistein), flavan-3-ols (catechins and gallocatechins isomers), and anthocyanins are under the category of flavonoids [14]. Phytoestrogens such as flavones and isoflavone influence signal transduction pathways which are regulated by sex hormones in cancer [15]. Estrogen receptors in human breast cancer cells and androgen receptors (AR) in prostate cancer are targeted by the flavonoids compounds [16, 17]. The risk of oesophageal cancer was dramatically reduced by the dietary intake of anthocyanidins, flavanones, flavonoids, and flavones [18]. Flavonoids are responsible for preventing invasion and metastasis of breast, lung, liver, prostate, ovarian, and colon cancer through epigenetic modifications [19–39] (Table 7.1).

7.3.1 Isothiocyanates (ITC)

Glucosinolates and sulforaphane are dietary isothiocyanates (ITC) which are commonly found in cruciferous vegetables such as broccoli and brussels sprouts. Sulforaphane hinders the proliferation of colon cancer cells through the cell cycle arrest and suppression of DNMT1 and HDAC3 activity. Sulforaphane also prevents the growth of cell lines of prostate cancer through reduced expression of DNMT1 and -3b and reprogramming of CCND2 expression [28]. It induces cell cycle arrest by overexpression of p21 and apoptosis by up-regulation of Bax expression in human prostate cancer epithelial cells such as LNCaP and PC-3 cells through the prevention of HDAC activity and acetylation of levels of histones [29].

7.3.2 Apigenin

Apigenin inhibits the growth of human prostate cancer cells such as PC-3 and 22Rv1 through the inhibition of class I HDACs activity and induction of programmed cell death [30].

7.3.3 Curcumin

Curcumin inhibits the activity of p300/CBP HAT in prostate cancer and is used for cancer chemoprevention. Curcumin showed global hypomethylation in leukemia cells [31, 38].

7.3.4 Epigallocatechin Gallate (EGCG)

EGCG treatment in oesophageal cancer cells reverses the expression of epigenetically silenced tumor suppressor genes including *p16, hMLH1, RARβ,* and *MGMT* [32]. EGCG treatment reduces expression levels of DNMT1 at mRNA and protein level which drives re-expression of p16 (INK4a) and p21/Cip1 mRNA [33]. EGCG reactivates the expression of estrogen receptor-α (ER-α) in triple-negative breast cancer cells such as MDAMB-231 [34]. EGCG reverses the expression of tumor suppressor gene RECK in oral carcinoma cells [35].

Table 7.1 Effect of phytochemicals on epigenetic alterations in cancer

Phytochemical	Source	Effect on epigenetic alterations in cancer	Anti-tumor activity	Reference
Epigallocatechin gallate	Green tea	Methylation of the promoter region of RARβ, suppression of acetylation level of H3K9, suppression of methylation of promoters of tumor suppressor genes such as p15 and p16.	Target epigenetic alteration of cancer cell, antioxidant activity; induction of apoptosis, inhibition of angiogenesis, and metastasis of human pancreatic adenocarcinoma cell line	[19]
Quercetin	Citrus fruits	Histone modification, induces expression of E-cadherin in triple-negative breast cancer cells	Antiproliferation and inhibition of epithelial to mesenchymal (EMT) transition	[20]
Apigenin	Citric fruit	Inhibits DNMT and HDAC activity through down-regulation of DNMT1, DNMT3a, and DNMT3b proteins	Antioxidant	[21]
Genistein	Soybeans	Demethylation of CpG island promoter regions of tumor suppressor genes such as GSTP1, EPHB2, RARβ, p16, and MGMT	Antioxidant; inhibition of cancer cell proliferation and invasion	[22]
Luteolin	Carrots, peppers, celery, olive oil, peppermint, thyme, rosemary, and oregano	Inhibition of DNMT enzymes	Antioxidant	[23]
Pelargonidin	Blueberries, cherries, strawberries, blackberries, raspberries,	Activation of tumor suppressor genes (p53) through inhibition of DNMT1 and DNMT3B	Anti-inflammation	[24]

(continued)

Table 7.1 (continued)

Phytochemical	Source	Effect on epigenetic alterations in cancer	Anti-tumor activity	Reference
Resveratrol	Blueberries, peanuts, raspberries, grapes	Inhibits methylation of the BRCA-1 gene and activity of DNMT 3b; reduces the level of RASSF-1α methylation and expression of androgen receptor	Induces apoptosis and inhibits cell proliferation, inflammation, and angiogenesis	[25]
Ellagic acid	Pomegranate	Induces cell cycle arrest of androgen-independent prostate cancer cells in S-phase through down-regulating the expression of cyclin B1 and cyclin D1 levels.	Antiproliferation	[26]
Isothiocyanates	Vegetables belong to Cruciferae family	Suppresses the expression of DNMT1 and -3b activity	Induces apoptosis and suppression of metastasis	[27]

7.3.5 Resveratrol

Resveratrol had demonstrated antiproliferative activities in breast, lung, liver, prostate, and colon cancer cells [36]. It prevents invasion and migration of colon cancer [37].

7.3.6 Genistein

Genistein reactivates the tumor suppressor genes such as *p16, RARβ, and MGMT* in oesophageal squamous cell carcinoma and prostate cancer cells [39].

7.4 Role of Natural Products in Controlling miRNAs Involved in Tumorigenesis

Dietary natural compounds such as curcumin, epigallocatechin-3-gallate, genistein, resveratrol, and indole-3-carbinol target carcinogenesis through the regulation of miRNAs (Table 7.2).

Table 7.2 Oncogenic and tumor-suppressive miRNAs regulated by dietary natural compounds (*↓ indicates down- and ↑indicates up-regulation)

Dietary natural compounds	Targeted miRNAs	Expression level*	Functional role in cancer
Resveratrol	miR-21	↓	Induces apoptosis of pancreatic cancer cells
	miR-17-92, miR-10ab	↓	Targets tumor suppressor gene PTEN in prostate cancer cells
Curcumin	miR-181b	↑	Inhibits migration of breast cancer through the inhibition of CXCL1, CXCL2, and MMP expression pattern.
	miR-21	↓	Targets PDCD4 in colon cancer
	miR-22	↑	Inhibits cell proliferation and migration in pancreatic cancer through the regulation of Erbb3.
EGCG	miR-16	↑	Inhibits the expression of apoptosis through the inhibition of Bcl-2 gene in hepatocellular carcinoma
	miR-98-5p	↓	Enhances the apoptosis by cisplatin in lung cancer cells
Genistein	miR-34a	↑	Induces programmed cell death and inhibits cell proliferation in pancreatic cancer
	miR-1296	↑	Arrest the prostate cancer cells at S-phase
	miR-23b-3p; miR-1260b	↓	Induces apoptosis and inhibits invasion of renal cell carcinoma through the regulation of PTEN and Smad4, respectively

7.4.1 Curcumin

Curcumin which is present in turmeric showed anti-inflammatory activity in different type of cancers. The expression of several genes involved in NFκB, Akt, and MAPK is regulated by curcumin [40]. Curcumin inhibits the proliferation of pancreatic cancer cells through the overexpression of miR-22 and lower expression of miR-199a. In gastric cancers, it reduces the expression of oncogenic miRNA (MiR-196) [41]. It reduces the expression of miR-200 and miR-21 which in turn drives the overexpression of tumor suppressor gene PTEN in pancreatic cancer [42]. It inhibits proliferation and induces apoptosis of triple-negative breast cancer cells MDA-MB-231 through the overexpression of miR-181b which prevents the expression of matrix metalloproteinases and release of chemokines [43]. Yang et al. reported that curcumin suppressed the expression of antiapoptotic gene Bcl-2 and over-expressed miR-15a and miR-16 in MCF-7 breast adenocarcinoma cell [44]. It induces apoptosis in A549 lung adenocarcinoma cells through the suppression of oncogenic miRNA miR-186 [45]. It induces cell cycle arrest and apoptosis by inhibiting oncogenic expression of miR-21 and miR-34a which targets Notch-1 and overexpression of tumor suppressor miRNA let-7a in human oesophageal cancer cells [46]. Curcumin induces apoptosis of hepatocellular carcinoma cells through the suppression of Bcl-2 level and overexpression of miR-200a- and miR-200b induces resistance in cells against curcumin [47].

7.4.2 EGCG

EGCG inhibits the expression of apoptosis inhibitory gene Bcl-2 through the up-regulation of miR-16 in HepG2 cells [40]. It enhances the apoptosis by cisplatin in lung cancer cells through the down-regulation of miR-98-5p [48]. Oncogenic miRNAs such as miR-92, miR-93, and miR-106b are reduced, and tumor-suppressive miRNA such as miR-7-1, miR-34a, and miR-99a are upregulated by EGCG in a neuroblastoma cell line [49].

7.4.3 Genistein

Genistein blocks the expression of tumor-inducing genes such as *EGFR, NFκB, IRAK-1,* and *MTA-2* through the overexpression of miR-146a in pancreatic cancer cells [50]. Genistein induces the expression of miR-1296 which arrests the prostate cancer cells at S-phase [51]. Down-regulation of miR-221 and miR-222 by genistein leads to the up-regulation of the ARH1 gene that inhibits proliferation and invasion of prostate cancer cells [52]. Genistein inhibits prostate cancer cells invasion and migration by miR-151 down-regulation [53]. Genistein induces apoptosis and reduces invasive potential of renal carcinoma through the inhibition of miR-23b-3p expression which targets PTEN expression [54]. miR-1260b, which is involved in proliferation and invasion of renal cancer, was inhibited by genistein [55]. Genistein up-regulates miR-34a which induces programmed cell death and inhibits cell proliferation in pancreatic cancer [56].

7.4.4 Sulforaphane and Indole-3-Carbinol (I3C)

Sulforaphane and indole-3-carbinol (I3C) are derived from the Cruciferous/Brassica vegetables group, and can be used as chemopreventive agents in cancer. I3C induces the expression of miR-34b, miR-26a, and miR-125a which targets p53, TGF-β, and ERBB2 genes regulation in cancer [57]. Phenethyl isothiocyanate (PEITC) induces the expression of miRNAs such as miR-99b, miR-123, miR-192, miR-146, let-7a, let-7c, and miR-222 which target activation of Ras gene, cell proliferation, angiogenesis, activation of NFκB, and apoptosis [58].

7.4.5 Quercetin

Quercetin drives apoptosis in hepatocellular carcinoma cells through the up-regulation of miR-34a [59]. Quercetin inhibits the expression of claudin-2 in lung adenocarcinoma cells through the overexpression of miR-16 [60]. Through the overexpression of miR-142-3p, it inhibits proliferation, and induces apoptosis of pancreatic ductal adenocarcinoma [61].

7.4.6 Resveratrol

Resveratrol prevents the proliferation of MCF-7 breast adenocarcinoma cells through the up-regulation of miR-663 and miR-774 which down-regulates mRNA and protein level of eukaryotic translation elongation factor 1A2 (eEF1A2) [62]. Resveratrol induces apoptosis of cancerous pancreatic cells through the inhibition of miR-21 expression [63]. It inhibits the expression of oncogenic miR-17-92 and miR-106ab clusters which targets tumor suppressor gene PTEN in prostate cancer cells [64]. It also inhibits the miR-21 in gastric cancer [65]. Tumor suppressive miRNAs such as miR-141 and miR-200c were up-regulated by resveratrol in triple-negative breast cancer cell lines such as MDA-MB-231 [66]. Resveratrol inhibits the expression of miRNAs such as miR-17, miR-21, miR- 25, miR-92a-2, and miR-196a which are involved in the progression of colorectal cancer [67].

7.5 Effect of Plant-Derived Phytochemicals on Genetic Alteration

Biological activity of plant-derived phytochemicals in a cancer cell is influenced by genetic alterations of cancer cells [68]. Evaluation of single nucleotide polymorphism (SNP) is used to study the effect of dietary habits in cancer [5]. Phytochemicals influence SNP of genes involved in the antioxidant defense system and phase I/II detoxification system in cancer. Natural compounds interact with SNPs of phase I detoxification enzymes such as cytochrome P450 (CYP) and phase II detoxification enzymes such as glutathione S-transferase P1 (GSTP1). The redox potential of APE1/Ref-1 is influenced by curcumin, resveratrol, and EGCG [69]. EGCG prevents transcriptional activation of mutant AR hotspot (T877A) in prostate cancer [70]. EGCG interacts with Gln54, Gly55, Asp57 ILe72, Cys73, and Lys96 residues of TRAF6 which is involved in the invasion of tumor cells [71].

7.6 Effect of Plant-Derived Phytochemicals on Gene Expression in Cancer

Plumbagin (5-hydroxy-2-methyl-1, 4-naphthoquinone) which is derived from plant *Plumbago zeylanica* induces apoptosis in NSCLC cell lines H460 through arresting the cells at G2/M-phase and regulating transcription of apoptosis inducing genes and cell survival genes [72]. Plumbagin prevents cell proliferation, invasion, angiogenesis, and metastasis through down-regulation of chemokine receptor CXCR4 in various tumor cell lines and animal models [73].

Cucurbitacin B (CuB) along with chemotherapeutic agents showed anti-tumor activity against head and neck cancer, breast, pancreas, liver, brain, and skin cancer without increasing cytotoxicity. Plumbagin induces programmed cell death in colon cancer cells through the release of Cytochrome C and activation of Caspase-3 through TNF-α mediated pathway [74]. MAPK, NFκB, COX-2, iNOS, STAT signaling pathways, which are involved in inflammation, are influenced by dietary nutrients [75]. EGCG induces apoptosis and cell cycle arrest in many cancer cells [76]. EGCG inhibits expression of antiapoptotic genes BAG3, XIAP, RIPK2 [77]. Genistein and resveratrol induce autophagy in ovarian tumor cells and endometrial cancer cells, respectively [5].

7.7 Role of the Potential Impact of Macro and Micronutrients on Cancer Therapy

Nutrients showed cancer inhibition through the inhibition of angiogenesis and inducing apoptosis (Table 7.3).

7.7.1 Vitamin A

All-trans-retinoic acid (RA) is a bioactive form of Vitamin A and it functions as a tumor suppressor in breast, lung, liver, pancreas, prostate, and bladder cancer models [78]. Dietary vitamin A is obtained from the carotenoids of plants. It reduces head, neck, and lung cancer in animal models through its antioxidant properties and DNA damage protection mechanism and modulation of DNA methylation [79]. Retinoids and lycopene showed inhibitory mechanisms against oral premalignant lesions such as leucoplakia through alterations of genes that drive cell proliferation and

Table 7.3 Role of nutrients in cancer chemoprevention

Nutrient	Role in cancer prevention
Vitamin A	Inhibits the growth of lung tumor through inducing of apoptosis and blocking of the JAK-STAT pathway
Vitamin C	Drives programmed cell death of tongue tumor cell
Vitamin D	Inhibits cell proliferation and angiogenesis through the regulation of p21 and p27
Vitamin E	Regulates the expression of miR-122 and miR-125b which are involved in inflammation and lipid metabolism
Folic acid	Inhibits the growth of hepatocellular carcinoma through the suppression of miR-122 expression
Selenium	Induces the expression of hypermethylated genes such as GSTP1, APC, and CSR1 in human prostate cancer cells through suppression of DNMT and HDAC activity
n-3-polyunsaturated fatty acids (n-3 PUFAs	Inhibits VEGF, PDGF, and MMP-2 driven tumor angiogenesis

differentiation [80]. 4-nitroquinoline 1-oxide induced oral carcinogenesis was inhibited by the combinatorial effect of bexarotene and retinoids through inhibition of ROS production [81]. Retinoic acid inhibits the growth of lung tumors through the induction of apoptosis and blocking the JAK-STAT pathway [82]. RAR promoter methylation has been reported in non-small cell lung cancer (NSCLC) [83].

7.7.2 Vitamin C

Concentrations of Vitamin C in cancer patients' plasma were significantly lowered as compared to healthy controls. A high dose of Vitamin C drives apoptosis of tumor cells through the production of hydrogen peroxide [84]. Vitamin C prevents the growth of tongue tumors through the production of hydrogen peroxide and superoxide anion radicals that drive programmed cell death of tongue tumor cells [85]. Vitamin C showed anti-tumor activity against laryngeal squamous cell carcinoma through the production of ROS (reactive oxygen species), activation of protein kinase C (PKC) and inducing the level of calcium in the cytoplasm that drives necrosis of laryngeal tumor [86].

7.7.3 Vitamin D

Vitamin D and one of its metabolites [25(OH) D or 1, 25$(OH)_2$ D] prevent colorectal and breast cancers [87]. Slattery et al. reported that VDR Fok I genotype (rs2228570) was associated with the risk of colorectal cancer. The ff genotype showed an association with colorectal cancer patients having obesity [88]. BsmI polymorphism, which is located in the intronic region of 3'end of the VDR gene, showed an association with the risk of prostate cancer [89]. Genes involved in inflammation, cell proliferation, and apoptosis have Vitamin D response elements (VDREs) in their promoters [90]. Growth of ovarian and breast cancer cells was inhibited at G1 to the S-phase of the cell cycle by 1, 25$(OH)_2$ D through transcriptional overexpression of cyclin A1 and cyclin D2 genes [91]. 1, 25(OH) 2 D induces the expression of insulin-like growth factor-binding protein 3 (IGFBP3) gene and p21 prostate epithelial cells [92]. Vitamin D is involved in prostaglandin metabolism, where the expression of COX-2 was inhibited by 1, 25$(OH)_2$ D in prostate cancer cell [93]. miRNAs are regulated by Vitamin D and its metabolic derivatives such as 1,25-dihydroxyvitamin D_3 (1, 25D_3), 25-hydroxyvitamin D_3 (25(OH)D_3 in different cancers. 1,25D_3 inhibits the expression of miR-181a and miR-181b in human myeloid leukemia cells which results in cell cycle arrest at the G1 phase through the overexpression of p27Kip1 and p21Cip1 [94]. Vitamin D showed anti-tumor activity against head and neck, breast, ovary, prostate, and colon tumor. Vitamin D receptor (VDR) is involved in cell proliferation, differentiation, inflammation, and metabolism (Fig. 7.2) [95]. Genetic polymorphisms of VDR genes and genes such as CYP27B1 and CYP24B1 involved in vitamin D metabolism pathway are involved in oral squamous cell carcinoma and prostate cancer. Fok I gene polymorphism in

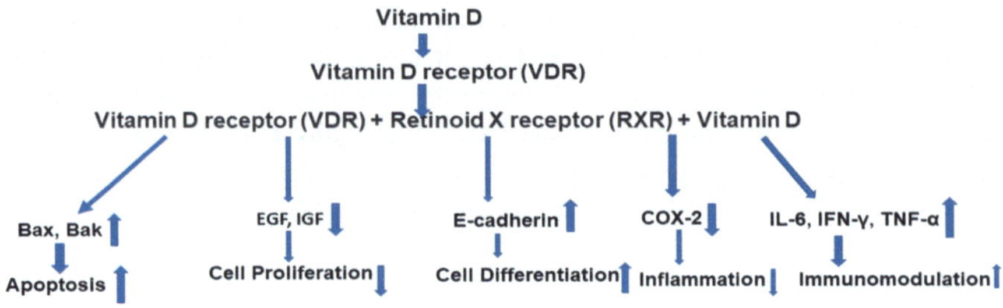

Fig. 7.2 Role of Vitamin D in cancer regulation

VDR was associated with the survival of oral cancer [96, 97]. Vitamin D induces autophagy through the sensitization of breast cancer cells to radiotherapy [98]. Vitamin D inhibits proliferation and angiogenesis of pancreatic cancer through the activation of cell cycle regulating genes such as p21 and p27 [99].

7.7.4 Vitamin E

Vitamin E succinate (VES) inhibits proliferation and induces apoptosis through the down-regulation of surviving and X-linked inhibitor of apoptosis proteins (XIAP) genes in pancreatic cancer cells [100]. Vitamin E is involved in the expression of miR-122 and miR-125b which are involved in inflammation and lipid metabolism [101].

7.7.5 Folic Acid

Folate inhibits the growth of hepatocellular carcinoma through the suppression of miR-122 expression [102]. People with MTHF reductase 677TT genotype and methylation of RASSF-1α have a higher risk of developing oral squamous cell carcinoma if they have a habit of ethanol consumption [103]. Reduced level of folate concentrations induces tumor-inducing mutation, breaking DNA double-strand by introducing uracil into DNA double helix. This is considerd as an important mechanism of tumorigenesis [104]. Methylene tetrahydrofolate reductase (MTHFR) is involved in folic acid metabolism. C677T polymorphism in MTHFR gene induces concentrations of homocysteine and DNA hypomethylation has been associated with the survival of lung, colon, breast, and blood cancer [105]. DNMT3B methylation enzyme polymorphism (C46359T) and SHMT1 C1420T) have been involved in folate metabolism which is related to head and neck cancer [106].

7.7.6 Selenium

Selenium enters into our body through food and drinking water [107]. Sodium selenite induces the expression of miR-34b and miR-34c in prostate cancer cells [108]. It has been reported that selenium prevents colorectal cancer of African Americans [109]. Selenocysteine is incorporated into glutathione peroxidases (GPXs) and thioredoxin reductases (TrxRs). Promoter methylation of GPX3 showed clinical association with oxaliplatin resistance colorectal cancer [110]. Selenium-containing proteins showed anti-inflammatory and antioxidant properties [111]. Oxidative forms of selenium such as selenium oxide, selenious acid, selenite salts inhibit the formation of DNA adducts [112]. Selenium binding protein 1 (SBP1) has been associated with lymph node metastasis in lung, nasopharyngeal, breast, and renal cell carcinoma [105]. Expression of selenoprotein induces the secretion of IL-6 and interferon-γ [113]. Overexpression of selenoprotein TXNRD1 has been reported in lung cancer [114]. Selenium induces expression of hypermethylated genes such as GSTP1, APC, and CSR1 in human prostate cancer cells through suppression of DNMT and HDAC activity [115].

7.7.7 Polyunsaturated Fatty Acids

n-3-Polyunsaturated fatty acids (n-3 PUFAs) which are present in green vegetables, seed oils, soybeans, walnuts, and fish-oil showed a protective role in colon cancer through up-regulation of tumor suppressor miRNAs [116]. PUFAs inhibit VEGF, PDGF, and MMP-2 driven tumor angiogenesis [117].

7.8 Role of Vitamins in Preventing Genomic Instability

Genomic instability triggers tumor initiation and progression. Selenium prevents the gain or loss of chromosomes, and damage of mitochondrial DNA (mtDNA). Vitamin D is involved in the protection of genome from chromosomal aberrations, shortening of telomere and telomerase activity. Vitamin B3 (niacin), folate, and vitamin B12 are involved in the protection of both nuclear and mitochondrial genomes [118].

7.9 Role of Phytochemicals as Adjuvants in Cancer Therapy

Genistein enhances the sensitivity of gentamicin through the prevention of miR-223 expression in pancreatic cancer models [119]. EGCG enhances response to cisplatin in non-small cell lung cancer patients through the up-regulation of Copper transporter 1 (CTR1) [120]. EGCG decreases resistance to Cisplatin in ovarian cancer cells through CTR1 and also reduces Cisplatin induced toxicity in nephron [121]. EGCG enhances doxorubicin induced autophagy in Hep3B cells [122].

7.10 Effect of Phytochemicals on Cancer Stem Cells (CSCs)

Phytochemicals target CSCs and stem cell signaling pathways such as Wnt/Frizzled/β-catenin, Notch signaling, Hippo signaling, Hedgehog signaling, JAK-STAT, and PI3K/Akt/mTOR signaling pathway which are involved in proliferation, differentiation, and maintenance of cancer stem cells. Curcumin inhibits the growth of CSCs in breast cancer through the down-regulation of Notch, Wnt-β-catenin, and Hedgehog pathways. Curcumin prevents translocation of β-catenin in the nucleus, transcriptional activation of the Slug transcription factor and inhibits the migration of breast CSCs. It enhances the sensitivity of breast cancer cell lines MCF-7 and MDA-MB-231 to mitomycin C, doxorubicin, cisplatin, and paclitaxel. Resveratrol induces autophagy in breast CSCs through the up-regulation of autophagic markers such as Atg-7, Beclin-1, and LC3-II.

Resveratrol reduces CSCs-like characteristics in breast cancer cells through the overexpression of Argonaute2 (Ago2) and tumor-suppressive microRNAs such as miR-16, −141, −143, and -200c. Green tea polyphenols such as epicatechin-3-gallate and epigallocatechin-3-gallate (EGCG) inhibit the growth of ALDH-positive inflammatory breast cancer cell lines such as SUM-149 through the down-regulation of VEGF-D. EGCG also inhibits the proliferation of CD44+/CD24- stem cell population in MDA-MB-231 human triple-negative breast cancer cells through the suppression of the mTOR pathway and overexpression of the AMPK pathway. Sulforaphane prevents proliferation, angiogenesis, and metastasis of MDA-MB-231 breast cancer cells through activating cell cycle arrest at S- and G2/M-phase by up-regulating p21WAF1 and p27KIP1 and by reducing cyclin A, cyclin B1, and CDC2. The synergistic effect of sulforaphane and HDAC inhibitor induces the expression of ER-α in MDA-MB-231 cells. Treatment of sulforaphane decreases the number of SOX9 and ALDH1 positive cells in ER-α-negative breast cancer cells. Sulforaphane reduces the secretion of IL-1β, IL-6, TNF-α, interferon-γ, IL-4, platelet-derived growth factor, and VEGF from MDA-MB-231 cells.

A combinatorial approach of Indole-3-carbinol (I3C) and its dimeric product 3, 3′-diindolylmethane (DIM) and Herceptin reduces clonogenicity of SKBR3 and MDA-MB-468 breast cancer cells through the overexpression of miR-200 and down-regulation of FoxM1 expression. Genistein inhibits mammospheres formation through the suppression of the Hedgehog-Gli1 signaling pathway. Quercetin enhances the inhibitory effects of geldanamycin on ALDH+ breast cancer cells. All-trans-retinoic acid (ATRA) which is the metabolic derivative of vitamin A inhibits the mammospheres formation capacity of MCF-7 cells through the suppression of expression of SLUG, Notch-3, and Jagged-1. ATRA enhances the sensitivity of ALDH+/CD44+ MDA-MB-468 cells to chemotherapy and radiotherapy through the up-regulation of CK8/18/19 expression [123].

Quercetin inhibits the proliferation of CSCs in pancreatic and head/neck cancer. Anthocyanin prevents the epithelial-mesenchymal transition of uterine cervical cancer cells. A combinatorial approach of green tea polyphenols, sulforaphane, and quercetin induces apoptosis of CSCs in pancreatic cancer. Sulforaphane inhibits the proliferation of green tea polyphenols through regulation of the Wnt/β-catenin signaling and hedgehog signaling pathway. Sulforaphane reduces the expression of the Akt/mTOR signaling pathway which is essential for survival and invasion of CSCs. Down-regulation of Akt induces apoptosis and reduces the motility of CSCs. Selenium induces apoptosis of CSCs in leukemia through the regulation of arachidonic acid metabolism and Akt/mTOR signaling pathways. Vitamin A reduces the proliferation of CSCs in glioblastoma through the reduced expression of Notch signaling pathways [124].

A combinatorial approach of phytochemicals such as tetrandrine, rhamnetin, and others along with paclitaxel, doxorubicin down-regulates the expression of P-glycoprotein (P-gp) in tumor cell lines such as U-2OS, Caco-2, MCF-7, CEM/ADR5000, HCC cells. Quercetin inhibited the growth of multidrug resistance AGS-cyr61 cells. Overexpression of CYR61 induces migration, invasion, and drug resistance in cancer cells. Isothiocyanate inhibited the growth of chemoresistance cancer stem cells in a colorectal cancer cell lines (HT29) through the down-regulation of LGR5 and PROM1 expression. ATRA enhances the sensitivity of radioresistant breast cancer cells (MCF7/C6) to Epirubicin. Gambogic acid (GA) enhances the chemosensitivity of colorectal cancer cells (HCT-15, HCT-15R) through the activation of apoptosis by up-regulating the JNK signaling pathway. Ellagic acid enhances the sensitivity of breast cancer cells (MCF-7) to radiotherapy. Resveratrol and berberine also enhance the sensitivity of nasopharyngeal carcinoma cells (CNE-1, CNE-2) to radiotherapy [125].

7.11 Future Perspectives

"Omic" technologies such as next-generation sequencing technology are used in nutrigenetics and nutrigenomics to identify the impact of nutrition in cancer patients. Dieticians, medical practitioners, and geneticists need knowledge about the field of nutrigenetics/nutrigenomics to integrate dietary natural compounds as personalized medicine in cancer treatments. Nutritional epigenetics may address the regulative role of bioactive food components on the epigenetic mechanisms in cancer. Nanomaterials encapsulation of bioactive natural compounds such as curcumin, resveratrol, EGCG, and genistein enhance bio-availability. Dietary natural compounds may not be useful for the advanced stage of the tumor due to the presence of multiple genetic alterations in the tumor cells. Epigenetically active dietary natural compounds may be used in early diagnosis of the tumor due to reverse the methylated cancer genome and possess multiple epigenetic targets. Dietary natural compounds may be used in the prevention of recurrent tumors by alterations of miRNAs. Nutrigenetics and nutrigenomics are the foundation to design the novel drug from natural products.

7.12 Conclusions

Dietary phytochemicals that alter DNA methylation of essential inflammatory genes and tumor suppressor genes, HDAC activity, histone modifications and alterations of oncogenic and tumor-suppressive miRNAs are used in cancer prevention. The doses of dietary nutrients used in in vitro and in vivo experiments need to apply in clinical studies. Dose, bio-availability, and routes of administration of bioactive dietary natural compounds are essential to determine chemopreventive strategies for the treatment of cancer patients. Nutrigenomics also addresses the impact of dietary components on the expression of genes involved in cell proliferation, apoptosis, DNA repair mechanism, angiogenesis, and metastasis in a tumor cell. Future studies are required to use dietary components as a personalized treatment of cancer.

Conflict of Interest The author declares that he has no conflict of interest.

References

1. Ardekani AM, Jabbari S (2009) Nutrigenomics and cancer. Avicenna J Med Biotechnol 1 (1):9–17
2. You JS, Jones PA (2012) Cancer genetics and epigenetics: two sides of the same coin? Cancer Cell 22:9–20
3. Waltenberger B, Mocan A, Smejkal K, Heiss EH, Atanasov AG (2016) Natural products to counteract the epidemic of cardiovascular and metabolic disorders. Molecules (Basel Swit) 21 (6):807
4. Irimie AI, Braicu C, Cojocneanu-Petric R, Berindan-Neagoe I, Campian RS (2015) Novel technologies for oral squamous carcinoma biomarkers in diagnostics and prognostics. Acta Odontol Scand 73:161–168
5. Braicu C, Mehterov N, Vladimirov B, Sarafian V, Nabavi SM, Atanasov AG, Berindan-Neagoe I (2017) Nutrigenomics in cancer: revisiting the effects of natural compounds. Semin Cancer Biol 46:84–106
6. Surh YJ (2003) Cancer chemoprevention with dietary phytochemicals. Nat Rev Cancer 3:768–780
7. Upadhyay S, Dixit M (2015) Role of polyphenols and other phytochemicals on molecular signaling. Oxidative Med Cell Longev 2015:504253
8. Irimie AI, Braicu C, Zanoaga O, Pileczki V, Gherman C, Berindan-Neagoe I, Campian RS (2015) Epigallocatechin-3-gallate suppresses cell proliferation and promotes apoptosis and autophagy in oral cancer SSC-4 cells. Onco Targets Ther 8:461–470
9. Mallery SR, Wang D, Santiago B, Pei P, Schwendeman SP, Nieto K, Spinney R, Tong M, Koutras G, Han B et al (2017) Benefits of multifaceted Chemopreventives in the suppression of the oral squamous cell carcinoma (OSCC) tumorigenic phenotype. Cancer Prev Res 10:76–88
10. Peregrin T (2001) The new frontier of nutrition science: nutrigenomics. J Am Diet Assoc 101:1306
11. Fenech M, El-Sohemy A, Cahill L, Ferguson LR, French T-AC, Tai ES, Milner J, Koh W-P, Xie L, Zucker M, Buckley M, Cosgrove L, Lockett T, Fung KYC, Head R (2011) Nutrigenetics and nutrigenomics: viewpoints on the current status and applications in nutrition research and practice. J Nutrigenet Nutrigenom 4:69–68
12. Farhud DD, Zarif Yeganeh M, Zarif Yeganeh M (2010) Nutrigenomics and nutrigenetics. Iran J Publ Health 39:1–14

13. Liu RH (2004) Potential synergy of phytochemicals in cancer prevention: mechanism of action. J Nutr 134(12 Suppl):3479–3485
14. Hardman WE (2014) Diet components can suppress inflammation and reduce cancer risk. Nutr Res Pract 8:233–240
15. He F-J, Chen J-Q (2013) Consumption of soybean, soy foods, soy isoflavones and breast cancer incidence: differences between Chinese women and women in Western countries and possible mechanisms. Food Sci Hum Wellness 2(3–4):146–161
16. Choi EJ, Kim GH (2013) Antiproliferative activity of daidzein and genistein may be related to ERalpha/c-erbB-2 expression in human breast cancer cells. Mol Med Rep 7:781–784
17. Adjakly M, Ngollo M, Boiteux JP, Bignon YJ, Guy L, Bernard-Gallon D (2013) Genistein and daidzein: different molecular effects on prostate cancer. Anticancer Res 33:39–44
18. Cui L, Liu X, Tian Y, Xie C, Li Q, Cui H, Sun C (2016) Flavonoids, flavonoid subclasses, and esophageal cancer risk: a meta-analysis of epidemiologic studies. Nutrients 8(6):850
19. Khan SI, Aumsuwan P, Khan IA, Walker LA, Dasmahapatra AK (2012) Epigenetic events associated with breast cancer and their prevention by dietary components targeting the epigenome. Chem Res Toxicol 25:61–73
20. Srinivasan A, Thangavel C, Liu Y, Shoyele S, Den RB, Selvakumar P, Lakshmikuttyamma A (2016) Quercetin regulates beta-catenin signaling and reduces the migration of triple negative breast cancer. Mol Carcinog 55:743–756
21. Paredes-Gonzalez X, Fuentes F, Su ZY, Kong AN (2014) Apigenin reactivates Nrf2 anti-oxidative stress signaling in mouse skin epidermal JB6 P+ cells through epigenetics modifications. AAPS J 16:727–735
22. Vardi A, Bosviel R, Rabiau N, Adjakly M, Satih S, Dechelotte P, Boiteux JP, Fontana L, Bignon YJ, Guy L, Bernard-Gallon DJ (2010) Soy phytoestrogens modify DNA methylation of GSTP1, RASSF1A, EPH2 and BRCA1 promoter in prostate cancer cells. In Vivo 24:393–400
23. Busch C, Burkard M, Leischner C, Lauer UM, Frank J, Venturelli S (2015) Epigenetic activities of flavonoids in the prevention and treatment of cancer. Clin Epigenetics 7(1):64
24. Wang LS, Kuo CT, Cho SJ, Seguin C, Siddiqui J, Stoner K, Weng YI, Huang TH, Tichelaar J, Yearsley M, Stoner GD, Huang YW (2013) Black raspberry derived anthocyanins demethylate tumor suppressor genes through the inhibition of DNMT1 and DNMT3B in colon cancer cells. Nutr Cancer 65:118–125
25. Papoutsis AJ, Lamore SD, Wondrak GT, Selmin OI, Romagnolo DF (2010) Resveratrol prevents epigenetic silencing of BRCA-1 by the aromatic hydrocarbon receptor in human breast cancer cells. J Nutr 140:1607–1614
26. Vicinanza R, Zhang Y, Henning SM, Heber D (2013) Pomegranate juice metabolites, ellagic acid and urolithin a, synergistically Inhibit androgen-independent prostate cancer cell growth via distinct effects on cell cycle control and apoptosis. Evid Based Complement Alternat Med 2013:247504
27. Yan H, Zhu Y, Liu B, Wu H, Li Y, Wu X, Zhou Q, Xu K (2011) Mitogen-activated protein kinase mediates the apoptosis of highly metastatic human non-small cell lung cancer cells induced by isothiocyanates. Br J Nutr 106:1779–1791
28. Bishop KS, Ferguson LR (2015) The interaction between epigenetics, nutrition and the development of cancer. Nutrients 7(2):922–947
29. Myzak MC, Hardin K, Wang R, Dashwood RH, Ho E (2006) Sulforaphane inhibits histone deacetylase activity in BPH-1, LnCaP and PC-3 prostate epithelial cells. Carcinogenesis 27:811–819
30. Pandey M, Kaur P, Shukla S, Abbas A, Fu P, Gupta S (2012) Plant flavone apigenin inhibits HDAC and remodels chromatin to induce growth arrest and apoptosis in human prostate cancer cells: in vitro and in vivo study. Mol Carcinog 51:952–962
31. Shankar E, Kanwal R, Candamo M, Gupta S (2016) Dietary phytochemicals as epigenetic modifiers in cancer: promise and challenges. Semin Cancer Biol 40-41:82–99

32. Fang MZ, Wang Y, Ai N et al (2003) Tea polyphenol (−)-epigallocatechin-3-gallate inhibits DNA methyltransferase and reactivates methylation-silenced genes in cancer cell lines. Cancer Res 63:7563–7570
33. Nandakumar V, Vaid M, Katiyar SK (2011) (−)-epigallocatechin-3-gallate reactivates silenced tumor suppressor genes, Cip1/p21 and p16INK4a, by reducing DNA methylation and increasing histones acetylation in human skin cancer cells. Carcinogenesis 32:537–544
34. Li Y, Yuan Y-Y, Meeran S, Tollefsbol T (2010) Synergistic epigenetic reactivation of estrogen receptor-α (ERα) by combined green tea polyphenol and histone deacetylase inhibitor in ERα-negative breast cancer cells. Mol Cancer 9:274
35. Kato K, Long NK, Makita H et al (2008) Effects of green tea polyphenol on methylation status of RECK gene and cancer cell invasion in oral squamous cell carcinoma cells. Br J Cancer 99:647–654
36. Mao Q-Q, Bai Y, Lin Y-W et al (2010) Resveratrol confers resistance against taxol via induction of cell cycle arrest in human cancer cell lines. Mol Nutr Food Res 54:1574–1584
37. Wu H, Liang X, Fang Y, Qin X, Zhang Y, Liu J (2008) Resveratrol inhibits hypoxia-induced metastasis potential enhancement by restricting hypoxia-induced factor-1[α] expression in colon carcinoma cells. Biomed Pharmacother 62:613–621
38. Liu H-L, Chen Y, Cui G-H, Zhou J-F (2005) Curcumin, a potent anti-tumor reagent, is a novel histone deacetylase inhibitor regulating B-NHL cell line Raji proliferation. Acta Pharmacol Sin 26:603–609
39. Fang MZ, Chen D, Sun Y, Jin Z, Christman K, Yang CS (2005) Reversal of hypermethylation and reactivation of p16INK4a, RARβ, and MGMT genes by genistein and other isoflavones from soy. Clin Cancer Res 11:7033–7041
40. Li Y, Kong D, Wang Z, Sarkar FH (2010) Regulation of microRNAs by natural agents: an emerging field in chemoprevention and chemotherapy research. Pharm Res 27:1027–1041
41. Sun M, Estrov Z, Ji Y, Coombes KR, Harris DH, Kurzrock R (2008) Curcumin (diferuloylmethane) alters the expression profiles of microRNAs in human pancreatic cancer cells. Mol Cancer Ther 7:464–473
42. Bao B, Ali S, Kong D, Sarkar SH, Wang Z, Banerjee S, Aboukameel A, Padhye S, Philip PA, Sarkar FH (2011) Anti-tumor activity of a novel compound-CDF is mediated by regulating miR-21, miR-200, and PTEN in pancreatic cancer. PLoS One 6(3):e17850
43. Kronski E, Fiori ME, Barbieri O et al (2014) miR181b is induced by the chemopreventive polyphenol curcumin and inhibits breast cancer metastasis via down-regulation of the inflammatory cytokines CXCL1 and -2. Mol Oncol 8:581–595
44. Yang J, Cao Y, Sun J, Zhang Y (2010) Curcumin reduces the expression of Bcl-2 by upregulating miR-15a and miR-16 in MCF-7 cells. Med Oncol 27:1114–1118
45. Zhang J, Du Y, Wu C et al (2010) Curcumin promotes apoptosis in human lung adenocarcinoma cells through miR-186* signalling pathway. Oncol Rep 24:1217–1223
46. Subramaniam D, Ponnurangam S, Ramamoorthy P et al (2012) Curcumin induces cell death in esophageal cancer cells through modulating Notch signaling. PLoS One 7(2):e30590
47. Liang HH, Wei PL, Hung CS et al (2013) MicroRNA-200a/b influenced the therapeutic effects of curcumin in hepatocellular carcinoma (HCC) cells. Tumor Biol 34:3209–3218
48. Zhou DH, Wang X, Feng Q (2014) EGCG enhances the efficacy of cisplatin by downregulating hsa-miR-98-5p in NSCLC A549 cells. Nutr Cancer 66:636–644
49. Chakrabarti M, Khandkar M, Banik NL, Ray SK (2012) Alterations in expression of specific microRNAs by combination of 4-HPR and EGCG inhibited growth of human malignant neuroblastoma cells. Brain Res 1454:1–13
50. Li Y, Vandenboom TG 2nd, Wang Z, Kong D, Ali S, Philip PA, Sarkar FH (2010) miR-146a suppresses invasion of pancreatic cancer cells. Cancer Res 70:1486–1495
51. Parker LP, Taylor DD, Kesterson J, Metzinger DS, Gercel-Taylor C (2009) Modulation of microRNA associated with ovarian cancer cells by genistein. Eur J Gynaecol Oncol 30:616–621

52. Chen Y, Zaman MS, Deng G et al (2011) MicroRNAs 221/222 and genistein-mediated regulation of ARHI tumor suppressor gene in prostate cancer. Cancer Prev Res 4:76–86
53. Chiyomaru T, Yamamura S, Zaman MS et al (2012) Genistein suppresses prostate cancer growth through inhibition of oncogenic MicroRNA-151. PLoS One 7(8):e43812
54. Zaman MS, Thamminana S, Shahryari V et al (2012) Inhibition of PTEN gene expression by oncogenic miR-23b-3p in renal cancer. PLoS One 7(11):e50203
55. Hirata H, Ueno K, Nakajima K et al (2013) Genistein downregulates onco-miR-1260b and inhibits Wnt-signalling in renal cancer cells. Br J Cancer 108:2070–2078
56. Xia J, Duan Q, Ahmad A et al (2012) Genistein inhibits cell growth and induces apoptosis through up-regulation of miR-34a in pancreatic cancer cells. Curr Drug Targets 13:1750–1756
57. Li Y, VandenBoom TG 2nd, Kong D, Wang Z, Ali S, Philip PA, Sarkar FH (2009) Up-regulation of miR-200 and let-7 by natural agents leads to the reversal of epithelial-to-mesenchymal transition in gemcitabine-resistant pancreatic cancer cells. Cancer Res 69:6704–6712
58. Izzotti A, Larghero P, Cartiglia C, Longobardi M, Pfeffer U, Steele VE, De Flora S (2010) Modulation of microRNA expression by budesonide, phenethyl isothiocyanate and cigarette smoke in mouse liver and lung. Carcinogenesis 31:894–901
59. Lou G, Liu Y, Wu S, Xue J, Yang F, Fu H et al (2015) The p53/miR-34a/SIRT1 positive feedback loop in quercetin induced apoptosis. Cell Physiol Biochem 35:2192–2202
60. Sonoki H, Sato T, Endo S, Matsunaga T, Yamaguchi M, Yamazaki Y et al (2015) Quercetin decreases Claudin-2 expression mediated by up-regulation of microRNA miR-16 in lung adenocarcinoma A549 cells. Nutrients 7:4578–4592
61. MacKenzie TN, Mujumdar N, Banerjee S et al (2013) Triptolide induces the expression of miR-142-3p: a negative regulator of heat shock protein 70 and pancreatic cancer cell proliferation. Mol Cancer Ther 12:1266–1275
62. Vislovukh A, Kratassiouk G, Porto E et al (2013) Proto-oncogenic isoform A2 of eukaryotic translation elongation factor eEF1 is a target of miR-663 and miR-744. Br J Cancer 108:2304–2311
63. Liu P, Liang H, Xia Q et al (2013) Resveratrol induces apoptosis of pancreatic cancers cells by inhibiting miR-21 regulation of BCL-2 expression. Clin Transl Oncol 15:741–746
64. Dhar S, Hicks C, Levenson AS (2011) Resveratrol and prostate cancer: promising role for microRNAs. Mol Nutr Food Res 55:1219–1229
65. Cao Z, Yoon JH, Nam SW, Lee JY, Park WS (2012) PDCD4 expression inversely correlated with miR-21 levels in gastric cancers. J Cancer Res Clin Oncol 138:611–619
66. Hagiwara K, Kosaka N, Yoshioka Y, Takahashi R, Takeshita F, Ochiya T (2012) Stilbene derivatives promote Ago2-dependent tumour-suppressive microRNA activity. Sci Rep 2:314
67. Tili E, Michaille JJ, Adair B, Alder H, Limagne E, Taccioli C, Ferracin M, Delmas D, Latruffe N, Croce CM (2010) Resveratrol decreases the levels of miR-155 by upregulating miR-663, a microRNA targeting JunB and JunD. Carcinogenesis 31:1561–1566
68. Harrison JG, Gompert Z, Fordyce JA, Buerkle CA, Grinstead R, Jahner JP, Mikel S, Nice CC, Santamaria A, Forister ML (2016) The many dimensions of diet breadth: phytochemical, genetic, behavioral, and physiological perspectives on the interaction between a native herbivore and an exotic host. PLoS One 11(2):e0147971
69. Preissner SC, Hoffmann MF, Preissner R, Dunkel M, Gewiess A, Preissner S (2013) Polymorphic cytochrome P450 enzymes (CYPs) and their role in personalized therapy. PLoS One 8(12):e82562
70. Siddiqui IA, Asim M, Hafeez BB, Adhami VM, Tarapore RS, Mukhtar H (2011) Green tea polyphenol EGCG blunts androgen receptor function in prostate cancer. FASEB J 25:1198–1207
71. Zhang J, Lei Z, Huang Z, Zhang X, Zhou Y, Luo Z, Zeng W, Su J, Peng C, Chen X (2016) Epigallocatechin-3-gallate(EGCG) suppresses melanoma cell growth and metastasis by targeting TRAF6 activity. Oncotarget 7:79557–79571

72. Gomathinayagam R, Sowmyalakshmi S, Mardhatillah F, Kumar R, Akbarsha MA, Damodaran C (2008) Anticancer mechanism of plumbagin, a natural compound, on non-small cell lung cancer cells. Anticancer Res 28(2A):785–792
73. Manu KA, Shanmugam MK, Rajendran P, Li F, Ramachandran L, Hay HS, Kannaiyan R, Swamy SN, Vali S, Kapoor S, Ramesh B, Bist P, Koay ES, Lim LH, Ahn KS, Kumar AP, Sethi G (2011) Plumbagin inhibits invasion and migration of breast and gastric cancer cells by downregulating the expression of chemokine receptor CXCR4. Mol Cancer 10:107
74. Subramaniya BR, Srinivasan G, Sadullah SS, Davis N, Subhadara LB, Halagowder D, Sivasitambaram ND (2011) Apoptosis inducing effect of plumbagin on colonic cancer cells depends on expression of COX-2. PLoS One 6(4):e18695
75. Kim M-K, Kim K, Han JY, Lim JM, Song YS (2011) Modulation of inflammatory signaling pathways by phytochemicals in ovarian cancer. Genes Nutr 6(2):109–115
76. Khan N, Afaq F, Saleem M, Ahmad N, Mukhtar H (2006) Targeting multiple signalling pathways by green tea polyphenol (−)-epigallocatechin-3-gallate. Cancer Res 66:2500–2505
77. Braicu C, Gherman C (2012) Epigallocatechin gallate induce cell death and apoptosis in triple negative breast cancer cells Hs578T. J Drug Target 21:250–256
78. Sun SY, Lotan R (2002) Retinoids and their receptors in cancer development and chemoprevention. Crit Rev Oncol Hematol 41:41–55
79. Contreras Vidaurre EG, Bagan Sebastian JV, Gavalda C, Torres Cifuentes EF (2001) Retinoids: application in premalignant lesions and oral cancer. Med Oral 6:114–123
80. Lodi G, Franchini R, Warnakulasuriya S, Varoni EM, Sardella A, Kerr AR, Carrassi A, MacDonald LC, Worthington HV (2016) Interventions for treating oral leukoplakia to prevent oral cancer. Cochrane Database Syst Rev 7:CD00182
81. Tang XH, Osei-Sarfo K, Urvalek AM, Zhang T, Scognamiglio T, Gudas LJ (2014) Combination of bexarotene and the retinoid CD1530 reduces murine oral-cavity carcinogenesis induced by the carcinogen 4-nitroquinoline 1-oxide. Proc Natl Acad Sci U S A 111:8907–8912
82. Li HX, Zhao W, Shi Y, Li YN, Zhang LS, Zhang HQ, Wang D (2015) Retinoic acid amide inhibits JAK/STAT pathway in lung cancer which leads to apoptosis. Tumor Boil 36:8671–8678
83. Feng H, Zhang Z, Qing X, Wang X, Liang C, Liu D (2016) Promoter methylation of APC and RAR-beta genes as prognostic markers in non-small cell lung cancer (NSCLC). Exp Mol Pathol 100:109–113
84. Chen Q, Polireddy K, Chen P, Dong R (2015) The unpaved journey of vitamin C in cancer treatment. Can J Physiol Pharmacol 93:1055–1063
85. Ohwada R, Ozeki Y, Saitoh Y (2017) High-dose ascorbic acid induces carcinostatic effects through hydrogen peroxide and superoxide anion radical generation-induced cell death and growth arrest in human tongue carcinoma cells. Free Radic Res 51:684–692
86. Baek MW, Cho HS, Kim SH, Kim WJ, Jung JY (2017) Ascorbic Acid induces necrosis in human laryngeal squamous cell carcinoma via ROS, PKC, and calcium signaling. J Cell Physiol 232:417–442
87. Davis CD, Milner JA (2011) Nutrigenomics, vitamin D and cancer prevention. J Nutrigenet Nutrigenomics 4:1–11
88. Slattery ML, Murtaugh M, Caan B, Ma KN, Wolff R, Samowitz W (2004) Associations between BMI, energy intake, energy expenditures, VDR genotype and colon and rectal cancers (United States). Cancer Causes Control 9:863–872
89. Raimondi S, Johansson H, Maisonneuve P, Gandini S (2009) Review and meta-analysis on vitamin D receptor polymorphisms and cancer risk. Carcinogenesis 30:1170–1180
90. Ebert R, Schutze N, Adamski J, Jakob F (2006) Vitamin D signaling is modulated on multiple levels in health and disease. Mol Cell Endocrinol 248:149–159
91. Townsend K, Trevino V, Falciani F, Stewart PM, Hewison M, Campbell MJ (2006) Identification of VDR-responsive gene signatures in breast cancer cells. Oncology 71:111–123

92. Boyle BJ, Zhao XY, Cohen P, Feldman D (2001) Insulin-like growth factor binding protein-3 mediates 1, 25-dihydroxyviatimin D growth inhibition in the LNCaP prostate cancer cell line through p21/WAF1. J Urol 165:1319–1324
93. Feldman D, Krishnan A, Moreno J, Swami S, Peehl DM, Srinivas S (2007) Vitamin D inhibition of the prostaglandin pathway as therapy for prostate cancer. Nutr Rev 65:S113–S115
94. Wang X, Gocek E, Liu CG, Studzinski GP (2009) MicroRNAs181 regulate the expression of p27Kip1 in human myeloid leukemia cells induced to differentiate by 1, 25-dihydroxyvitamin D3. Cell Cycle 8:736–741
95. Peterlik M, Grant WB, Cross HS (2009) Calcium, vitamin D and cancer. Anticancer Res 29:3687–3369
96. Zeljic K, Supic G, Stamenkovic Radak M, Jovic N, Kozomara R, Magic Z (2012) Vitamin D receptor, CYP27B1 and CYP24A1 genes polymorphisms association with oral cancer risk and survival. J Oral Pathol Med 41:779–787
97. Holick CN, Stanford JL, Kwon EM, Ostrander EA, Nejentsev S, Peters U (2007) Comprehensive association analysis of the vitamin D pathway genes, VDR, CYP27B1, and CYP24A1, in prostate cancer. Cancer Epidemiol Prev Biomark 16:1990–1999
98. Gewirtz DA, Hilliker ML, Wilson EN (2009) Promotion of autophagy as a mechanism for radiation sensitization of breast tumor cells. Radiother Oncol 92:323–328
99. Chiang KC, Chen TC (2009) Vitamin D for the prevention and treatment of pancreatic cancer. World J. Gastroenterol 15:3349–3354
100. Patacsil D, Osayi S, Tran AT, Saenz F, Yimer L, Shajahan AN, Gokhale PC, Verma M, Clarke R, Chauhan SC, Kumar D (2012) Vitamin E succinate inhibits survivin and induces apoptosis in pancreatic cancer cells. Genes Nutr 7:83–89
101. Gaedicke S, Zhang X, Schmelzer C, Lou Y, Doering F, Frank J, Rimbach G (2008) Vitamin E dependent microRNA regulation in rat liver. FEBS Lett 582:3542–3546
102. Kutay H, Bai S, Datta J, Motiwala T, Pogribny I, Frankel W, Jacob ST, Ghoshal K (2006) Down regulation of miR-122 in the rodent and human hepatocellular carcinomas. J Cell Biochem 99:671–678
103. Supic G, Jovic N, Kozomara R, Zeljic K, Magic Z (2011) Interaction between the MTHFR C677T polymorphism and alcohol--impact on oral cancer risk and multiple DNA methylation of tumor-related genes. J. Dent. Res 90:65–70
104. Berger SH, Pittman DL, Wyatt MD (2008) Uracil in DNA: consequences for carcinogenesis and chemotherapy. Biochem Pharmacol 76:697–706
105. Irimie AI, Braicu C, Pasca S, Magdo L, Gulei D, Cojocneanu R, Ciocan C, Olariu A, Coza O, Berindan-Neagoe I (2019) Role of key micronutrients from nutrigenetic and nutrigenomic perspectives in cancer prevention. Medicina (Kaunas) 55:6
106. Succi M, de Castro TB, Galbiatti AL, Arantes LM, da Silva JN, Maniglia JV, Raposo LS, Pavarino EC, Goloni-Bertollo EM (2014) DNMT3B C46359T and SHMT1 C1420T polymorphisms in the folate pathway in carcinogenesis of head and neck. Mol Biol Rep 41:581–589
107. Meplan C (2018) Association of single nucleotide polymorphisms in selenoprotein genes with cancer risk. Methods Mol Biol 1661:313–324
108. Parasramka MA, Ho E, Williams DE, Dashwood RH (2012) MicroRNAs, diet, and cancer: new mechanistic insights on the epigenetic actions of phytochemicals. Mol Carcinog 51:213–230
109. Williams CD, Satia JA, Adair LS et al (2010) Antioxidant and DNA methylation related nutrients and risk of distal colorectal cancer. Cancer Causes Control 21:1171–1181
110. Pelosof L, Yerram S, Armstrong T, Chu N, Danilova L, Yanagisawa B, Hidalgo M, Azad N, Herman JG (2016) GPX3 promoter methylation predicts platinum sensitivity in colorectal cancer. Epigenetics 12:540–550

111. Tsavachidou D, McDonnell TJ, Wen S, Wang X, Vakar-Lopez F, Pisters LL, Pettaway CA, Wood CG, Do K-A, Thall PF et al (2009) Selenium and vitamin E: cell type- and intervention-specific tissue effects in prostate cancer. J Natl Cancer Inst 101:306–320
112. Ferguson LR, Karunasinghe N, Zhu S, Wang AH (2012) Selenium and its' role in the maintenance of genomic stability. Mutat Res Mol Mech Mutagen 733:100–110
113. Tsuji PA, Carlson BA, Anderson CB, Seifried HE, Hatfield DL, Howard MT (2015) Dietary selenium levels affect selenoprotein expression and support the interferon-gamma and IL-6 immune response pathways in mice. Nutrients 7:6529–6549
114. Tobe R, Carlson BA, Tsuji PA, Lee BJ, Gladyshev VN, Hatfield DL (2015) Differences in redox regulatory systems in human lung and liver tumors suggest different avenues for therapy. Cancers 7:2262–2276
115. Dolinoy DC (2008) The agouti mouse model: an epigenetic biosensor for nutritional and environmental alterations on the fetal epigenome. Nutr Rev 66:S7–S11
116. Davidson LA, Wang N, Shah MS, Lupton JR, Ivanov I, Chapkin RS (2009) n-3 polyunsaturated fatty acids modulate carcinogen-directed non-coding microRNA signatures in rat colon. Carcinogenesis 30:2077–2084
117. Kang JX (2013) Nutrigenomics and cancer therapy. J. Nutr. Nutr. 6:I–II
118. Ferguson LR, Chen H, Collins AR, Connell M, Damia G, Dasgupta S, Malhotra M, Meeker AK, Amedei A, Amin A, Ashraf SS, Aquilano K, Azmi AS, Bhakta D, Bilsland A, Boosani CS, Chen S, Ciriolo MR, Fujii H, Guha G, Halicka D, Helferich WG, Keith WN, Mohammed SI, Niccolai E, Yang X, Honoki K, Parslow VR, Prakash S, Rezazadeh S, Shackelford RE, Sidransky D, Tran PT, Yang ES, Maxwell CA (2015) Genomic instability in human cancer: molecular insights and opportunities for therapeutic attack and prevention through diet and nutrition. Semin Cancer Biol 35(Suppl):S5–S24
119. Ma J, Zeng F, Ma C, Pang H, Fang B, Lian C, Yin B, Zhang X, Wang Z, Xia J (2016) Synergistic reversal effect of epithelial-to-mesenchymal transition by miR-223 inhibitor and genistein in gemcitabine-resistant pancreatic cancer cells. Am J Cancer Res 6:1384–1395
120. Jiang P, Wu X, Wang X, Huang W, Feng Q (2016) NEAT1 upregulates EGCG-induced CTR1 to enhance cisplatin sensitivity in lung cancer cells. Oncotarget 7:43337–43351
121. Wang X, Jiang P, Wang P, Yang CS, Wang X, Feng Q (2015) EGCG enhances cisplatin sensitivity by regulating expression of the copper and cisplatin influx transporter CTR1 in ovary cancer. PLoS One 10:e0125402
122. Chen L, Ye HL, Zhang G, Yao WM, Chen XZ, Zhang FC, Liang G (2014) Autophagy inhibition contributes to the synergistic interaction between EGCG and doxorubicin to kill the hepatoma Hep3B cells. PLoS One 9:e85771
123. Dandawate PR, Subramaniam D, Jensen RA, Anant S (2016) Targeting cancer stem cells and signaling pathways by phytochemicals: novel approach for breast cancer therapy. Semin Cancer Biol 40-41:192–208
124. Oh J, Hlatky L, Jeong YS, Kim D (2016) Therapeutic effectiveness of anticancer phytochemicals on Cancer stem cells. Toxins (Basel) 8:7
125. Kapinova A, Kubatka P, Liskova A, Baranenko D, Kruzliak P, Matta M, Büsselberg D, Malicherova B, Zulli A, Kwon TK, Jezkova E, Blahutova D, Zubor P, Danko J (2019) Controlling metastatic cancer: the role of phytochemicals in cell signaling. J Cancer Res Clin Oncol 145:1087–1109

8 Nano-Delivery Carriers for Enhanced Bioavailability of Antitumor Phytochemicals

Anu Kalia and Gagandeep Kaur

Abstract

Phytochemicals, the plant derived natural products or bioactive compounds, exhibit immense diversity regarding origin and mechanism of action. The recent two decades have witnessed renaissance in anti-cancer therapeutics stressing identification of anti-neoplastic or anti-cancer properties of different phytochemicals/plant nutraceuticals. However, the available formulations of these phytochemicals exhibit pharmacological limitations such as low water solubility and reduced bioavailability. These aspects can be possibly improved by developing nano-enabled formulations of these phytochemicals. Various nano-scale delivery vehicles which include branched globular polymeric particles (dendrimers), unilamellar micelles, double layered liposomes, and other zero-dimensional nanomaterials have been developed to address the low water solubility and poor uptake issues. The phytochemical of interest can be encapsulated, embedded, or adsorbed on these nano-scale carriers. The nano-scale dimensions of these engineered delivery vehicles could help enhance the stability of water dispersed formulations. Further, the nano-size enables easy infiltration to cancer cells at rates higher than the non-nano-formulations of the same anti-cancer phytochemical thereby reducing the dosage required to achieve effective anti-cancer action. The amenability to multiple surface functionalization of these nano-delivery vehicles can ensure decoration with ligands that can lead to targeted delivery of the phytochemical to cancer cells leading to decreased

A. Kalia (✉)
Electron Microscopy and Nanoscience Laboratory, Department of Soil Science, College of Agriculture, Punjab Agricultural University, Ludhiana, Punjab, India
e-mail: kaliaanu@pau.edu

G. Kaur
Department of Microbiology, College of Basic Sciences and Humanities, Punjab Agricultural University, Ludhiana, Punjab, India

M. Kumar et al. (eds.), *Pharmacotherapeutic Botanicals for Cancer Chemoprevention*, https://doi.org/10.1007/978-981-15-5999-0_8

cyto-toxicity to normal tissues. This manuscript describes the various types of phytochemical payloaded nano-carriers for targeted or site-specific delivery of active anti-cancer chemical(s) to reduce undesirable side effects of chemotherapeutic agents on application in cancer subjects.

Keywords

Anti-cancer drug · Bioactive compounds · Nano-carriers · Nanoparticles · Natural products

8.1 Introduction

Cancer has emerged as the second most dreaded cause of human deaths claiming every one in six individuals globally. Cancer is characterized by rapid and uncontrolled growth of abnormal cells exhibiting omission of the contact inhibition phenomena followed by invasion of adjoining tissues and other visceral organs of the body leading to metastasis and ultimately death [1]. International Agency for Research on Cancer (IARC) through its World Cancer Report (2014) has predicted an upsurge in the reported cancer cases to 25 million by the year 2035. In 2018, 9.6 million deaths have been reported due to cancer with the under-developed and developing countries being the prima foci accounting for nearly 70% of the cancer related deaths. Due to expensive treatment options and post-treatment patient care costs, the economic impact of cancer is immense. In 2010, approximately US$ 1.16 trillion was the estimated gross annual economic cost of cancer [2]. Cancer Prevention and Control through an Integrated Approach (WHA70.12), a resolution passed by WHO in 2017 aims to accelerate designing strategies; conventional, alternative, and advanced, to reduce premature mortality due to cancer.

8.2 Use of Phytochemicals for Treatment of Cancer

Use of products of natural origin as remedies has been practiced since ancient time. Various ancient scriptures like *Charaka Samhita* from India and *Wu Shi Er Bing Fang* from China include illustrative documentation of more than 200 natural products of therapeutic relevance. *Ben Cao Gang Mu* published in sixteenth century documented more than 1000 natural bioactive agents [3]. Early nineteenth century marked the isolation of morphine from *Opium* plant by a German pharmacist, Friedrich Serturner, and since then it has been widely applied in the medical field as anesthesia [4]. A plethora of natural bioactive compounds have since been isolated including alkaloids, glycosides, and flavonoids. Plant based pharmaceuticals have been extensively studied as treatment options for cancer and currently over 60% of the total anti-cancer drugs in use have been derived from plants [5]. For example, curcumin, a polyphenol extracted from *Curcuma longa* (turmeric), possesses remarkable anti-cancer properties [6]. Similar anti-cancer potential has also been reported for secondary metabolites derived from actinobacteria [7]. However, it is earier to obtain scaled up quantities of the plant based anticancer [8]. The non-edible or even edible tissues/organs of diverse plant genera including fruits, vegetables, and medicinal plants contain anti-neoplastic phytochemicals. Relative safety of these drugs and their easy availability makes these phytochemicals a promising alternative treatment for cancer.

Common anti-neoplastic phytochemicals include original compounds or their derivatives namely taxanes, podophyllotoxin, vinca alkaloids, camptothecin, anthracyclines, and many others. The action mechanism of these drugs is not just limited to anti-oxidative and immune-modulatory roles, but they have been observed to actively target cancer related metabolic pathways [7]. Genistein, a plant based estrogen homolog obtained from legumes, is being tested for its therapeutic effects against cancers of pancreas, kidneys, rectum, and ovaries [9]. Likewise, lycopene (beta carotenoid), a lipophilic hydrocarbon isolated from tomatoes, papaya, watermelon, and carrots can prevent Reactive Oxygen Species (ROS) and Reactive Nitrogen Species (RNS) mediated DNA damage. Apart from the preventative effects, lycopene also directly affects cancer related metabolic pathways in prostate cancer [10]. The phytochemicals with more targeted effect on cancer such as paclitaxel isolated from Pacific yew, *Taxus brevifolia* acts as a mitotic inhibitor and thus can prevent cancer cell division by stabilizing the microtubules. Both paclitaxel and its derivative compound, docetaxel, have been approved by FDA to be used as anti-cancer drugs against breast, lung, and ovarian cancers [11].

Plant based anti-cancer drugs, therefore, show a range of possibilities, from their anti-oxidative and immune-modulating activities to directly acting as anti-mitotic agents. Their low toxicity, higher tolerance, and easy availability signify their role and potential in cancer therapy. However, their use has been limited by some factors discussed below.

8.3 Phytochemical Use: Possible Limitations

Plant based drugs encounter many limitations in terms of their pharmacological activities. One of the major challenges is low solubility of phytochemicals which hinders development of drug formulation. Such poorly formulated drugs exhibit inferior blood circulation. Some of these drugs may show low gastrointestinal absorption, while others get quickly metabolized and does not remain effective for longer durations. When the drugs have faster clearance rate, higher doses have to be applied at frequent intervals, making the process very tedious (Table 8.1).

Table 8.1 Lists of the some common phytochemicals and their limitations when applied as anti-cancer drugs

Phytochemical or its derivative	Limitation	Reference
Paclitaxel, docetaxel	Low solubility and poor penetration	[12, 13]
Taxol and taxotere	Hypersensitivity, toxicity	[14]
Cremophor®EL	Hepatic disposition	[15]
Vincristine	Short terminal elimination half-life	[16]
Docetaxel and paclitaxel	Multi-drug resistance	[17]

8.4 Nano-Enabled Approaches for Phytochemical Application: Can Nanotechnology Help Circumvent Possible Pharmacological Limitations of Phytochemicals?

As discussed in Sect. 8.3, the pharmacological limitations of phytochemicals need to be addressed by use of appropriate delivery systems. One of the convergent disciplines, Nanomedicine—the nanotechnological applications for biomedicine, has gained tremendous interest involving use of nanomaterials as novel drug delivery vehicles. Likewise, nano-enabled devices can ensure improved targeting, high efficacy, and decreased side effects to patient under treatment. The phytochemicals can be loaded on to nano-carriers to achieve improved solubility, better stability, bioavailability, and target specificity on application. The bio-safety of the nano-vehicles can be ensured if these carriers can be fabricated or encapsulated using biodegradable compounds [18, 19].

One of the main characters of cancerous tissue is angiogenesis. Tumors generate defective and leaky blood vessels that continuously leak fluids into interstitial spaces. From here, the fluids are drained by poorly formed lymphatic system. By targeting this phenomenon, a nano-sized drug carrier can be applied that can easily invade the tumor and accumulate there. This is called Enhanced Permeability and Retention effect or EPR effect [14]. The challenge of multi-drug resistance in tumor cells can also be averted by modifying the surface of the nano-carriers through chemical functionalization. These modifications will alter the interactions between drug carriers and cell membranes [20]. Some of the FDA approved nano-carrier based delivery systems are discussed in the following text.

8.4.1 Antibodies Drug Conjugation (ADC)

The conventional unconjugated anti-cancer drugs exhibit high cellular toxicity causing death of normal body cells besides the tumor cells due to low specificity to target the cancerous tissue alone. Therefore, one of the novel biopharmaceutical drugs, the Antibody drug conjugation (ADC), which combine the specific immunogenicity feature of monoclonal antibodies with the higher toxicity of the drug molecule can effectively ensure improved targeting of the cancer cells [21]. The monoclonal antibody(ies) is/are raised against specific surface antigens produced and secreted by the tumor cells. These antigenic moieties exist on the outer end of the cell membrane of tumor cell. Monoclonal antibodies against these specific tumor cell surface antigens are then attached to highly potent anti-cancer agents or drug molecules via a chemical linker domain. The size of the final bioconjugate remains in nano-scale dimensions; however, ADCs are considered to be non-nano anti-cancer agents. The common anti-cancer agents used include two types of cytotoxins viz., DNA damaging agents and microtubule inhibitors. Microtubule inhibitors can be further of two types: maytansinoids and auristatins. Maytansinoids are derivatives of maytansine, a phytochemical isolated from African shrub *Maytenus serrata* [22, 23]. Selection of suitable cytotoxin, linkers, and target antibodies are key components to be considered while designing ADCs.

8.4.2 Nanoparticles

Nanoparticle is a term used in nanopharmaceutics to identify zero-dimensional particles falling in size range of 10–1000 nm. The drug of interest can either be embedded or coated in these nanoparticles. Chemically, the origin of nanoparticles is diverse and include metal/ non-metal oxides/sulfides/nitrides, polymers, and carbon nanomaterials. Among these nano-delivery vehicles, the polymeric nanoparticles represent a very diverse group including the nanoparticles derived or formulated from biodegradable high molecular weight polymer compounds (e.g., poly(lactic-co-glycolic acid (PLGA) nanoparticles), and natural polymers (e.g., albumin nanoparticles, gelatin cellulose nanoparticles, and chitosan nanoparticles) [24]. These nanoparticles have high storage stability. PGLA and poly lactic acid (PLA) based nanoparticles have been synthesized using formulation of ginsenoside and luteolin, and have been found to be effective against lung cancer cells [25, 26].

8.4.2.1 Liposomes

Liposome-polymeric nanoparticles were the first well-explored and prudently commercialized nano-based drug delivery system used in cancer therapeutics [27]. Structurally, liposomes are comprised of a hydrophobic shell and hydrophilic core. Therefore, the polar drugs can be loaded in the core while the non-polar drugs can be loaded in the shell. The properties of a liposome can be suitably altered by simply modifying the composition of the phospholipid bilayer. A commercial preparation under the name Lipusu® has been developed as a more stable and less toxic substitute for Taxol® using liposome based delivery. Lipusu® showed higher retention in tumor tissues in mice compared to Taxol® [28].

8.4.2.2 Micelles

Micelles are the smallest nano-vehicles ($\sim$10 to 400 nm) that can be utilized for efficient drug delivery. These are formed when the concentration of a surfactant gets higher than critical micelle concentration (CMC). Polymeric micelles are being increasingly used for the development of several drug formulations. These micelles can be prepared by polymerization of the natural organic or synthetic compounds (monomer). Most commonly used compounds include the hydrophilic monomer PEG while the core-forming compounds include poly(propyleneoxide), poly (caprolactone), poly(D, L-lactic acid), and poly(L-aspartic acid) compounds [29]. An example of polymeric micelle formulation of a known anti-cancer drug containing Paclitaxel, Genexol-PM, has a size dimension in nano-regime (24 nm). In clinical trials for cancer, this drug has shown higher inhibition of tumor cell growth compared to Taxol [24].

8.4.2.3 Dendrimers

Dendrimers are nanoscopic, radially symmetric large molecular weight polymer molecules that exhibit extensive branched structure [30]. The peripheral groups present on the branched structure can be easily modified to help in binding of hydrophobic drug molecules [30]. The distinct physical-chemical and structural properties of dendrimers can be very useful for their role as 'Excipients' [31]. Dendrimers of gallic acid with polyamidoamine (PAMAM) have been formulated and found to be active against breast cancer cell lines [32].

8.4.2.4 Metal Nanoparticles

Metal nanoparticles can be synthesized using green nanotechnology approaches where solutions of desired metals are treated with suitable phytochemical solution. Phenolic compound rich extracts of plant *Albizia adianthifolia* were used to synthesize silver nanoparticles (AgNPs). Similarly, stem latex of *Euphorbia nivulia* plant have been used to generate AgNPs. Both of these nanoparticles showed anti-cancer effects against A-549 cancer cells [33]. Gold, copper, and titanium based nanoparticles have been synthesized and tried at pre-clinical levels for their effects against cancer cells.

Shape of nanoparticles is also an important aspect ensuring efficacy and uptake of the loaded anti-cancer drug. It has been demonstrated that nano-rods have 1.6-fold higher uptake by cancer cells compared to nano-spheres [34]. While larger particles are cleared quickly from the body, smaller ones are harder to filter. For EPR, 5–100 nm size has been found to be most suitable.

Application of nanoparticles is limited due to unavailability of substantial information regarding their toxic effects. The small size that makes nanoparticles suitable for therapeutic use can also create problems inside the body. The nanoparticles can cross barriers and get accumulated in vital organs like heart, lungs, brain, and liver. This can lead to complications such as systemic failure or inactivation of immune cells, inflammation, edema, etc. [35].

8.4.2.5 Carbon Nanotubes

Carbon nanotubes are hollow, rolled cylindrical tubes derived from single C-atom sheets, graphene. These long and thin cylinders exhibit remarkable mechanical and physical properties. The walls of these nanotubes can be functionalized with desired drug [36]. Multi-walled carbon nanotubes functionalized by PTX have been shown to be effective against HeLa cell lines [37].

8.5 Conclusion

Plant based anti-cancer drugs have a unique potential for use in cancer therapy. However, the efficacy of these phytochemicals get limited due to issues regarding the stability of the formulation and its bioavailability. Therefore, among the other alternatives that have been explored for addressing these issues, use of nano-inspired or -enabled phytochemical delivery systems can possibly circumvent the pharmacological limitations of these natural products. Nano-carriers can improve the solubility due to decrease in size dimensions, bioavailability, and uptake owing to larger surface area on nano-scaling, target specificity, and kinetics of these drugs besides lowering the toxicity of these drugs [38]. Further, analysis of advanced toxicity studies can ensure the safer administration of these drugs.

Acknowledgements The authors are thankful to the Head, Department of Microbiology, Punjab Agricultural University, Ludhiana for providing the necessary infrastructural facilities.

Conflict of Interest The authors declare no conflict of interest.

References

1. Siegel RL, Miller KD, Jemal A (2019) Cancer statistics, 2019. CA Cancer J Clin 69(1):7–34
2. IARC (2014) World Cancer Report 2014. Stewart BW and Wild CW (Ed). https://www.who.int/cancer/publications/WRC_2014/en/. Accessed 27 February 2020
3. Ji HF, Li XJ, Zhang HY (2009) Natural products and drug discovery. Can thousands of years of ancient medical knowledge lead us to new and powerful drug combinations in the fight against cancer and dementia? EMBO Rep 10(3):194–200
4. Ba GRH (2000) History of anesthesia: in the arms of morpheus: the development of morphine for postoperative pain relief. Can J Anesth 47(4):367–374
5. Newman DJ, Cragg GM (2016) Natural products as sources of new drugs from 1981 to 2014. J Nat Prod 79(3):629–661
6. Kuttan R, Bhanumathy P, Nirmala K et al (1985) Potential anticancer activity of turmeric (*Curcuma longa*). Cancer Lett 29(2):197–202
7. Wang H, Khor TO, Shu L et al (2012) Plants vs. cancer: a review on natural phytochemicals in preventing and treating cancers and their druggability. Anti Cancer Agents Med Chem 12:1281–1305
8. Kaur S, Kalia A, Gangwar M (2019) Bioprospecting endophytic actinobacteria of medicinal plants as potential anticancer therapeutic agents. In: Sharma A, Kunar M (eds.) Pollutants and protectants: valuation and assessment techniques. I.K International, New Delhi, pp 196–214.
9. Hwang KA, Park MA, Kang NH et al (2013) Anticancer effect of genistein on BG-1 ovarian cancer growth induced by 17 beta-estradiol or bisphenol A via the suppression of the crosstalk between estrogen receptor alpha and insulin-like growth factor-1 receptor signaling pathways. Toxicol Appl Pharmacol 272:637–646
10. Holzapfel NP, Holzapfel BM, Champ S et al (2013) The potential role of lycopene for the prevention and therapy of prostate cancer: From molecular mechanisms to clinical evidence. Int J Mol Sci 14:14620–14646
11. Skwarczynski M, Hayashi Y, Kiso Y (2006) Paclitaxel prodrugs: toward smarter delivery of anticancer agents. J Med Chem 49(25):7253–7269
12. Mastropaolo D, Camerman A, Luo Y et al (1995) Crystal and molecular structure of paclitaxel (taxol). Proc Natl Acad Sci U S A 92:6920–6924
13. Eisenhauer EA, ten Bokkel Huinink WW, Swenerton KD et al (1994) European-Canadian randomized trial of paclitaxel in relapsed ovarian cancer: high-dose versus low-dose and long versus short infusion. J Clin Oncol 12:2654–2666
14. Fang J, Nakamura H, Maeda H (2011) The EPR effect: unique features of tumor blood vessels for drug delivery, factors involved, and limitations and augmentation of the effect. Adv Drug Deliv Rev 63:136–151
15. Gelderblom H, Verweij J, Nooter K et al (2001) Cremophor EL: the drawbacks and advantages of vehicle selection for drug formulation. Eur J Cancer 37(13):1590–1598. https://doi.org/10.1016/s0959-8049(01)00171-x
16. Gidding CE, Meeuwsen-de Boer GJ, Koopmans P et al (1999) Vincristine pharmacokinetics after repetitive dosing in children. Cancer Chemother Pharmacol 44(3):203–209

17. Liu B, Staren ED, Iwamura T et al (2001) Mechanisms of taxotere-related drug resistance in pancreatic carcinoma. J Surg Res 99(2):179–186
18. Manuja A, Raguvaran R, Kumar B, Kalia A, Tripathi BN (2020) Accelerated healing of full thickness excised skin wound in rabbits using single application of alginate/acacia based nanocomposites of ZnO nanoparticles. Intl J Biol Macromol, 155:823–833
19. Raguvaran R, Manuja BK, Chopra M, Thakur R, Anand T, Kalia A, Manuja A (2017) Sodium alginate and gum acacia hydrogels of ZnO nanoparticles show wound healing effect on fibroblast cells. Intl J Biol Macromol, 96:185–191
20. Patel NR, Pattni BS, Abouzeid AH et al (2013) Nanopreparations to overcome multidrug resistance in cancer. Adv Drug Deliv Rev 65:1748–1762
21. Nejadmoghaddam MR, Minai-Tehrani A, Ghahremanzadeh R et al (2019) Antibody-drug conjugates: possibilities and challenges. Avicenna J Med Biotechnol 11(1):3–23
22. Kupchan SM, Komoda Y, Court WA et al (1972) Maytansine, a novel antileukemic ansa macrolide from *Maytenus ovatus*. J Am Chem Soc 94:1354–1356
23. Lopus M, Oroudjev E, Wilson L et al (2010) Maytansine and cellular metabolites of antibody-maytansinoid conjugates strongly suppress microtubule dynamics by binding to microtubules. Mol Cancer Ther 9:2689–2699
24. Liang Y, Xiao L, Li Y et al (2011) Poly(ester anhydride)/mPEG amphiphilic block co-polymer nanoparticles as delivery devices for paclitaxel. J Biomater Sci Polym Ed 22:701–715
25. Geng L, Fan J, Gao QL et al (2016) Preliminary study for the roles and mechanisms of 20(R)-ginsenoside Rg3 and PEG-PLGA-Rg3 NPs in the Lewis lung cancer mice. Beijing Da Xue Xue Bao 48(3):496–501
26. Majumdar D, Jung KH, Zhang H, Nannapaneni S, Wang X, Amin ARMR et al (2014) Luteolin nanoparticle in chemoprevention: in vitro and in vivo anticancer activity. Cancer Prev Res 7 (1):65–73
27. Bulbake U, Doppalapudi S, Kommineni N et al (2017) Liposomal formulations in clinical use: an updated review. Pharmaceutics 9(2):12
28. Ye L, He J, Hu Z et al (2013) Antitumor effect and toxicity of Lipusu in rat ovarian cancer xenografts. Food Chem Toxicol 52:200–206
29. Torchilin VP (2001) Structure and design of polymeric surfactant-based drug delivery systems. J Control Release 73:137–172
30. de Araújo RV, Santos SS, Ferreira EI et al (2018) New advances in general biomedical applications of PAMAM dendrimers. Molecules 23:2849. https://doi.org/10.3390/molecules23112849
31. Santos A, Veiga F, Figueiras A (2020) Dendrimers as pharmaceutical excipients: synthesis, properties, toxicity and biomedical applications. Materials 13:65. https://doi.org/10.3390/ma13010065
32. Sharma A, Gautam SP, Gupta AK (2011) Surface modified dendrimers. Bioorg Med Chem 19:3341–3346. https://doi.org/10.1016/j.bmc.2011.04.046
33. Valodkar M, Jadeja RN, Thounaojam MC et al (2011) *In vitro* toxicity study of plant latex capped silver NPs in human lung carcinoma cells. Mater Sci Eng C 31(8):1723–1728
34. Barua S, Yoo JW, Kolhar P et al (2013) Particle shape enhances specificity of antibody-displaying nanoparticles. Proc Natl Acad Sci U S A 110:3270–3275
35. Akhter S, Ahmad J, Rizwanullah M et al (2015) Nanotechnology-based inhalation treatments for lung cancer: state of the art. Nanotechnol Sci Appl 8:55–66
36. Jogi H, Maheshwari R, Raval N et al (2018) Carbon nanotubes in the delivery of anticancer herbal drugs. Nanomedicine 13(10):1187–1220. https://doi.org/10.2217/nnm-2017-0397
37. Tian Z, Shi Y, Yin M et al (2011) Functionalized multiwalled carbon nanotubes anticancer drug carriers: synthesis, targeting ability and antitumor activity. Nano Biomed Eng 3:157–162. https://doi.org/10.5101/nbe.v3i3.p157-16
38. Lagoa R, Silva J, Rodrigues JR et al (2020) Advances in phytochemical delivery systems for improved anticancer activity. Biotechnol Adv. https://doi.org/10.1016/j.biotechadv.2019.04.004

Biphasic Effects of Phytochemicals and their Relevance to Cancer Therapeutics

9

Urvashi Vijayvergia, Shreetama Bandyopadhayaya, and Chandi C. Mandal

Abstract

Cancer is one of the most deadly diseases caused worldwide. Enormous active anti-cancer agents that are available in nature play an efficient role in therapies associated to cancer for different organ-specific cancers like breast, skin, pancreas, and thyroid. These anti-cancer agents which occur naturally are derived mostly from various sources such as herbs, plants, and microorganisms. However, spontaneous uses of these therapies have no proper data on their therapeutic benefits available, which might induce side effects or off-target toxicity on due course. This book chapter throws light on the effect of these natural compounds in a dose-dependent manner on various human cancer cells and how these compounds show biphasic effects at different concentrations. The naturally occurring compounds exhibit a biphasic nature depending on their dose, i.e., they show both stimulatory and inhibitory effects on tumor cells depending on their concentrations. There has been no study conducted which explains the reason behind the biphasic nature of these naturally occurring compounds. This book chapter summarizes the role of different naturally occurring compounds such as resveratrol, quercetin, falcarinol, genistein, kaempferol, berberine, daidzein, indole-3-carbinol, umbelliprenin, glabridin, and 12-O-tetradecanoylphorbol-13-acetate (TPA) used as chemotherapeutic agents based on their toxicity, dosage, and its exhibition of biphasic effects. This chapter also highlights the major signaling pathways involved in showing anti-cancer effects of the above compounds having a biphasic effect.

Keywords

Natural compounds · Biphasic effects · Cancer · Therapeutics

U. Vijayvergia · S. Bandyopadhayaya · C. C. Mandal (✉)
Department of Biochemistry, Central University of Rajasthan, Kishangarh, Rajasthan, India

M. Kumar et al. (eds.), *Pharmacotherapeutic Botanicals for Cancer Chemoprevention*, https://doi.org/10.1007/978-981-15-5999-0_9

Abbreviations

COX-2	Cyclooxygenase-2
CYP1A1	Cytochrome P450 family 1 subfamily A member 1
EGFR	Epidermal growth factor receptor
EMT	Epithelial mesenchymal transition
ERK	Extracellular receptor kinase
EROD	7-ethoxyresorufin-O-deethylase
GSK	Glycogen synthesis kinase
GST-P	Glutathione S-transferase placental form
HCT-116	Human colorectal carcinoma
HIF-1α	Hypoxia-inducible factor-alpha
HL-60	Human leukemia cell line
JAK/STAT	Janus kinase/signal transducer and activator of transcription
LNCaP	Lymph node carcinoma of the prostate
MMP	Matrix metalloproteinase
NF-kB	Nuclear factor kappa-light-chain-enhancer of activated B cells
PGE2	Prostaglandin E2
PKB/AKT	Protein kinase B
PTEN	Phosphatase and tensin homolog deleted on chromosome 10
ROS	Reactive oxygen species
SCC-25	Squamous cell carcinoma
TGF-B	Transforming growth factor-beta
TPA	12-O-tetradecanoylphorbol-13-acetate

9.1 Introduction

Cancer is the most complex and majorly occurring disease worldwide. At least 60% of anti-cancer compounds are derived naturally from plants, i.e., about three-quarters of anti-cancer compounds are natural products [1]. Some anti-tumor plant products derived from alkaloids and flavonoids. Dietary phytoestrogens like flavonoids (quercetin), isoflavonoids (genistein, daidzein), and stilbenes (resveratrol) prevent estrogen-related diseases like post-menopausal symptoms and reduce the risk of estrogen-dependent cancer. Different approaches to cancer management have been proved ineffective because of various reasons like drug resistance, adverse reactions, insufficient and inadequate target specificity of a single anti-cancer agent [2]. Therefore, an approach developed which involves the application of one or more anti-cancer agents combined, which thereby produced a synergistic effect boosting the cytotoxicity of cancer cells.

Natural compounds, herbs, and plant derivatives or extracts used to make drugs for cancer therapy [3]. Although these compounds are assumed to be safe at times, these can be toxic, too, if not prescribed properly. Many compounds exhibit a biphasic effect, which means the compound at low and high doses will produce opposite effects towards cancer cells. The process by which an organism or cell exhibits a biphasic effect with the response to increasing amounts of any compound or a drug is known as hormesis. In 2005, the first assessment and documentation of the biphasic dose-response relationship in tumorigenesis came to the limelight [4]. The biphasic dose-response relationship is highly generalized, i.e., the responses not restricted by the biological model [4]. Some natural compounds specific for the treatment of cancer might become not effective for the same cell at any other concentration, either high or low dose. The sources and common uses of some of the natural compounds having biphasic effects in tumorigenesis are mentioned in Table 9.1.

This book chapter describes various natural compounds which have a biphasic effect in different cancer cells (Table 9.2). For example, daidzein from soybean possesses estrogen-like and estrogen-dependent activity [26]. Resveratrol from blueberries, mulberry, and raspberry is a potent tumorigenic inhibitor [27]. Falcarinol found in carrots is known to reduce the cancer risk [28]. Kaempferol found in onions, tomatoes, cucumbers, and spinach is a effectual anti-oxidant showing anti-cancer effects on human cancer cells [29]. Quercetin from red onions regulates cell signaling, suppression of growth and proliferation, pro-apoptotic, and anti-oxidant effects [30]. Berberine from Oregon grape and poppy exerts anti-tumor potential in multiple ways, including the suppression of angiogenesis, cell proliferation, and metastasis in various types of cancers such as breast cancer, melanoma, liver cancer, and gastric cancer [31]. Indole-3-carbinol also exerts anti-cancer potential [32]. Similarly, an anti-cancer effect of umbelliprenin has been depicted in the case of colorectal cancer in a dose-dependent manner [33]. Genistein is used as a chemotherapeutic agent, which alters cell cycle, apoptosis, angiogenesis, and metastasis, also is inhibited by it [34]. A phytochemical glabridin from *Glycyrrhiza glabra* exhibits anti-tumor activity [35]. 12-O-tetradecanoylphorbol-13-acetate (TPA) causes many tumor cells to die [36] (ChemBioDraw Ultra draws all the structures of mentioned compounds in this book chapter). This book chapter has described the biphasic effects of the above mentioned natural products on various human cancer cells (Tables 9.1, 9.2, and 9.3).

9.2 Natural Compounds Showing Biphasic Effects in Cancer

Resveratrol

Resveratrol, also known as stilbenoid (3, 4, 5-trihydroxy-trans-stilbene) produced by the plants which are attacked by the bacteria and fungi, grapes skin, and seed, peanuts, blueberries, mulberries, and raspberries (Table 9.1). It is identified as a PAN assay interference compound that directly affects the cell membranes and also has the ability of varied interactions. Resveratrol inhibits carcinogenesis in mouse-skin cancer [27], and it also exhibits cytotoxic effects in vitro in cancers like breast, skin, cervix, stomach, prostate, ovary, pancreas, and thyroid. Studies conducted in vitro and in vivo report the inhibition of all the stages of carcinogenesis by resveratrol, e.g., initiation, promotion, and progression [56]. It also affects the signaling pathways like Akt/GSK and ERK pathways, EMT and SMAD signaling, ROS-notch1/PTEN/Akt, and HIF-1α/ROS/p53 signaling pathways as described in

Table 9.1 Sources and uses of natural compounds having a biphasic effect

Sr. no.	Compound	Source	Uses	Ref.
1.	Resveratrol (polyphenol)	Grapes skin, peanuts, blueberries, mulberries, and raspberries	High cholesterol, cancer, heart disease	[5]
2.	Falcarinol (fatty alcohol)	Carrots, ivy, and red ginseng	Protects roots from fungal diseases, such as licorice root, colon cancer	[6]
3.	Kaempferol (flavonoid)	Onions, cucumbers, tomatoes, potatoes, spinach, raspberries, plants like *Moringa oleifera*, Cuscuta, Chinensis, cocaine grandis, and *Glycine max*	Reduces oxidative stress, cancer treatment	[7]
4.	Quercetin (flavonol)	Red onions, tomatoes, and honey	Arthritis, bladder infections, and diabetes, cancer treatment	[8]
5.	Berberine (alkaloid)	Barberry, Oregon grape, goldenseal, yellow root, Amur cork, and prickly poppy	High blood pressure, high levels of cholesterol or other fats (lipids) in the blood (hyperlipidemia), and diabetes	[9]
6.	Indole-3-carbinol	Brussels sprouts, collards, broccoli, cabbage, mustard greens, kale, cauliflower, turnips, and rutabagas	Prevents various cancers like colon cancer, breast cancer, etc.	[10]
7.	Daidzein (isoflavone)	Soybeans (hypocotyl) and legume	Blood cholesterol, menopausal relief, osteoporosis, heart disease, and lowers the risk of some cancers caused due to hormonal imbalance	[11]
8.	Umbelliprenin	*Anethum graveolens*, *Pimpinella anisum*, *Ferulago campestris*, and *grecescu*	Acts as skin whitening agent	[12]
9.	Genistein	Tofu, fava beans, soybeans, kudzu, and lupin	Prostate cancer, bladder cancer, and breast cancer	[13]
10.	Glabridin	The root extract of licorice (*Glycyrrhiza glabra*)	Acts as anti-inflammatory, anti-oxidant, and skin whitening agent	[14]
11.	12-O-tetradecanoylphorbol-13-acetate (TPA)	Croton plant (Euphorbiaceae)	Tumor promoter	[15]

Table 9.2 Summarizes biphasic role of natural compounds

Sr. no.	Compound name	Biphasic effects	Type of cancer cells	Ref.
1.	Resveratrol	• Both stimulatory and inhibitory effects in SCC-25 cell line	Oral squamous carcinoma cells, prostate cancer	[16]
2.	Falcarinol	• Decreases DNA strand breakage in Caco2 cells • Increase in caspase-3 activity in Caco2 cell line	Colon cancer	[17]
3.	Kaempferol	• Biphasic response on the concentration of E2 via ER-dependent or independent pathway in MCF-7	Breast cancer	[18]
4.	Quercetin	• Stimulates and inhibits cell proliferation in MCF-7, HT-29, and HCT-116	Breast cancer, colon cancer	[19]
5.	Berberine	• Induces CYP1A1 mRNA expression and elevates 7-ethoxyresorufin-O-demethylase (EROD) activity in HEPG2 cells	Liver cancer cell	[20]
6.	Indole-3-carbinol	• Exhibits enhanced effect on post-initiation stage and inhibitory effect on the pre-initiation stage in the medium-term liver bioassay in glutathione S-transferase placental form (GST-P) positive liver cell foci	Hepatocarcinoma	[21]
7.	Daidzein	• Stimulate cell growth, estrogen-like effects and inhibits cell growth, cells got arrested at G0/G1 phase, anti-estrogen effect in LoVo cell line	Colon cancer	[22]
8.	Umbelliprenin	• Increases the percentage of apoptosis at <10 μM in Jurkat T-cell line • Decreases the cell-death in the Jurkat T-cell line at >100 μM	Myeloid leukemia	[23]
9.	Genistein	• In PC3 cell line, the number of cells increased to 1.5 fold at 500–1000 nmol/L • A threefold decrease in cell number at 50,000 nmol/L in the PC3 cell line	Prostate cancer, breast cancer	[24]
10.	Glabridin	• Inhibits the proliferation of ER- and ER+ at □10 μM and increases proliferation at 100 nM-10 μM in PC3 cell line	Prostate cancer	[25]
11.	TPA	• Before 24 h rapid cell–cell contacts and after 24 h loss of cell–cell contact seen in human SCC cell line	Human epidermal squamous carcinoma	[25]

Table 9.3 Signaling pathways affected by various phytochemicals exhibiting biphasic activity

Sr. no.	Compound	Signaling pathway	Type of cancer/cell line
1.	Resveratrol	• Downregulates AKT/GSK and ERK pathways [37]	OVCAR-3 ovarian cancer cells,
		• Suppresses EMT and SMAD dependent signaling [38],	Glioblastoma cells
		• ROS dependent downregulation of notch1/PTEN/Akt signaling [39]	Human ovarian cancer cell line A2780 and SKOV3
		• Induction of apoptosis via HIF-1α/ ROS/p53 signaling [40]	Prostate cancer
2.	Falcarinol	• Downregulates NF-kB [41]	Rat colorectal cancer
3.	Kaempferol	• Impairs MAPK pathway [42],	Hepatocellular carcinoma
		• Downregulates PI3K/AKT signaling [43]	Skin, colon, liver cancer
		• Blocks the EGFR-related pathway [29]	Pancreatic cell cancer
4.	Quercetin	• Inhibits NF-kB and MMP-2/9 pathway [44] • Inhibits PI3K/AKT/mTOR • Regulates Wnt/β-catenin, MAPK/ ERK1/2 pathway [45]	SAS human oral cancer cells
5.	Berberine	• Downregulates COX2/PGE2-JAK2/STAT3 pathway [46]	Colorectal cancer cells
6.	Indole-3-carbinol	• Represses ER-alpha signaling [47]	Breast cancer cells
7.	Daidzein	• Inhibits Hedgehog/gli1 signaling [48]	Breast cancer,
		• Inhibits RAF/MEK/ERK cascade [49]	SKOV3 human ovarian cancer cells,
		• Inhibits FGFR3 pathway [50]	Bladder cancer
8.	Umbelliprenin	• Activates intrinsic and extrinsic apoptosis pathways [51]	Jurkat T-cell line
9.	Genistein	• Inhibits Wnt signaling via ERS dependent pathway • Inhibits NF-kB [52]	Prostate cancer
10.	Glabridin	• Inhibits FAK/Rho signaling [53]	Human lung cancer A549,
		• Inhibits TGF-β/SMAD2 signaling pathway [54]	Hepatocellular carcinoma
11.	TPA	• Activates MEK-ERK [55]	Oesophageal squamous cell carcinoma

Table 9.3. Resveratrol also exhibits a dual effect in some conditions. For example, in human oral squamous carcinoma cells SCC-25, literature represents that, at 1 μM, resveratrol had stimulatory effect while at 10 μM resveratrol exhibits inhibitory effect. At 100 μM concentration, resveratrol shows inhibitory effect on both DNA synthesis and cell growth. HL-60 when treated with TPA (12-o-tetra-hydrodecanoylphorbol-13-acetate), resveratrol inhibits in a dose-dependent manner

[16]. In cultured mouse hepatoma cells (Hepa 1c1c7), resveratrol induces quinone reductase activity. A combination of quercetin with resveratrol has proved to be a potent inhibitor of oral cancer cell growth. In LNCaP cells, resveratrol at 20 μM concentration inhibited 50% of DNA synthesis (Table 9.2) [16].

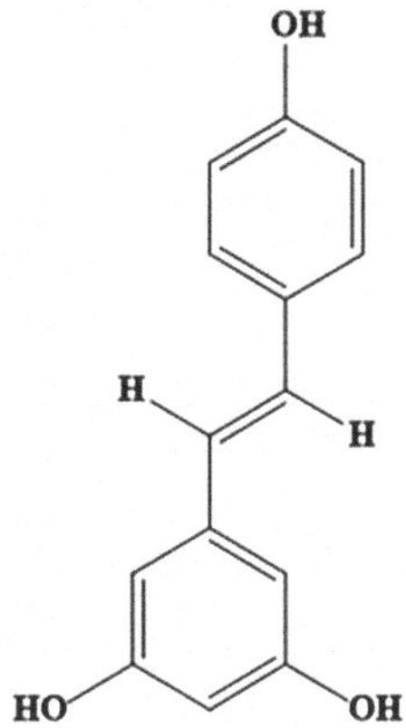

Resveratrol

Falcarinol

Falcarinol [(3R)–(9Z)-heptadeca-1, 3-diene-4, 6-diyn-3-ol], also known as panaxynol or carotatoxin is a natural pesticide present in carrots, ivy, and red ginseng (Table 9.1). It guards roots from fungal diseases like licorice root and acts as an antagonist for covalent cannabinoid receptor inducing pro-allergic effects in the skin [57]. The NF-kB pathway is affected in rat colorectal cancer by it (Table 9.3). Hormesis of biphasic effects has recently observed in falcarinol in the Caco2 cell line [17]. At 0.5–10 μM conc., basal DNA strand breakage decreases with the decrease in the expression of apoptosis indicator caspase-3, while at a concentration more than 20 μM, the detachment of the CaCo2 cells from the surface of cultured flask increases within 24–48 h. At a concentration of 1–20 μM falcarinol, there was a decrease in DNA strand breakage in colon cells, and at 100 μM concentration, active caspase-3 concentration increases in CaCo2 cells. By SCGE (single-cell gel electrophoresis), falcarinol at a concentration of 50–100 μM induces comet remnants, and this reflects nuclear degradation taking place during apoptosis. Therefore, the integrity and proliferation of CaCo2 cells were induced dose-dependently by falcarinol (Table 9.2) [17].

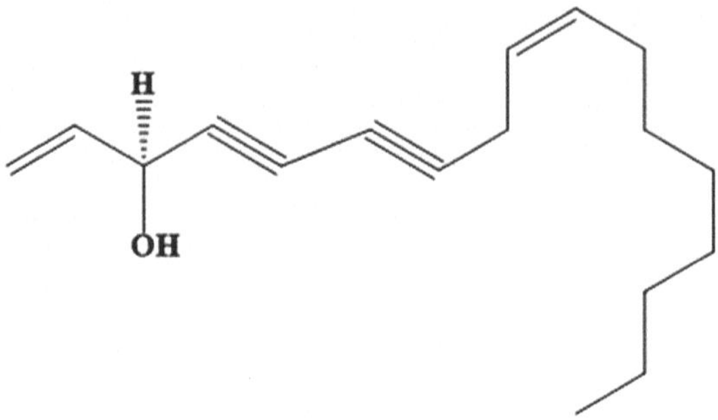

Falcarinol

Kaempferol

Kaempferol is a class of flavonol present in pteridophytes, angiosperms, pinophytes, vegetables, and fruits like Brussels sprouts, blackberries, cucumbers, tomatoes, onions, potatoes, spinach, and raspberries. Moreover, some plants like *Moringa oleifera*, Cuscuta, Chinensis, cocaine grandis, glycine max (Table 9.1) contain kaempferol [7]. It reduces oxidative stress by acting as an anti-oxidant and the most important dietary phytoestrogen. It also impairs MAPK, PI3K/PKB, and EGFR-related signaling pathways, as mentioned in Table 9.3. However, it has a response of hormesis on estrogen receptor, i.e., it has both estrogenic and anti-estrogenic activity. In the absence of E2, in the MCF-7 cell line, kaempferol induces the estrogen agonist activity. It activated the E2 responsive reporter gene dose-dependently and showed maximal effect at 10^{-5} M, whereas at higher concentration, at $>10^{-5}$, it decreases the luciferase activity (Table 9.2). When a focus assay performed to investigate the malignancy, it is seen that kaempferol inhibited the cell foci formation by E2. In the absence of estrogen, at concentration $> 10^{-5}$ M kaempferol, cytotoxicity not restoring the anti-proliferation caused by excess E2. So, kaempferol regulates the estrogenic levels in the body and might also prevent estrogen imbalance diseases like osteoporosis and breast cancer [18].

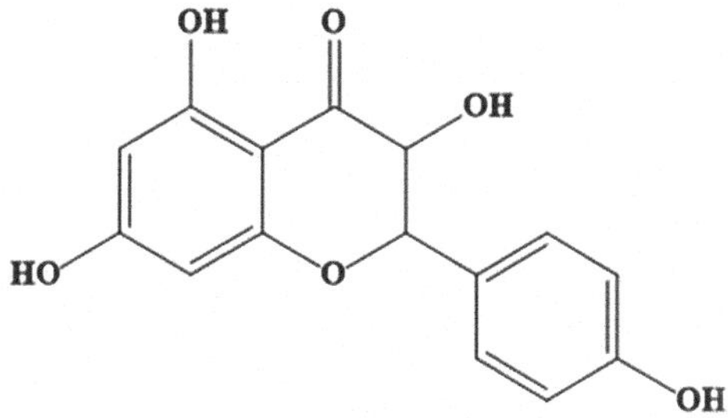

Kaempferol

Quercetin

Quercetin, plant flavonol derived from Quercetum (oak forest), can be found in red onions, organically grown tomatoes, and honey (Table 9.1) [58]. It regulates cell signaling, growth suppression, and shows anti-proliferative, pro-apoptotic, and anti-oxidant effects [30]. It inhibits COX (cyclooxygenase) and lipooxygenase which thereby, decreases the inflammation, inhibits the platelet aggregation, and improves the endothelium, has potent anti-cancer properties, and also reduces the number and size of rectal adenomas [59]. It alters the NF-kB and MMP2/9 signaling pathways and also inhibits the PI3K/Akt/mTOR, MAPK/ERK1/2, and Wnt/B-catenin pathways as summarized in Table 9.3. However, quercetin shows different stabilities at different conditions. It is unstable at 7.4pH; in 0.1 M potassium phosphate, it degrades in 10 hr., whereas, at the same pH in McCoy's 5A culture, it degrades in 2 hr. It modulates cell proliferation as a biphasic response, which is also known as growth hormesis, and in both the conditions, quercetin gets stabilized by ascorbic acid. Quercetin shows a stimulating effect in MCF7 cell lines, whereas a biphasic effect seen in cell proliferation in colon cancer cell lines. Generally, quercetin inhibits the proliferation in colon cancer cell lines, but only at high concentrations (above 30 μM for HCT-116 cells and 80 μM HT29 cells) (Table 9.2). On the contrary, quercetin shows a significant increase in cell proliferation even at low concentrations [19], which suggests that quercetin in a concentration-dependent manner balances between the pro-oxidant and anti-oxidant activity. Quercetin shows a dual effect in SCC-25 (oral squamous carcinoma cell line), at a concentration of 1–10 μM, its stimulatory effects on growth, whereas at 100 μM it causes growth inhibition. Ranges of quercetin concentrations can have both promoting as well as inhibiting effect, depending upon the cell type. For example, human blood serum levels have enhanced cell proliferation after quercetin ingestion from onions, wine, etc. (around 1 μM), and higher concentration of quercetin has been found in intestine (250–500 mg).

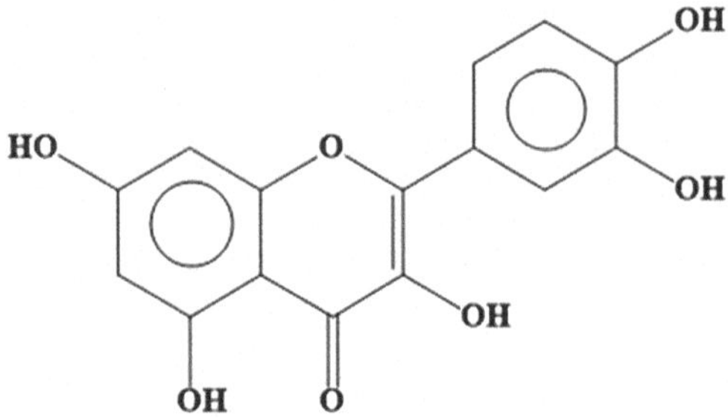

Quercetin

Berberine

Berberine, derived from Oregon grape and poppy (Table 9.1), exerts anti-tumor activity by multiple routes such as by suppressing the cell proliferation, angiogenesis, and metastasis in breast cancer, melanoma, gastric cancer, and hepatoma [31]. It is a yellow alkaloid which was first isolated from Xanthoxylon cava in 1826 by Chevallier and Pelletan [60] found in plants like barberry, Oregon grape, goldenseal, yellow root, amur cork, prickly poppy [61]. Berberine is a quaternary ammonium salt derived from the protoberberine group having a strong yellow fluorescence and thus is used in histology. It is considered as an anti-malarial drug also effective against diabetes, hypertension, obesity, artery diseases, Alzheimer's, and polycystic ovary. It alters the COX2/PGE2-JAK2/STAT3 pathway, as discussed in Table 9.3. The biphasic effect of berberine is seen on CYP1A1 in HepG2 cells [20]. It induces CYP1A1 mRNA expression at 1uM concentration, whereas, at 50 μM, the induction attained efficiency. It inhibits the enzyme catalytic activity and induces the gene expression at high doses of berberine for 24–48 h (Table 9.2). On treatment, it significantly elevated the EROD (7-ethoxyresorufin-O-demethylase) activity, whereas after 6 h of incubation with high dose of berberine, the EROD activity was significantly decreased. The PI3K/AKT/Bcl-2 cell survival and Nrf2/HO-1 anti-oxidative signaling pathway were upregulated which helped in the mediation of the hormetic and neuroprotective effects of berberine [62].

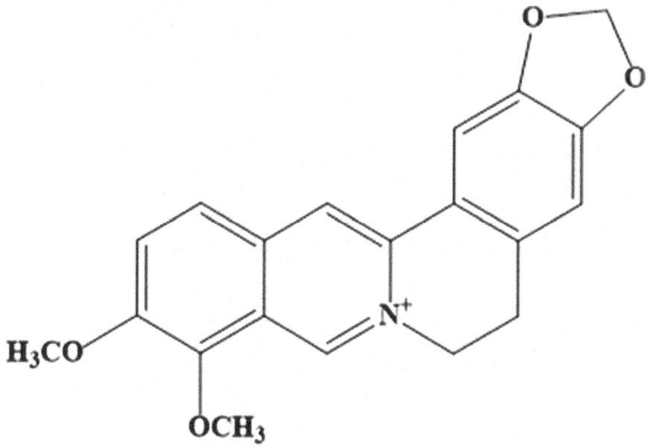

Berberine

Indole-3-Carbinol

Indole-3-carbinol is commonly present in vegetables like cabbage, broccoli, cauliflower, sprouts, and kale and collard greens. It is derived from glucobrassicin of the Brassica vegetables [63] (Table 9.1), possesses anti-oxidant, atherogenic and anti-carcinogenic properties. It is a phytochemical and also exerts anti-cancer properties [32]. It works by repressing ER-α signaling, as mentioned in Table 9.3. Indole-3-carbinol shows biphasic modifying effects on glutathione S-transferase placental form (GST-P) positive liver foci development and on the post-initiation stage, it exerts a promoting effect while on the pre-initiation stage; it exerts an inhibitory effect in the liver bioassay. On the contrary, AFB (alpha toxin B1) induced liver tumor of rainbow trout, indole-3-carbinol shows biphasic effects on hepatocarcinogenesis. Pre-initiation exposure to indole-3-carbinol reduces the AFB initiated hepatocarcinoma, whereas indole-3-carbinol given 3 weeks before diethylnitrosamine (DEN) exposure inhibits the development of GST-P positive liver cell foci in rats (Table 9.2) [21].

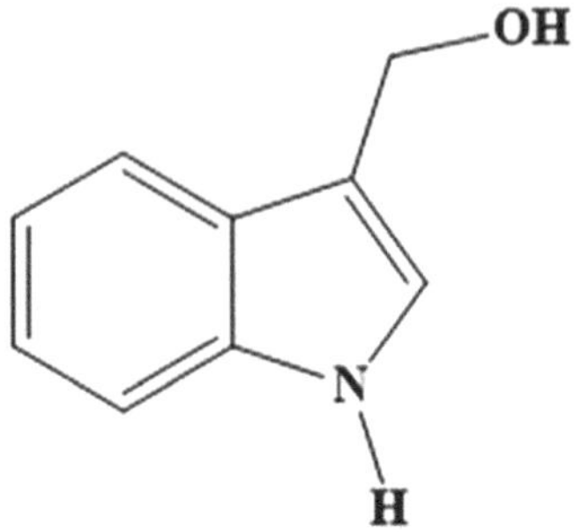

Indole-3-carbinol

Daidzein

Daidzein (7-hydroxy-3-(4-hydroxyphenyl)-4H-chromen-4-one) belonging to the class of isoflavone is found in soybeans [26] and legumes [64] (Table 9.1) and it possesses estrogen-like and estrogen-dependent properties derived from the shikimate pathway. Epidemiologic studies evidenced some association between soy exposure and a decrease in breast cancer, whereas few other studies report no association between these. It modulates Hedgehog/Gli1 signaling in breast cancer, inhibits RAF/MEK/ERK cascade, and inhibits FGFR3 signaling in bladder cancer cells as depicted in Table 9.3. The effect of daidzein has proved to be biphasic [65]. For instance, in LoVo cell lines, during MTT assay, when cells incubated in a medium of different concentrations of daidzein, it enhanced cell growth at lower concentration (0.1 and 1 μM), whereas inhibited the cell growth at higher doses (10–100 μM). Daidzein induces apoptosis in LoVo cells dose-dependently. At higher concentrations of daidzein, cells got arrested at G0/G1 phase (Table 9.2). Low concentration of daidzein causes estrogen-like effects which stimulates cancer cells growth, while at higher concentration, the cell growth is inhibited by anti-estrogen effect. Daidzein treatment induces DNA fragmentation due to apoptosis via caspase activation. Daidzein reduces the MNU (N-methyl-N-nitrosourea) induced mammary carcinogenesis by 20%. Therefore, about 600 g of soybean must be taken in diet per day to benefit from the anti-tumorigenic effects of daidzein [65]. No effect on differentiation is seen as proved by the ALP activity in daidzein treated colon cancer cells where ALP is the marker used to determine the differentiation in colon cancer cells [22].

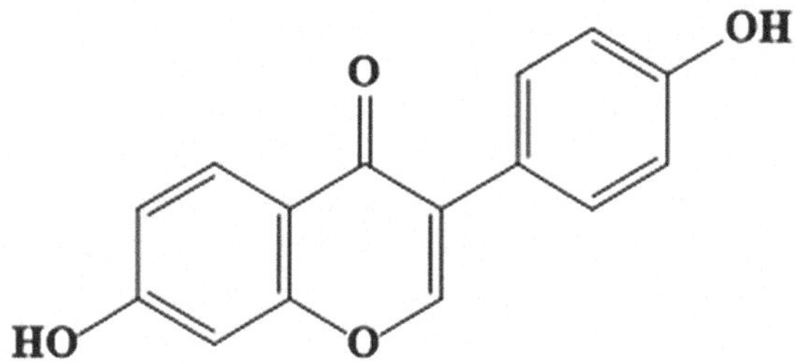

Daidzein

Umbelliprenin

Found in plants belonging to the Apiaceae-*Anethum graveolens* (dill an herb in celery) L. *Pimpinella anisum* L. (aniseed) and *Ferulago campestris* (Table 9.1) [66]. It shows anti-inflammatory, immunomodulatory, anti-melanogenic, and pro-apoptotic effects as well as promising skin whitening agent [66]. In Jurkat T-cell line, it activates the intrinsic and extrinsic apoptosis pathways, as shown in Table 9.3. It initiates apoptosis by activating the caspase-8 following the intrinsic pathway and caspase-9 of the extrinsic pathway which is mediated by the tumor necrosis factor receptor (Table 9.2) [23].

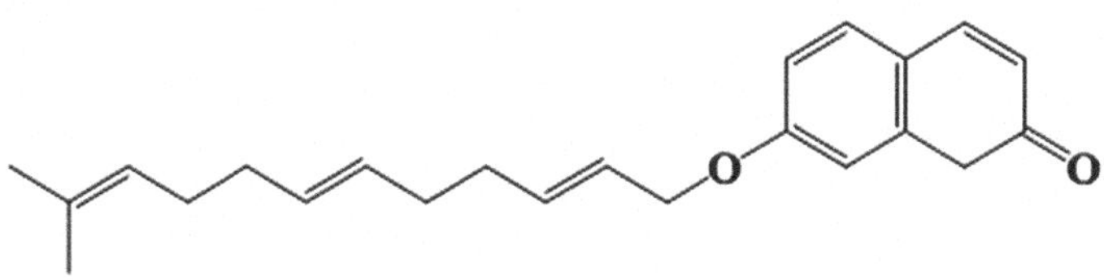

Umbelliprennin

Genistein

Genistein is an isoflavone, best known as a phytoestrogen and an angiogenesis inhibitor found in a variety of plants like kudzu, lupin, soybeans, fava beans, and Psoralea (Table 9.1) [67]. Genistein, a chemotherapeutic agent alters cell cycle, apoptosis, angiogenesis, and also inhibits metastasis [34]. It inhibits NF-kB signaling, as mentioned in Table 9.3. Regulation of cell proliferation in PC3 cells and in vitro invasion is carried out by genistein in a biphasic manner. A concentration of 500–1000 nmol/L of genistein for 72 h induces a significant 1.5 fold increase in several cells while >threefold decrease in cell number occurs with 50,000 nmol/L genistein. To assess the invasion, cells were subjected in the Boyden chamber with Matrigel assay and with 500 nmol/L genistein, an increase in the number of invaded cells was observed, whereas a twofold reduced invasion was observed with 50,000 nmol/L genistein (Table 9.2) [24]. In a 250 mg/kg diet group of TRAMP-FVB mice (transgenic adenocarcinoma mouse prostate) consuming diet supplemented with genistein, a significant twofold upregulation was found while no significant decrease was observed in 1000 mg/kg diet in the prostatic lysates. These studies show that genistein exposure affects the progression in TRAMP-FVB mice dose-dependently.

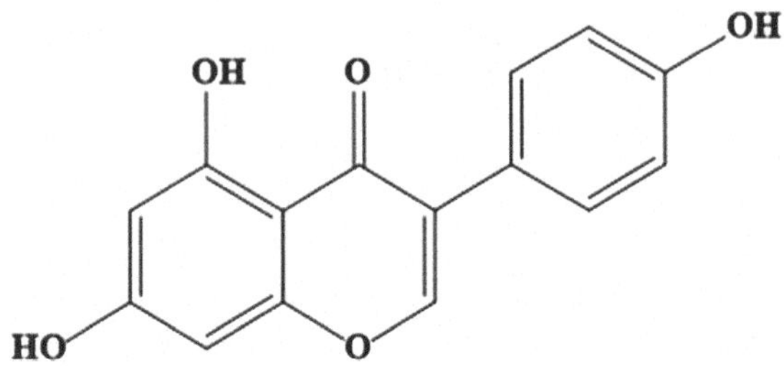

Genistein

Glabridin

Glabridin, present in the root extract of licorice (*Glycyrrhiza glabra*) (Table 9.1), is a phytoestrogen that binds to the human estrogen receptor and enhances creatine kinase activity in the ventricle, bone, heart, and cartilage. It also exhibits anti-tumor properties [35]. It works by inhibiting FAK/Rho signaling and TGF-β/SMAD2 signaling pathway in hepatocellular carcinoma, as shown in Table 9.3. Zava et al. first showed biphasic effects in glabridin [68]. At 0.1–10 nM conc., glabridin showed equal and maximum growth stimulation like estradiol. At <10 μM, it promotes cell growth, and at >15 μM, glabridin inhibits the cell growth. With 0.1–10 μM glabridin, in ER+ breast cancer cell lines, cell growth increased above control, and then abruptly, it inhibited at approximately 25 μM glabridin. The cell growth was not increased in the ER-cell line, but at 25 μM, it exhibited an inhibitory effect. When glabridin (10 μM) was added to the tamoxifen-treated cells (1 μM), it inhibited the maximum growth of ER+ breast cancer cells by 50% while an increase in tamoxifen up to 5 μM inhibited the proliferative effect of glabridin (Table 9.2) [69].

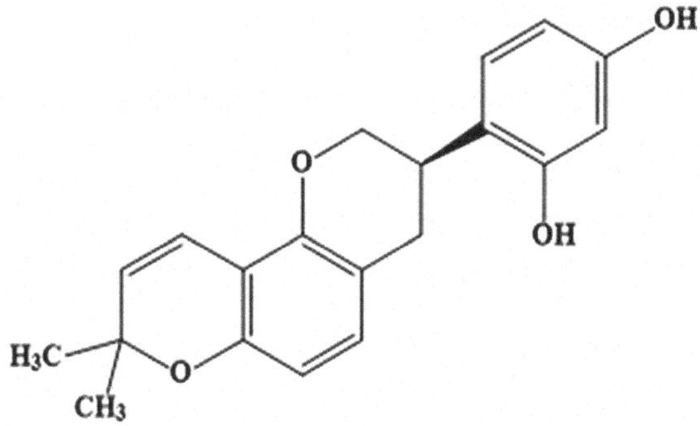

Glabridin

12-O-Tetradecanoylphorbol-13-Acetate (TPA)
12-O-Tetradecanoylphorbol-13-acetate (TPA), a phorbol diester (compound isolated from croton seeds) (Table 9.1), is the most effective tumor promoter and modulator of differentiation in epidermal keratinocytes, which induces a rapid formation of cell–cell contact with an increase in protein kinase C on low calcium grew cells. TPA can cause many tumor cells to die [36]. It can modulate the MEK-ERK pathway in oesophageal squamous cell carcinoma, as discussed in Table 9.2. It is used for cytogenetic testing as a B-cell specific mitogen in cancer diagnostics and is also applied along with ionomycin to stimulate T-cell activation, cytokine production, and cell proliferation [70]. However, it exhibits a biphasic effect on treatment more than 24 h, causing loss of cell–cell contact and a decrease in membrane-bound protein kinase C activity [25].

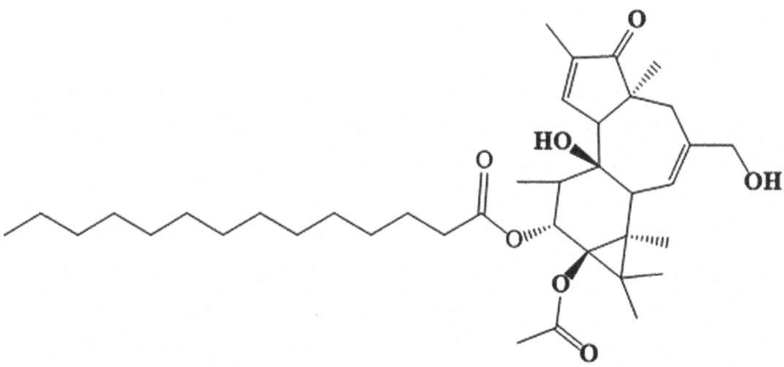

12-0-Tetradecanoylphorbol-13-acetate (TPA)

9.3 Metabolomics Studies

Metabolomics deals with the evaluation of the mechanism of action of various types of anti-cancer agents. It is applicable to understand the tumor cell response to nutritional agents. It throws light on the quantitative analysis of all the metabolites present in an organism to examine the relationship between the metabolites and the physiological and pathological changes. New inflammatory drugs can be discovered using metabolomics as it acts as an effective tool to understand the mechanism of inflammation [71]. Metabolomics of resveratrol showed that it modulated the polyamine biosynthesis in human breast cancer cell lines, MDA-MB-231 and MCF-7 [72].

9.4 Quality of the Product

The medicines from natural products or their derivatives often puzzle consumers by being labeled as "natural," and thereby, implicating them as "completely safe" products simply based on perception. These natural products might affect the body in unwanted ways and might cause severe side effects or even death in certain cases. Therefore, there is a need to understand the detection, separation, purification, quantification, and identification of the best suitable compounds for therapeutics of various diseases, especially cancer. However, the main issue comes from a drug development perspective as the question arises as to how to expand scientific research into pre-clinical studies? Depending on the upcoming rapid and sensitive techniques, novel approaches developed.

9.5 Different Types of Drug Toxicity

Different types of drug toxicities reported like cardiotoxicity, nephrotoxicity, neurotoxicity, skin toxicity, and hepatotoxicity. Even the natural compounds show a different type of toxicity or the side effects on different doses of varying concentrations. For example, resveratrol causes nephrotoxicity when given in high doses when studied in in vivo animal studies. Resveratrol can also lead to DNA damage and cell cycle interruption by its pro-oxidant effect [56]. Some substances interfere with the normal reproduction and affect the normal sexual functionality of humans, known as reproductive toxicity shown mainly by genistein [73]. Daidzein causes changes in the fertility and developmental toxicity in the female rat's reproductive tract. High dose of daidzein reduces the body weight and circulating progesterone levels [74]. Kaempferol exhibits genotoxic effects [75] as it possesses in vitro pro-oxidant activity. The activity of quercetin was evaluated and it was found, after 2 years, toxic and neoplastic lesions appeared in the kidney of male rat, also increased hyperplasia, neoplasia, nephropathy, benign tumor in the renal tubular epithelium was caused [76]. Some adverse effects of indol-3-carbinol (I3C) are also reported, where a patient taking 400 mg of I3C twice a day had tremor and imbalance [77].

9.6 Possible Reasons for Drug Toxicity

As natural products are easily available in the market, people take it easily without any prescription. These products advertised without side effects; thereby, a large population gets attracted to it, and hence, an influence of self-treatment starts. Medical practitioners who possess a vast knowledge about the drugs, its mechanism of action, the human body, case studies, and pharmacology are allowed to do the practice. However, 50% of practitioners are not qualified enough to practice medicine. These under-qualified and untrained practitioners are unaware of the toxic side effects of natural products which are used as medicine and, therefore, are unable to treat their patients leading to non-rectified mistakes causing serious health issues [78]. Many sub-standard natural products are available in the market, not tested for the quality before marketing and use. Some products do not contain the active ingredients at all, and some might have a small number of active ingredients. Therefore, using such adulterated products instead of original loses its efficacy [78]. Some of the natural compounds taken as dietary supplements, but the proper dose not mentioned. The same dose is taken by different persons of different ages and weight without any mentioned period for it and, thus, might cause many harmful side effects [78].

9.7 Limitations

Multidisciplinary collaboration will facilitate the natural product discovery which requires proper escalation through the application of both combinatorial and medicinal chemistry and biochemistry. There is still so much to explore, and with the new emerging technology, new prototypes are getting developed, which are proved to be efficient candidates for pharmacologically active compounds and hence, increase the exploitation of the elements present in nature. The biggest limitation of clinical translation of resveratrol is its rapid metabolism, which leads to limited bioavailability in vivo [79]. The major limitation of genistein is its low bioavailability and high pleiotropy [34]. Reducing the side effect, the toxicity of quercetin nanoparticles, and the cost-effectiveness of the Nano formulations considered as the limitation of quercetin [80]. Kaempferol has a limitation of poor bioavailability [43]. Question arises whether these agents produce biphasic response and immune response at different stages of carcinogenesis and/or simultaneously. These naturally occurring agents might either enhance or diminish the cancer cell proliferation at one particular concentration while at some other concentrations, it may enhance or diminish the cell-mediated immunity [4]. Also, it is found that in certain clinical circumstances, some agents with low concentration generate a stimulatory response thereby, enhancing the tumor proliferation [4]. It is still not known why the compounds show biphasic effect or the opposite effect at low and high concentrations, but dose-response relationship mainly depends on the exposure time and route of administration, i.e., different routes and time exposure leads to different relationship and a different conclusion. This limitation is major because of the complex biosystems and bioprocesses between the cells and tissues and external exposure.

9.8 Conclusion

The different natural compounds described in this book chapter have shown to produce opposite effects at different concentrations and conditions (Table 9.3). For example, the LoVo cells, daidzein can both stimulate the cell growth at low concentrations, whereas it can also arrest the cell cycle at the G0/G1 phase at high concentrations. In SCC-25 and LNCaP cells, resveratrol can show both inhibitory and stimulatory effects on the cells. In colon cancer, falcarinol decreases the DNA strand breakage as well as can also increase the caspase-3 activity of Caco2 cell lines. In the MCF-7 cells, kaempferol can show biphasic response on different concentrations of estradiol via ER-dependent or independent pathways. In HT-29 and HCT-116 cell lines, quercetin can both inhibit and stimulate cell proliferation at different concentrations. In the HEPG2 cell line, berberine increases as well as decreases the EROD activity at a different time of exposure. At post-initiation and pre-initiation, indole-3-carbinol in the glutathione S-transferase placental form liver cell foci shows both promoting and inhibiting effects, respectively. In Jurkat T-cell line, Umbelliprenin can increase the percentage of apoptosis at <10 μM as well as decrease the apoptosis. Genistein in PC3 and MCF-7 cells changes the number of fold increase in cell number. Glabridin when treated in mouse adenocarcinoma prostate cell line, PC3 and MCF-7, both lead to the increment as well as decrement of cellular proliferation at different concentrations. In human epidermal squamous cell carcinoma consisting of both ER+ and ER− cells, when treated with 12-O-tetradecanoylphorbol-13-acetate (TPA), increases and decreases the cell-cell contact at a different time of exposure. The basic signaling pathways, biphasic effects, and toxicity issues of these natural compounds summarized in Fig. 9.1.

9.9 Future Perspective

High-throughput screening along with combinatorial chemistry drives the future success of the pharmaceutical industry. Also, natural product discovery, proteomics, genomics, metabolomics, metagenomics, semi-synthesis, recombinant DNA methodology, genome mining, structure–function drug design, and combinatorial biosynthesis are associated with it. Intelligent screening methods like screening by a robot or robotic separation alongside metabolic engineering and structural analysis offer dimensions to technologies for new natural product discovery for the origination of anti-tumor compounds [81]. Another promising approach to develop effective chemotherapeutic anti-cancer agents is, by conjugating the potent cytotoxic

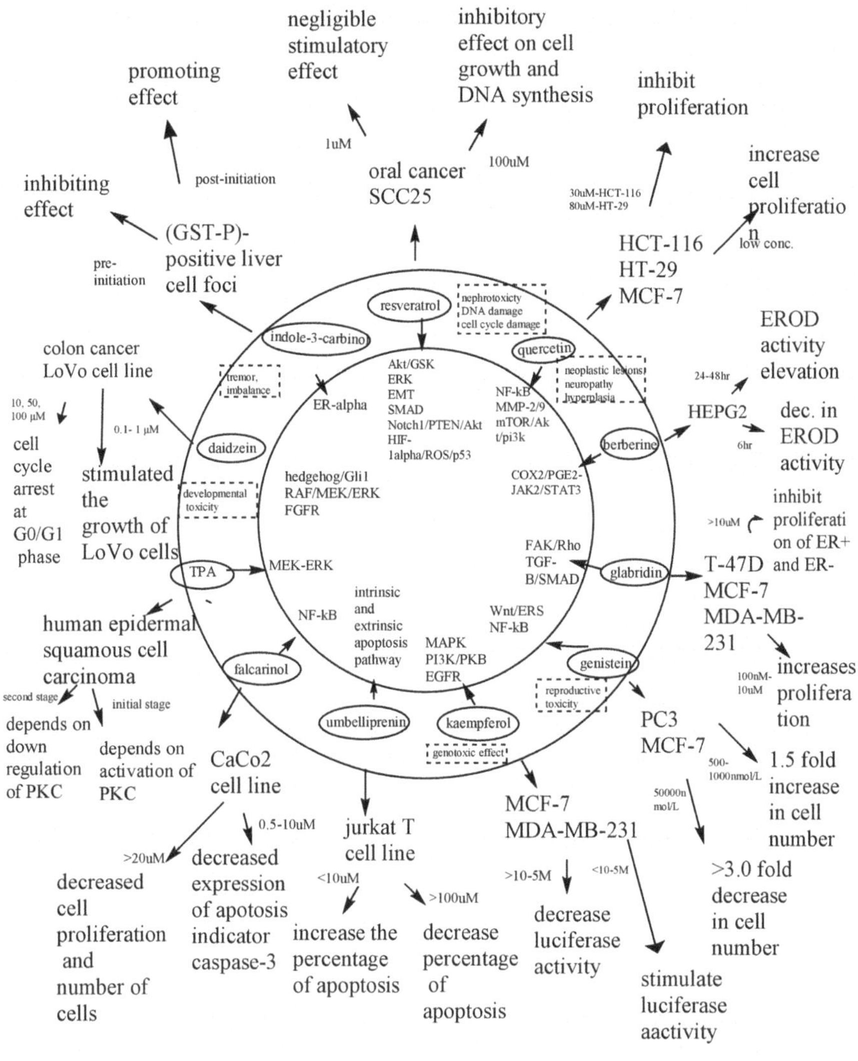

Fig. 9.1 Figure depicting different signaling pathways, toxicity, and biphasic effects of natural compounds. The natural compounds are depicted here by an oval inside the circular ring. The inner circle depicts the signaling pathways exhibited by the natural compounds. The dotted rectangular box inside the ring represents the toxicity or the side effects by the respective natural compound. The portion outside the ring represents the biphasic effects shown by the natural compounds in various cancer cell lines at different concentrations

natural products to monoclonal antibodies by targeting epitopes present on tumors of interest. Many natural compounds act by more than one mechanism as some have a greater degree of efficacy, whereas many works against multi-drug resistance [3].

Moreover, in-depth research is required for the better understanding of the compounds derived from the natural products and their therapeutic applications, which will help to meet the major challenges in treating cancer. Of course, a broad-spectrum study is recommended in various cell culture and/or animal models to verify and affirm the biphasic effect of the same drug/natural compound. Understanding the differential molecular mechanisms and signaling transduction pathways involved in the biphasic effect of the same natural compound and drug is necessary to increase the effectiveness of the cancer treatment. This detailed study might lead to identifying some target molecules/other drugs used in combination with the compound having a biphasic effect to prevent the tumorigenic potential of this natural compound.

Conflict of Interest The author confirms that there is no conflict of interest in this book chapter.

References

1. Cragg GM, Newman DJ (2005) Plants as a source of anti-cancer agents. J Ethnopharmacol 100 (1–2):72–79
2. Holohan C, Van Schaeybroeck S, Longley DB et al (2013) Cancer drug resistance: an evolving paradigm. Nat Rev Cancer 13(10):714–726
3. Nobili S, Lippi D, Witort E et al (2009) Natural compounds for cancer treatment and prevention. Pharmacol Res 59(6):365–378
4. Calabrese EJ (2005) Cancer biology and hormesis: human tumor cell lines commonly display hormetic (biphasic) dose responses. Crit Rev Toxicol 35(6):463–582
5. Frémont L (2000) Biological effects of resveratrol. Life Sci 66(8):663–673
6. Hansen SL, Purup S, Christensen LP (2003) Bioactivity of falcarinol and the influenceof processing and storage on its content in carrots (Daucus carota L). J Sci Food Agric 83 (10):1010–1017
7. Calderón-Montaño JM, Burgos-Morón E, Pérez-Guerrero C et al (2011) A review on the dietary flavonoid kaempferol. Mini Rev Med Chem 11(4):298–344
8. Lamson DW, Brignall MS (2000) Antioxidants and cancer, part 3: quercetin. Altern Med Rev 5 (3):196–208
9. Tillhon M, Guamán Ortiz LM, Lombardi P et al (2012) Berberine: new perspectives for old remedies. Biochem Pharmacol 84(10):1260–1267
10. Weng JR, Tsai CH, Kulp SK et al (2008) Indole-3-carbinol as a chemopreventive and anti-cancer agent. Cancer Lett 262(2):153–163
11. Liggins J, Bluck LJ, Runswick S et al (2000) Daidzein and genistein contents of vegetables. Br J Nutr 84(5):717–725
12. Iranshahi M, Sahebkar A, Takasaki M et al (2009) Cancer chemopreventive activity of the prenylated coumarin, umbelliprenin, in vivo. Eur J Cancer Prev 18(5):412–415
13. Dixon RA, Ferreira D (2002) Genistein. Phytochemistry 60(3):205–211
14. Simmler C, Pauli GF, Chen SN (2013) Phytochemistry and biological properties of glabridin. Fitoterapia 90:160–184
15. Abb J, Bayliss GJ, Deinhardt F (1979) Lymphocyte activation by the tumor-promoting agent 12-O-tetradecanoylphorbol-13-acetate (TPA). J Immunol 122(5):1639–1642
16. ElAttar T, Virji AS (1999) Modulating effect of resveratrol and quercetin on oral cancer cell growth and proliferation. Anti-Cancer Drugs 10(2):187–193
17. Young JF, Duthie SJ, Milne L et al (2007) Biphasic effect of falcarinol on CaCo-2 cell proliferation, DNA damage, and apoptosis. J Agric Food Chem 55(3):618–623
18. Oh SM, Kim YP, Chung KH (2006) Biphasic effects of kaempferol on the estrogenicity in human breast cancer cells. Arch Pharm Res 29(5):354

19. van der Woude H, Gliszczyńska-Swigło A, Struijs K et al (2003) Biphasic modulation of cell proliferation by quercetin at concentrations physiologically relevant in humans. Cancer Lett 200 (1):41–47
20. Vrzal R, Zdarilová A, Ulrichová J et al (2005) Activation of the aryl hydrocarbon receptor by berberine in HepG2 and H4IIE cells: biphasic effect on CYP1A1. Biochem Pharmacol 70 (6):925–936
21. Kim DJ, Lee KK, Han BS et al (1994) Biphasic modifying effect of indole-3-carbinol on diethylnitrosamine-induced preneoplastic glutathione S-transferase placental form-positive liver cell foci in Sprague-Dawley rats. Jpn J Cancer Res 85(6):578–583
22. Murata M, Midorikawa K, Koh M et al (2004) Genistein and daidzein induce cell proliferation and their metabolites cause oxidative DNA damage in relation to isoflavone-induced cancer of estrogen-sensitive organs. Biochemistry 43(9):2569–2577
23. Gholami O (2017) Umbelliprenin mediates its apoptotic effect by hormesis: a commentary. Dose-Response 15(2):1559325817710035
24. El Touny LH, Banerjee PP (2009) Identification of a biphasic role for genistein in the regulation of prostate cancer growth and metastasis. Cancer Res 69(8):3695–3703
25. Kitajima Y, Inoue S, Nagao S et al (1988) Biphasic effects of 12-O-tetradecanoylphorbol-13-acetate on the cell morphology of low calcium-grown human epidermal carcinoma cells: involvement of translocation and down regulation of protein kinase C. Cancer Res 48 (4):964–970
26. Reinli K, Block G (1996) Phytoestrogen content of foods-a compendium of literature values. Nutr Cancer 26(2):123–148
27. Jang M, Cai L, Udeani GO et al (1997) Cancer chemopreventive activity of resveratrol, a natural product derived from grapes. Science 275(5297):218–220
28. Tan KW, Killeen DP, Li Y et al (2014) Dietary polyacetylenes of the falcarinol type are inhibitors of breast cancer resistance protein (BCRP/ABCG2). Eur J Pharmacol 723:346–352
29. Lee J, Kim JH (2016) Kaempferol inhibits pancreatic cancer cell growth and migration through the blockade of EGFR-related pathway in vitro. PLoS One 11(5):e0155264
30. Brito AF, Ribeiro M, Abrantes AM et al (2015) Quercetin in cancer treatment, alone or in combination with conventional therapeutics? Curr Med Chem 22(26):3025–3039
31. Wang H, Zhu C, Ying Y et al (2018) Metformin and berberine, two versatile drugs in treatment of common metabolic diseases. Oncotarget 9(11):10135
32. Katz E, Nisani S, Chamovitz DA (2018) Indole-3-carbinol: a plant hormone combatting cancer. F1000Res 7:F1000 Faculty Rev-689
33. Alizadeh MN, Rashidi M, Muhammadnejad A et al (2018) Antitumor effects of umbelliprenin in a mouse model of colorectal cancer. Iran J Pharm Res 17(3):976
34. Spagnuolo C, Russo GL, Orhan IE et al (2015) Genistein and cancer: current status, challenges, and future directions. Adv Nutr 6(4):408–419
35. Jiang F, Li Y, Mu J et al (2016) Glabridin inhibits cancer stem cell-like properties of human breast cancer cells: an epigenetic regulation of miR-148a/SMAd2 signaling. Mol Carcinog 55 (5):929–940
36. Han ZT, Tong YK, He LM et al (1998) 12-O-Tetradecanoylphorbol-13-acetate (TPA)-induced increase in depressed white blood cell counts in patients treated with cytotoxic cancer chemotherapeutic drugs. Proc Natl Acad Sci U S A 95(9):5362–5365
37. Vergara D, Simeone P, Toraldo D et al (2012) Resveratrol downregulates Akt/GSK and ERK signalling pathways in OVCAR-3 ovarian cancer cells. Mol BioSyst 8(4):1078–1087
38. Song Y, Chen Y, Li Y et al (2019) Resveratrol suppresses epithelial-mesenchymal transition in GBM by regulating smad-dependent signaling. Biomed Res Int 2019:1321973
39. Kim TH, Park JH, Woo JS (2019) Resveratrol induces cell death through ROS-dependent downregulation of Notch1/PTEN/Akt signaling in ovarian cancer cells. Mol Med Rep 19 (4):3353–3360

40. Wang D, Gao Z, Zhang X (2018) Resveratrol induces apoptosis in murine prostate cancer cells via hypoxia-inducible factor 1-alpha (HIF-1α)/reactive oxygen species (ROS)/P53 signaling. Med Sci Monit 24:8970
41. Kobaek-Larsen M, Baatrup G, Notabi MK et al (2019) Dietary polyacetylenic oxylipins falcarinol and falcarindiol prevent inflammation and colorectal neoplastic transformation: a mechanistic and dose-response study in a rat model. Nutrients 11(9):2223
42. Mylonis I, Lakka A, Tsakalof A et al (2010) The dietary flavonoid kaempferol effectively inhibits HIF-1 activity and hepatoma cancer cell viability under hypoxic conditions. Biochem Biophys Res Commun 398(1):74–78
43. Imran M, Salehi B, Sharifi-Rad J et al (2019) Kaempferol: a key emphasis to its anticancer potential. Molecules 24(12):2277
44. Lai WW, Hsu SC, Chueh FS et al (2013) Quercetin inhibits migration and invasion of SAS human oral cancer cells through inhibition of NF-κB and matrix metalloproteinase-2/-9 signaling pathways. Anticancer Res 33(5):1941–1950
45. Reyes-Farias M, Carrasco-Pozo C (2019) The anti-cancer effect of quercetin: molecular implications in cancer metabolism. Int J Mol Sci 20(13):3177
46. Liu X, Ji Q, Ye N et al (2015) Berberine inhibits invasion and metastasis of colorectal cancer cells via COX-2/PGE2 mediated JAK2/STAT3 signaling pathway. PLoS One 10(5):e0123478
47. Meng Q, Yuan F, Goldberg ID et al (2000) Indole-3-carbinol is a negative regulator of estrogen receptor-α signaling in human tumor cells. J Nutr 130(12):2927–2931
48. Bao C, Namgung H, Lee J et al (2014) Daidzein suppresses tumor necrosis factor-α induced migration and invasion by inhibiting Hedgehog/Gli1 signaling in human breast cancer cells. J Agric Food Chem 62(17):3759–3767
49. Hua F, Li CH, Chen XG et al (2018) Daidzein exerts anticancer activity towards SKOV3 human ovarian cancer cells by inducing apoptosis and cell cycle arrest, and inhibiting the Raf/MEK/ERK cascade. Int J Mol Med 41(6):3485–3492
50. He Y, Wu X, Cao Y et al (2016) Daidzein exerts anti-tumor activity against bladder cancer cells via inhibition of FGFR3 pathway. Neoplasma 63(4):523–531
51. Gholami O, Jeddi-Tehrani M, Iranshahi M et al (2013) Umbelliprenin from Ferula szowitsiana activates both intrinsic and extrinsic pathways of apoptosis in Jurkat T-CLL cell line. Iran J Pharm Res 12(3):371
52. Adjakly M, Ngollo M, Boiteux JP et al (2013) Genistein and daidzein: different molecular effects on prostate cancer. Anticancer Res 33(1):39–44
53. Tsai YM, Yang CJ, Hsu YL et al (2011) Glabridin inhibits migration, invasion, and angiogenesis of human non–small cell lung cancer A549 cells by inhibiting the FAK/Rho signaling pathway. Integr Cancer Ther 10(4):341–349
54. Jiang F, Mu J, Wang X et al (2014) The repressive effect of miR-148a on TGF beta-SMADs signal pathway is involved in the glabridin-induced inhibition of the cancer stem cells-like properties in hepatocellular carcinoma cells. PLoS One 9(5):e96698
55. Meng L, Wang M, Du Z et al (2017) Cell signaling pathway in 12-O-tetradecanoylphorbol-13-acetate-induced LCN2 gene transcription in esophageal squamous cell carcinoma. Biomed Res Int 2017:9592501
56. Salehi B, Mishra AP, Nigam M et al (2018) Resveratrol: a double-edged sword in health benefits. Biomedicine 6(3):91
57. Leonti M, Casu L, Raduner S et al (2010) Falcarinol is a covalent cannabinoid CB1 receptor antagonist and induces pro-allergic effects in skin. Biochem Pharmacol 79(12):1815–1826
58. Li Y, Yao J, Han C et al (2016) Quercetin, inflammation and immunity. Nutrients 8(3):167
59. David AVA, Arulmoli R, Parasuraman S (2016) Overviews of biological importance of quercetin: a bioactive flavonoid. Pharmacogn Rev 10(20):84
60. Hahn FE, Ciak J (1975) Berberine. In: Mechanism of action of antimicrobial and antitumor agents. Springer, Berlin, pp 577–584
61. Preeti S, Prabhat U, Shardendu M et al (2015) Berberine a potent substance for researcher: a review. World J Pharm Pharm Sci 4(07):547–573

62. Kafi Z, Cheshomi H, Gholami O (2018) 7-Isopenthenyloxycoumarin, arctigenin, and hesperidin modify myeloid cell leukemia Type-1 (Mcl-1) gene expression by Hormesis in K562 cell line. Dose Response 16(3):1559325818796014
63. Brew CT, Aronchik I, Kosco K et al (2009) Indole-3-carbinol inhibits MDA-MB-231 breast cancer cell motility and induces stress fibers and focal adhesion formation by activation of Rho kinase activity. Int J Cancer 124(10):2294–2302
64. Wiseman H (2000) The therapeutic potential of phytoestrogens. Expert Opin Investig Drugs 9 (8):1829–1840
65. Guo J, Xiao BX, Liu DH et al (2004) Biphasic effect of daidzein on cell growth of human colon cancer cells. Food Chem Toxicol 42(10):1641–1646
66. Taddeo VA, Epifano F, Preziuso F et al (2019) HPLC analysis and skin whitening effects of umbelliprenin-containing extracts of anethum graveolens, pimpinella anisum, and ferulago campestris. Molecules 24(3):501
67. Saha S, Sadhukhan P, Sil PC (2014) Genistein: a phytoestrogen with multifaceted therapeutic properties. Mini Rev Med Chem 14(11):920–940
68. Zava DT, Blen M, Duwe G (1997) Estrogenic activity of natural and synthetic estrogens in human breast cancer cells in culture. Environ Health Perspect 105(suppl 3):637–645
69. Tamir S, Eizenberg M, Somjen D et al (2001) Estrogen-like activity of glabrene and other constituents isolated from licorice root. J Steroid Biochem Mol Biol 78(3):291–298
70. Barch MJ, Knutsen T, Spurbeck JL (1997) The AGT cytogenetics laboratory manual. 3rd edn Philadelphia, PA Lippincott-Raven 355–359
71. Beger RD, Dunn W, Schmidt MA et al (2016) Metabolomics enables precision medicine:"a white paper, community perspective". Metabolomics 12(9):149
72. Jäger W, Gruber A, Giessrigl B et al (2011) Metabolomic analysis of resveratrol-induced effects in the human breast cancer cell lines MCF-7 and MDA-MB-231. OMICS 15(1–2):9–14
73. National Toxicology Program (2008) Toxicology and carcinogenesis studies of genistein (Cas no. 446–72-0) in Sprague-Dawley rats (feed study). National Toxicology Program technical report series (545). NIH, Bethesda, MD, pp 1–240
74. Lamartiniere CA, Wang J, Smith-Johnson M et al (2002) Daidzein: bioavailability, potential for reproductive toxicity, and breast cancer chemoprevention in female rats. Toxicol Sci 65 (2):228–238
75. Galati G, O'Brien PJ (2004) Potential toxicity of flavonoids and other dietary phenolics: significance for their chemopreventive and anticancer properties. Free Radic Biol Med 37 (3):287–303
76. Dunnick JK, Hailey JR (1992) Toxicity and carcinogenicity studies of quercetin, a natural component of foods. Toxicol Sci 19(3):423–431
77. Karimabad MN, Falahati-Pour SK, Hassanshahi G et al (2017) WITHDRAWN: the anti-cancer properties in parallel with toxic effects of indole-3-carbinol derivatives. Immunol Lett S0165-2478(17):30138–30134
78. Fatima N, Nayeem N (2016) Toxic effects as a result of herbal medicine intake. In: Toxicology - new aspects to this scientific conundrum. IntechOpen, London, pp 193–204
79. Singh CK, Ndiaye MA, Ahmad N (2015) Resveratrol and cancer: challenges for clinical translation. Biochim Biophys Acta Mol basis Dis 1852(6):1178–1185
80. Nam JS, Sharma AR, Nguyen LT et al (2016) Application of bioactive quercetin in oncotherapy: from nutrition to nanomedicine. Molecules 21(1):108
81. Bishayee A (2017) Natural products for cancer prevention and therapy: progress, pitfalls and promise. Multidiscip Digit Publ Instit Proc 1(10):982

Potential Pharmacotherapeutic Phytochemicals from Zingiberaceae for Cancer Prevention

10

Indrani Manna, Debalina Das, Sejuty Mondal, and Maumita Bandyopadhyay

Abstract

Cancer, one of the most nefarious maladies, is set to affect one in five of the global population soon. Aberrant uncontrolled cell divisions, proliferation and metastasis are hallmarks of cancer. For quite some time now, numerous cancer prevention and treatment strategies have been formulated with a capricious investment of wealth and resources. The state-of-the-art treatment procedures rely on surgeries, radiation therapies, stem cell induction in conjunction with chemotherapy, immunotherapy and hormonal therapeutics. Yet, these combined treatments are not foolproof often leading to secondary health risks, unspecific outcomes and toxicity. Plant extracts have been used to prevent and cure cancerous growth since times immemorial. In the traditional Indian pharmacopoeia, many phytochemical extracts are listed as potent pharmacotherapeutics against cancer. Zingiberaceae, one of the largest monocot families with a centre of diversity in India, is a promising source of many anti-cancerous, anti-proliferative compounds, attributable to its high polyphenol and flavonoid contents. Principal phytochemicals include curcumin, curcumol, kaempferol, zerumbone, apigenin, galangin, 6-gingerol and 8-gingerol. These compounds are reportedly effective against human colorectal, cervical, breast, lung, ovarian, gastric and liver cancers. Interestingly, the modus-operandi of each compound against cancer cells is unique: curcumin and curcumol reportedly induced apoptosis via p53 regulation and accumulation of ROS/oxidative stress or by modulation of MAPK pathway and inhibition of NF-κB; kaempferol inhibited angiogenesis by suppressing ERK-NFκB-cMyc-p21-VEGF pathway, while apigenin modulated signalling pathways that include PI3K/AKT, MAPK/ERK, JAK/STAT, NF-κB and

I. Manna · D. Das · S. Mondal · M. Bandyopadhyay (✉)
Plant Molecular Cytogenetics Laboratory, Centre of Advanced Study, Department of Botany, Ballygunge Science College, University of Calcutta, Kolkata, India
e-mail: mbbot@caluniv.ac.in

M. Kumar et al. (eds.), *Pharmacotherapeutic Botanicals for Cancer Chemoprevention*, https://doi.org/10.1007/978-981-15-5999-0_10

Wnt/β-catenin pathways and zerumbone caused apoptosis by expression of pro-apoptotic proteins like Bax via cytochrome-c dependent caspase activation, simultaneously decreasing levels of anti-apoptotic proteins like Bcl2. These phytochemicals are effective in cancer cell lines resistant to chemotherapeutic drugs like cisplatin and 5-fluorouracil. Plant-based compounds offer flexibility of usage and diversity of action, affording recourse to most of the woes left behind by systematic and commercial chemical drugs. In this context, the present chapter will thoroughly look into the pros and cons of using phytochemicals of Zingiberaceae on various cancer cell lines, delving into their mode of action, potential side effects, discussing how far research has progressed and what the immediate future holds for us.

Keywords

Cancer · Zingiberaceae · Phytochemicals · Pharmacotherapeutics · Treatment

Abbreviations

AP 1	Activator protein 1
cAMP	Cyclic adenosine monophosphate
COX	Cyclooxygenase
CREB	Cyclic AMP response element binding
EGCG	Epigallocatechin gallate
EGFR	Epidermal growth factor receptor
ERE	Oestrogen responsive element
ERK	Extracellular signal-regulated kinases
FAK	Focal adhesion kinase
FDA	Food and Drug Administration
FOXO	Forkhead box O
FU	Fluorouracil
GSK	Glycogen synthase kinase
HDAC	Histone deacetylases
HIF	Hypoxia-inducible factor
HNSCC	Head and neck squamous cell carcinoma
IL	Interleukin
JAK	Janus kinase
MAPK	Mitogen-activated protein kinase
MMP	Matrix metalloproteinase
mTOR	Mammalian target of rapamycin
NF κB	Nuclear factor-κB
NRF	Nuclear factor erythroid 2-related factor
NSCLC	Non-small cell lung carcinoma
PARP	Poly ADP ribose polymerase
PI3K	Phosphatidylinositol 3-kinase

PTEN	Phosphatase and tensin homolog gene
ROCK	Rho-associated protein kinase
ROS	Reactive Oxygen Species
STAT	Signal transducer and activator of transcription
TGF	Transforming growth factor
TNF	Tumour necrosis factor
uPA	Urokinase plasminogen activator
VEGF	Vascular endothelial growth factor

10.1 Introduction

Cancer is the second largest non-communicable killer worldwide after cardiovascular diseases [1]. In developed and developing countries cancer has emerged as a major concern to public health with high incidence of morbidity [2]. With the third highest number of deaths by cancer registered globally per year, number of cancer patients in India is predicted to increase by a further 25% by the year 2020 (https://economictimes.indiatimes.com/industry/healthcare/biotech/pharmaceuticals/indias-share-in-global-herbal-medicinal-market-just-0-5-government/articleshow/55498419.cms).The National Cancer Prevention and Research Institute confirmed that around 1.7 million new patients would be reported in India by early 2020s, killing at least 0.8 million of those affected, at the rate of around 1300 lives lost per day according to a recent report (http://www.vims.ac.in/blog/cancer-treatment-in-india/. Accessed 07 Jan 2020). While breast cancer is most common in women, oral and lung cancer kills most cancer afflicted men in India (http://ncdirindia.org/NCRP/ALL_NCRP_REPORTS/PBCR_REPORT_2012_2014/ALL_CONTENT/PDF_Printed_Version/Chapter1_Printed.pdf. Accessed 07 Jan 2020). Against the worldwide trend, in India more women are affected by cancer than men (https://www.bbc.com/news/world-asia-india-43539369. Accessed 07 Jan 2020). Recent advancements in terms of cancer drugs and therapeutics have seen a ginormous growth and development of target oriented drugs have opened novel dimensions in cancer treatment but even the most advanced strategies are often rendered ineffective during metastasis [3]. For the last five decades, chemotherapy has been the most common treatment module for cancer patients, though more often than not, these agents cause multiple toxicities in the already cancer-ravaged weakened patients [4], inciting a range of problems like renal failures, cardiovascular problems, myelotoxicity and vasospasm, pulmonary problems, immunosuppression and alopecia [5]. Most of the common and effective chemotherapeutic drugs, like 5-fluorouracil, doxorubicin, bleomycin and cyclophosphamide, cause one or more of these above problems [6]. As recourse to such plight, multiple plant-based products have gained favour of doctors and patients alike for their promise in controlling cancerous growth, as evident in many recent studies. Interestingly, though people of different ethnicities around the world, especially in Asia, Africa and America, have used plants for curing different diseases since times immemorial, in recent times plant-based remedies have found newer proclivity, their usages having increased manifold [7, 8].

Plants have retained prominent status in human civilizations for thousands of years as chief sources of economic, industrial and commercial products. Ayurveda, the ancient Indian system of healing and well-being, is based entirely on plant products [8]. Similar plant-based medicines are prevalent in many Asian and African cultures too. Compounds extracted from terrestrial plants had long been known for their anti-cancerous properties, especially those which are now known to be rich in polyphenols, flavonoids, brassinosteroids, etc. [9]. According to a recent report, approximately 50–60% patients suffering from various cancers in the USA rely on and use plant-based compounds along with prescription drugs and conventional therapies for efficient disease mitigation [10, 11].

The most common mechanism of action of these plant-based compounds includes their effects on expression of P53 protein, NF-κB expression and beginning of apoptosis cascades, reduction of cyclin-dependent proteins like P21 and P27 expression, interfering with or inhibiting pathways like PI3K/Akt/mTOR and associated biochemical changes like reduced acid phosphatase levels and lipid peroxidation, culminating in checking cell cycle proliferation of different cancers [12].

Compounds like curcumin from turmeric, polyphenols from tea, gingerol from ginger, genistein from soybean, resveratrol from red grapes, sulforaphane and isothiocyanates from cruciferous vegetables, lycopene from tomato as well as rosmarinic acid and apigenin from members of Umbelliferae are the most sought after for anti-cancer drug development [11, 13, 14]. While traditional medicinal systems have used plant extracts for cancer treatment for a long time, the first plant metabolite to be adopted for cancer therapeutics was taxol [15].

The search for safer, selectively toxic anti-cancer plant-based drug has been a long and exhaustive one. While many metabolites, like taxol, campothecin, vinca alkaloids, were approved by FDA, USA, for human use and are routinely used as chemotherapeutic agents, they too can induce severe side effects ranging from cardiomyopathy to neurotoxicity. In fact, most anti-cancer drugs developed so far have potent cytotoxicity that would hamper a wide range of cells. In the recent years, with multiple developments, specific agents have been developed that target only tumorigenic pathways and cellular checkpoints. However, presently, many precision drugs have been designed to meet the particular needs of a patient [16]. Chemically, the secondary metabolites exuded by plants have been a wholesome source of anti-tumorigenic, anti-neoplastic compounds. The most common chemotherapeutics, namely taxanes, vinca alkaloids and anti-microtubule compounds like camptothecin and podophyllotoxins, are all originally extracted from various plants [17].

Chemopreventive compounds have been recommended by health care professionals in the recent years because of their manifold health benefits. Figure 10.1 shows a simplified diagram showing types of cancer therapies that can be achieved through plant products. With the rise in so many lifestyle disorders, use of phyto-nutraceuticals has gained much impetus. Compounds like resveratrol, genistein, eriocitrin, apigenin, rosmarine, piperine, andrographolide are some plant derived preventive compounds found mostly in leafy greens and plant-based food items that are being used extensively all across the globe [18–20].Through detailed analyses and trials, many new plant-based compounds have gained popularity, and

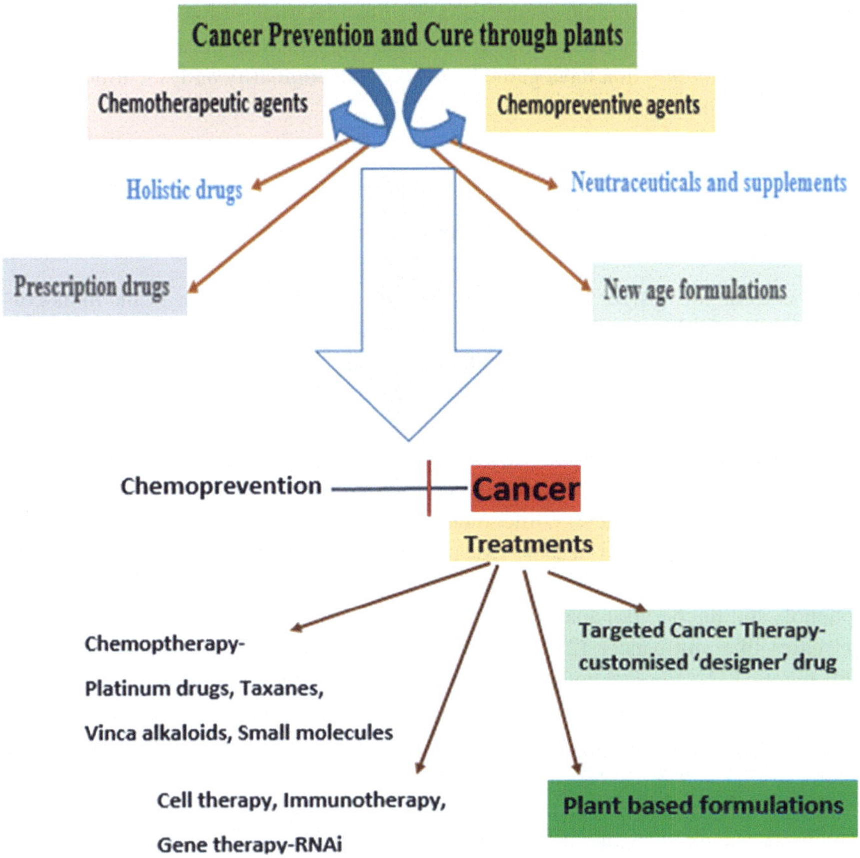

Fig. 10.1 Schematic diagram showing types of cancer therapies based on plant-based products

the global market is around $70 billion; India's share in it is, however, minute, only around 0.5% [3]. Many novel anti-cancer products are now being used in conjunction with other known drugs to better their effectiveness [21, 22]. India has a rich diverse reserve of endemic flora that has long been used as a good repertoire of many indigenous compounds with widespread health benefits.

Zingiberaceae, perhaps the most diverse family in India, has a plush variety of various chemical compounds that could potentially be used as miracle drugs to cure multiple ailments and health conditions. From diabetes to coronary dysfunction to cancer prevention and therapy, this family can provide an elixir for all. Yet it is one of the most underutilized, overlooked family, the main incentive has been put on the use of few rhizomes as herbs and spices. Few compounds like curcumin, gingerol, zerumbone and kaempferol have received both praises and flack for their astounding health benefits, but were soon relegated to obscurity. India already fares poorly when it comes to reaching out, popularizing and marketing its plant products for holistic purposes. Compared to the wide market availability and popularity of Chinese traditional medicines, Indian products have remained underutilized. In this review, the authors would try to expound and illustrate such underutilized plants of Zingiberaceae family that have gained considerable popularity and scientific interest in the recent years.

10.2 Zingiberaceae: The Wonder Family

Zingiberaceae the 'ginger family' belongs to the order Zingiberales of the monocots and is one of the largest plant families [23], comprising of 52 genera and more than 1200 species worldwide [24]. India has one of the richest diversities of Zingiberaceae, with 20 genera and more than 200 species, including a host of endemic taxa [25]. Members of Zingiberaceae are well known for their medicinal value since times immemorial. A large variety of phytochemicals obtained from different parts of these plants possess potent anti-diabetic, anti-tumour, antioxidant and anti-cancer properties. While members of this family are characterized by presence of rhizomes and in most cases aroma in their leaves [26], chemically they are distinguished by high phenolic and flavonoid contents [27]. Phenolic compounds (phenols and flavonoids) apart from being potent antioxidants have other biological activities like anti-cancer [28, 29], anti-inflammatory [30], anti-diabetic [31] and anti-pyretic [32]. Very recently, anti-obese property of phenolic compounds was also established [33]. The most important genera of medicinal importance from this family are *Alpinia*, *Curcuma* and *Zingiber* [34, 35]. Figure 10.2 shows the diverse nature of this family via few distinct representatives.

Medicinally important phytochemicals from rhizomes and leaves of Zingiberaceae include galangin, apigenin, kaempferol, acacetin, quercetin, alpinetin, rutin, oxyphyllacinol, luteolin [36], curcumin, curcumenol, cineole, pinocembrin, cinnamic-acid, coumaric acid, eugenol, curdione, limonene, cuminyl-alcohol, turmerone, arturmerone, germacrone, ar-curcumene [37], 6-gingerol, 8-gingerol, 10-gingerol, shogaols like 6-shogaol, 8-shogaol, kaempferol, zerumbone, zingerone, zingiberene, cardamonin, α-zingiberene, β-bisabolene, β-sesquiphellandrene, β-bisabolene, β-phellandrene, kaempferol 3-glucuronide, quercetin, sabiene, 3-glucuronide, quercetin 3-glucoside, myricetin, beta-sitosterol, proglumide, convallatoxin, osthol [38]. Out of this array, the principal phytochemicals with reported anti-cancer properties are curcumin, apigenin, galangin, alpinetin, 6-gingerol, 8-gingerol, 10-gingerol, 6-shogaol, 8-shogaol, zerumbone, kaempferol, zingiberene, zingerone and cardamonin. However, no other secondary metabolite of Zingiberaceae can even come close to the scientific attention that curcumin has merited till date. Figure 10.2 shows some of representatives depicting the morphological diversity in this family.

The following section focuses on the different phenolics and flavonoids found in various members of Zingiberaceae that have potent anti-cancer activities.

Fig. 10.2 Diversity in Zingiberaceae family—few representatives

10.3 Chemical Nature of Medicinally Important Secondary Metabolites of Zingiberaceae

Curcuma longa (L.), the golden spice of India, is one of the most important species of the genus for its utilities as medicine, spice, food, cosmetics, dye, along with myriad cultural and spiritual importance in Asian countries. Active constituents of turmeric are Curcuminoids, diarylheptanoid or diphenylheptanoid flavonoids and include curcumin, demethoxycurcumin, bisdemethoxycurcumin and cyclocurcumin [39, 40]. Curcumin ($C_{21}H_{20}O_5$), chemically a diferuloyl methane or 1,6-heptadiene-3,5-dione-1,7-bis(4-hydroxy-3-methoxyphenyl)-(1E,6E), was isolated in 1815, and is the principle compound of curcuminoids [41]. Polyphenolic curcuminoids impart the characteristic yellow colour to rhizomes and extracts typically contain 71.5% curcumin (curcumin I), 19.4% demethoxycurcumin (curcumin II) and 9.1% bisdemethoxycurcumin (curcumin III) [42–44]. Apigenin (4′,5,7-trihydroxyflavone) is a dietary flavonoid usually synthesized naturally in rhizomes and leaves of different *Alpinia* species [45]. Galangin (3,5,7-trihydroxyflavone) is a naturally active flavonoid, present in high concentrations in propolis and roots of *Alpinia officinarum* [45]. Alpinetin (ALP,7-hydroxy-5-methoxyflavanone) is a medicinally important plant flavonoid isolated from *Alpinia katsumadai* [46]. The pungent phenolic substances from *Zingiber* species, gingerols and shogaols are generally extracted from *Zingiber officinale.* 6-gingerol (1-[4′-hydroxy-3′-methoxyphenyl]-5-hydroxy-3-decanone) is the most potent anti-cancer compound, while 8-gingerol [5S)-5-hydroxy-1-(4-hydroxy-3-methoxyphenyl) dodecan-3-one], 10-gingerol [(5S)-5-ethoxy-1-(4-hydroxy-3-methoxyphenyl)tetradecan-3-one], 6-shogaol [1-(3,4-dimethoxyphenyl)dec-4-en-3-one], 8-shogaol [1-(4-hydroxy-3-methoxyphenyl)dodec-4-en-3-one] are also promising with respect to cancer inhibition [47]. Zerumbone (2,6,9,9-tetramethylcycloundeca-2,6,10-trien-1-one), a cyclic sesquiterpene, is the main constituent of *Zingiber zerumbet* [48], which has garnered extensive attention in the recent decade for anti-cancer activities. Zingiberene (2-methyl-5-[(2S)-6-methylhept-5-en-2-yl]cyclohexa-1,3-diene) is a sesquiterpene hydrocarbon, isolated from *Zingiber officinale* [49], as is Zingerone (4-(4-hydroxy-3-methoxyphenyl)butan-2-one), a phenolic alkanone in nature [50]. Kaempferol [3,5,7-trihydroxy-2-(4-hydroxyphenyl)chromen-4-one] is a natural dietary flavonol derived from the rhizome of *Kaempferia galanga* L. [51]. Cardamonin [1-(2,4-dihydroxy-6-methoxyphenyl)-3-phenylprop-2-en-1-one], a chalcone isolated from *Alpinia katsumadai,* is also an emerging anti-cancer phytochemical from Zingiberaceae [52].

The list of such medicinally important secondary metabolites from Zingiberaceae is literally unending with newer ones being characterized regularly. However, an exhaustive table detailing all the reported anti-cancer metabolites of Zingiberaceae, their chemical structures as well as reported activities has been compiled (Table 10.1). The following section focuses on the mode of action of some of the most important compounds that have been elucidated with rigorous scientific research spanning over decades as depicted in Fig. 10.3.

Table 10.1 Different phytochemicals from Zingiberaceae family with their potent anti-cancer activity

Name of phytochemical and source plant	Type of cancer and cell line/ mouse model	Therapeutic effect	Mechanism	Reference
Curcumin ($C_{21}H_{20}O_6$) 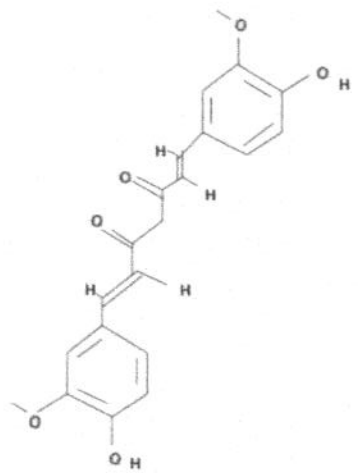 *Curcuma longa*	Gastric cancer BGC823 and SGC7901 MGC803	Decreased proliferation and inducing apoptosis Inhibition of proliferation	Modulating expression of microRNA 33b (mir-33b) Induction and apoptosis via the mitochondrial pathway	[60, 267]
	Human papillary thyroid carcinoma BCPAP cell line	Papillary thyroid cancer cell invasion and metastasis arrested	Suppression of the TGF/Smad2/3 pathway	[315]
	Head neck squamous cell carcinoma (HNSCC) Fadu cell line (hypopharyngeal cancer) and Cal27 cell line (tongue cancer)	Reduction of cell proliferation and imposes cytotoxic effect	Pro-apoptotic Bik cascade over-expression, decreased pro-survival signalling by AKT and NF-κb	[75]
	Breast cancer cell lines MCF-7 and MDAMB231	Inducing breast cancer apoptosis Inducing cell cycle arrest at the G2/M phase and late S-phase	Regulate the apoptosis-related proteins Regulating Spindle-related signalling pathways	[53, 125]
	Acute myeloid leukaemia Human trial	Attenuated the proliferation and clonogenicity of leukemic cells	Down-regulation of wt1 protein through over-expression of mir-15a/16-1	[61]
	Colorectal cancer Orthotopic mouse model	Inhibition of tumour growth and angiogenesis	Inhibition of NF-κb activation	[79]
	Bone marrow cancer Human multiple myeloma cells U266	Modulated tumorigenic proteins	Inhibition of both constitutive and inducible STAT3 activation	[109]
	Brain cancer LN229	Inhibition of growth, invasion and metastasis	Elevation of either cytostatic and/or cytotoxic effects of some EGFR kinase inhibitors through EGFR pathway	[176]

(continued)

Table 10.1 (continued)

Name of phytochemical and source plant	Type of cancer and cell line/ mouse model	Therapeutic effect	Mechanism	Reference
	Hepatocellular carcinoma Mouse model	concentration- and time-dependent inhibition of migration and invasion	Signalling cascades like EGFR and CAV-1 pathways repressed	[181]
	Oesophageal cancer Raji cells	Inhibition of growth of cancer cells	Notch 1 down-regulation through repression of vital complex proteins	[226]
	Lung 95D and A549	Inhibition of NSCLC proliferation and invasion	Induction of G0/G1cell cycle arrest through MTA1-mediated deactivation of Wnt/β-catenin pathway	[259]
	Prostrate LNCaP	Decreased proliferation, colony formation and cellular motility; enhanced cell–cell aggregation	Interference between the Wnt/β-catenin signalling pathway and androgen receptors	[257]
	Ovarian OVAR3, SKOV3, HO-8910 and A2780	Reduction in mean tumour growth, decreased proliferation and lessened micro-vessel density; significant increase in apoptosis	Induction of apoptosis by down-regulation of HSP27 and HSP70 pathways	[66]
	Oesophageal squamous cell carcinoma cell line EC 109	Apoptosis	Suppression of the PTEN/PI3K/AKT pathway	[132]
	Human leiomyosarcoma	Cancer cell repression	Modulation cross talk between signal cascades of autophagy and apoptosis	[137]
	Hepatocellular carcinoma/liver cancer SK-Hep-1 Huh7, Hep3B, HepG2 Bel7402, SGC7901, HL60 HA22T	Inhibition of both migration as well as invasion, increase in apoptosis	Decrease in Nf-kβ activation, as well as a lowering of cox-2 levels	[87]
	Pancreatic cancer (Xenograft study) Female mice model	Reduced tumour size, and inhibited tumour angiogenesis	Decrease in the expression of CD31 in addition to that of VEGF and IL-8 indicating suppression of pancreatic carcinoma growth	[83]
	Blood cancer Chronic leukaemia K562	Results in growth arrest and apoptosis	Down-regulation of JAK-STAT3 pathway by suppression of JAK2, cyclin d1 and v-src gene expression	[117]

Curcumol ($C_{15}H_{24}O_2$) 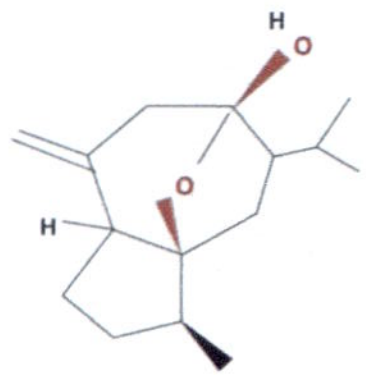 *Curcuma longa*	Nasopharyngeal—CNE-2, NPC 5-8F Breast—MDA-MB-231 Bladder—EJ, T24 Ovarian—SKOV3 Liver—HSC-T6, HepG2 Gastric—AGS MGC—803 Colorectal—LoVo, HCT-116, SW-480	Cell cycle arrest and apoptosis, inhibition of cell proliferation and invasion	Down-regulation of various signalling pathways, lowering of mitochondrial membrane potential, decrease in levels of MMP-9, decrease in levels of cyclins and CDKs	[300]
6-gingerol ($C_{19}H_{30}O_4$) 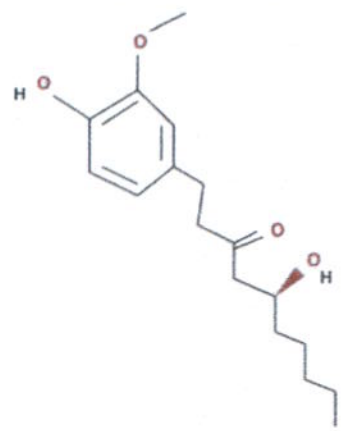 *Zingiber officinale*	Pancreatic cancer cell BxPC-3 and HPAC cell lines	Cell cycle arrest and cancer cell death	Cyclin A and Cyclin-dependent kinase (Cdk) expression were decreased p53 expression was decreased down-regulation of Rb phosphorylation up regulation of p21	[306]
	Liver cancer AH109A Hep3B PLC/PRF/5 Hep G2	Cell cycle arrest at S-phase Anti-invasive action Anti-invasive action Inhibition of metastasis Improvement of tumourmicroenvironment Apoptosis of cancer cells	Suppresses the production of Hepatocyte Growth Factor (HGF) Inhibition of phosphorylation of mitogen-activated protein kinase (MAPK) Down-regulation of PI3K/Akt signalling Inhibition of translocation of NF-κB Down-regulation of STAT3 pathway Decrease in Microvascular Structural Entropy (MSE) via p-VEGFR2/VEcadherin/β-catenin/actin pathway Inhibition of de novo fatty acid synthesis via inhibition of carnitine palmitoyltransferase-1 activity	[100, 205, 304, 334]

(continued)

Table 10.1 (continued)

Name of phytochemical and source plant	Type of cancer and cell line/ mouse model	Therapeutic effect	Mechanism	Reference
	Colorectal cancer LoVo HCT-116, SW480, HT-29, LoVo and Caco-2	Cell cycle arrest at the G2/M phase Cell cycle arrest and apoptosis	Trigger the G2/M cell cycle Arrest via down-regulation of cyclin A, CDK2, cyclin B1 and CDK1 Suppressed cyclin D1 expression activation of PKCε and GSK3 pathways Suppresses β-catenin signalling pathway	[100, 303, 305]
	Cervical cancer HeLa, CaSki, SiHa HeLa	Decreased proliferation and apoptosis; cell cycle arrest at G2/M phase Apoptosis (through shrinkage, detachment and membrane blebbing) and cell cycle arrest	ROS generation and p53 reactivation via proteasome inhibition Up regulation of p21-p53/p21 mediated cell cycle arrest Suppression of cyclin D1 Increase in caspase 3/7 activity Down-regulation of PIK3/Akt/ mTOR	[151, 207]
	Oral cancer OSCC, KB, SCC4	Apoptosis (through shrinkage, detachment and membrane blebbing) and cell cycle arrest	Suppression of cyclin D1 Increase in caspase 3/7 activity Down-regulation of PIK3/Akt/ mTOR	[151]
	Retinoblastoma cancer cells RB355 Cell line	Apoptosis and cell cycle arrest	Upregulation of PI3K/Akt signalling pathway	[150]
	Breast cancer MCF-7	Inhibition of metastasis Improvement of tumour microenvironment	Decrease in micro vascular Structural Entropy (MSE) via p-VEGFR2/VEcadherin/β-catenin/ actin pathway	[334]

	Skin cancer Mouse skin Mouse skin	Inhibited skin tumour promotion Inhibits mouse skin tumour promotion and anchorage-independent growth of epidermal growth factor stimulated cultured mouse epidermal cells	Down-regulation of TPA-induced epidermal ornithine decarboxylase activity suppressed NF-κB DNA binding activity in mouse skin Activation of p38 MAPK	[102, 103]
	Bone cancer 143B, MG63	Inhibition of cell proliferation and apoptosis	Activation of AMP-activated protein kinase (AMPK) Activation of caspase cascades Alteration of BCl2 protein	[275]
6-shogaol ($C_{17}H_{24}O_3$) 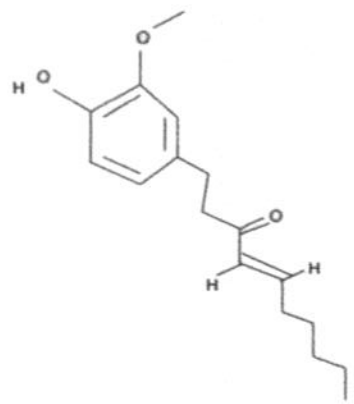 *Zingiber officinale*	Lung cancer A549 A549	Inhibition of proliferation of cancer cells Pro-apoptotic activity induced Induction of autophagy	Modulation of STAT 3 and MAPK signalling cascade Inhibition of the AKT/mTOR pathway	[118, 154]
	Breast cancer MCF-7 and MDA-MB-231 And cancer stem cells (CSC) MDA-MB-468, MDA-MB-231, MCF-7, and T47D	Induction of autophagy in cancer cells Inhibition of cell proliferation	Modulation of notch signalling pathway Activation of peroxisomal proliferator activated receptor c (PPARc) and suppression of NF-κB signalling	[230, 281]
	Liver cancer Hep3B HepG2	Anti-invasive action Induction of apoptosis	Inhibition of phosphorylation of mitogen-activated protein kinase (MAPK) Down-regulation of PI3K/Akt signalling Inhibition of translocation of NF-κB Down-regulation of STAT3 pathway Caspase activation and endoplasmic reticulum (ER) stress signalling	[100, 280]

(continued)

Table 10.1 (continued)

Name of phytochemical and source plant	Type of cancer and cell line/ mouse model	Therapeutic effect	Mechanism	Reference
	Squamous carcinoma SCC4	Inhibition of proliferation of cancer cells Pro-apoptotic activity induced	Modulation of STAT 3 and MAPK signalling cascade	[118]
	Colorectal cancer and colon cancer COLO 205 HT29 and HCT116	Induction of apoptosis Inhibition of cell proliferation	Fas activation, and modulation of GADD153 gene expression Mitochondrial dysfunction (loss of mitochondrial membrane potential), resulting in cytochrome-c release, Caspases activation and apoptotic death Activation of peroxisomal proliferator activated receptor c (PPARc) and suppression of NF-κB signalling	[106, 281]
	Prostate cancer DU145	Inhibition of proliferation of cancer cells Pro-apoptotic activity induced	Modulation of STAT 3 and MAPK signalling cascade	[118]
8-shogaol ($C_{19}H_{28}O_3$) 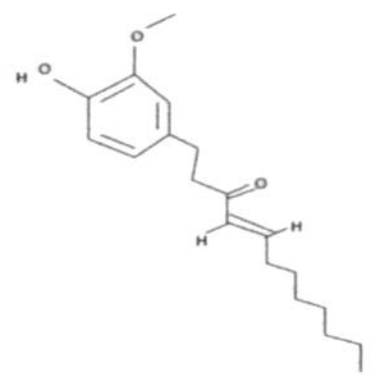 *Zingiber officinale*	Blood cancer Leukaemia cells	Induction of apoptosis	Reactive oxygen species (ROS) generation, glutathione depletion and caspase activation	[223]

10-gingerol ($C_{21}H_{34}O_4$) 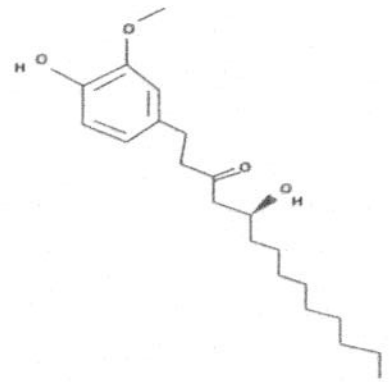 *Zingiber officinale*	Breast cancer MDA-MB-231TNBC	Inhibition of cell proliferation and cell invasion Induction of apoptosis	Suppression of Akt and p38MAPK activity Mitochondrial dysfunction (loss of mitochondrial membrane potential), resulting in cytochrome-c release, Caspases activation	[152, 277]
	Colon cancer HCT 116	Induction of mitochondrial apoptosis	Down-regulation of MAPK signalling pathway	[209]
	Cervical cancer HeLa	Cell cycle arrest and apoptosis	Decrease in phosphorylation of Akt and down-regulation of PIK3/Akt/mTOR pathway and suppression of NF-κB activation	[104]
	Ovarian cancer HEY, OVCAR3, and SKOV-3	Cell cycle arrest at G2/M	Combined effects of decreased expression of cyclin B1 and D3	[307]
Alpinetin ($C_{16}H_{14}O_4$) 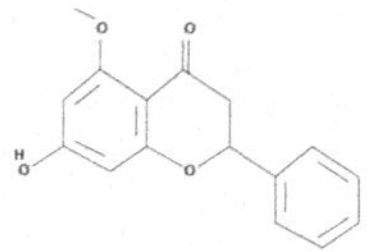 *Alpinia katsumadai*	Oral cancer TCA-8113 and CAL-27	Inhibition of cell proliferation	Through miR-211-5p (micro-RNA) upregulation and Notch Pathway Deactivation	[231]
	Gastric cancer AGS N87	Induction of apoptosis and cell cycle arrest at G2/M phase	By translocation of Bax and triggering mitochondrial pathway of apoptosis	[274]
	Pancreatic cancer BxPC-3	Anti-proliferative activity		[298]
	Lung cancer A549, SK-MES-1, NCI-H292, and A549/cis-diammined dichloridoplatium (CDDP)	Inhibition of lung cancer progression and increased sensitization towards chemotherapeutic drug	Inhibition of the PI3K/Akt signalling pathway	[146]
	Colorectal cancer HT-29	Cell cycle arrest at G0/G1 phase and S-phase and apoptosis	p53 mediated cell cycle arrest and Uridine-cytidine kinase 2 (UCK 2) enzyme inhibition	[314]

(continued)

Table 10.1 (continued)

Name of phytochemical and source plant	Type of cancer and cell line/ mouse model	Therapeutic effect	Mechanism	Reference
	Ovarian cancer SKOV3	Inhibition in proliferation and migration of cancer cells, anti-invasive property	Suppression of STAT 3 signalling pathway	[120]
	Brain cancer Glioma stem cells (GSC) U-87, U-251, and U-373	Inhibition of proliferation of cancer stem cells	By suppression of notch signalling pathway	[232]
	Liver cancer	Supresses proliferation of cancer cells and sensitizes those towards chemotherapeutic agents	By the activation of Mitogen-activated protein kinase kinase-7 (MKK7)	[322]
Galangin ($C_{15}H_{10}O_5$) 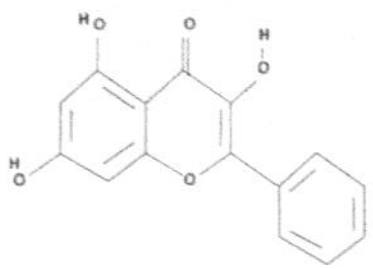 *Alpinia officinarum*	Kidney cancer or renal cell carcinoma 786-0 and Caki-1 A498	Suppresses EMT and induces apoptosis Inhibition of cell migration and invasion	Increased levels of ROS Down-regulation of PI3K/AKT/ mTOR signalling pathways	[317, 318]
	Hepatocellular carcinoma	Induction of apoptosis	Mitochondrial dysfunction	[268]
	Oesophageal carcinoma Eca9706, TE-1, and EC109 nude mice with xenograft tumours	Inhibition of proliferation	Increased intracellular reactive oxygen species (ROS) levels	[289]
	Cervical cancer HeLa	Induces cell death of cancer cells	Modulating the expression of glyoxalase-1 and Nrf-2	[203]
	Laryngeal carcinoma TU212 and M4e	Suppressed proliferation	Modulation of caspase 3 and AKT signalling pathways	[249]
	Ovarian cancer OVCAR-3 A2780/CP70	Inhibition of angiogenesis	Decrease in levels of AKT	[250]
	Human nasopharyngeal carcinoma	Cell cycle arrest at S-phase	Inhibition of PI3K/AKT signalling pathway	[148]

	Breast cancer MCF-7 and T47D	Induces apoptosis	TRAIL pathway by activating AMPK	[168]
	Human fibrosarcoma HT-1080	Inhibited proliferation	Decrease in MMP-9 expression	[316]
	Brain cancer glioma cells U87, U251 and U87-luciferase and HUVECs	Suppresses EMT and inhibits angiogenesis	Down-regulation of CD44	[96, 251]
	Murine melanoma mice bearing B16F1melanoma tumour	Inhibits cancer	–	[45]
Apigenin ($C_{15}H_{10}O_5$) 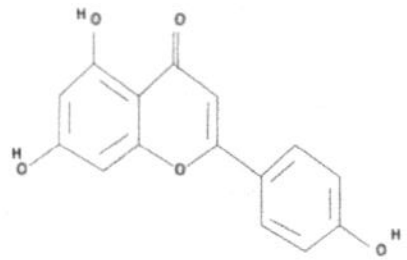 *Alpinia galanga*	Breast cancer MDA-MB-231	Cell cycle arrest at G2/M phase	Histone H3 acetylation leading to p21WAF1/CIP1 expression	[302]
	Thyroid cancer (papillary thyroid cancer) BCPAP cells	Induces autophagy	Increase in acidic vesicular organelles (AVOs) formation with respect to control	[204]
	Colorectal cancer and colon cancer SW480 and HCT15 HCT-116 Cisplatin resistant colon cancer cells	Inhibition of migration and cancer cell proliferation Induces apoptosis Induction of autophagy and programmed cell death Inhibits epithelial-mesenchymal transition	Wnt/β-catenin signalling pathway repression extrinsic and intrinsic pathways of apoptosis stimulated Targeting m-TOR/PI3K/Akt signalling pathway targeting NF-κB/ Snail signalling pathway	[143, 145, 263, 272, 292, 293]
	Liver cancer HepG2	Induction of autophagy and induction of apoptosis	Inhibition of PI3K/AKT/mTOR signalling pathways	[144]
	Brain cancer U87MG and U373MG	Inhibits cancer stem cell like phenotype	Suppression of c-met signalling	[232, 339]
	Bone cancer U2OS and MG63	Inhibition of proliferation and invasion of cancer cells	Suppression of Wnt/β-catenin signalling pathway	[262]
	Skin cancer B16F10 cells in mouse model	Anti-metastasis	Inhibition of STAT 3 signalling pathway	[291]

(continued)

Table 10.1 (continued)

Name of phytochemical and source plant	Type of cancer and cell line/ mouse model	Therapeutic effect	Mechanism	Reference
	Cervical cancer HeLa	Inhibits formation of sphere-like cancer stem cells	Inactivation of casein kinase 2α	[338]
	Oesophageal cancer	Inhibits proliferation of tumour	Inhibits IL-6 transcription	[343]
	Prostate cancer	Inhibits metastasis	Targeting the SPOCK1-snail/slug axis-mediated epithelial-to-mesenchymal transition	[333]
Zerumbone ($C_{15}H_{22}O$) 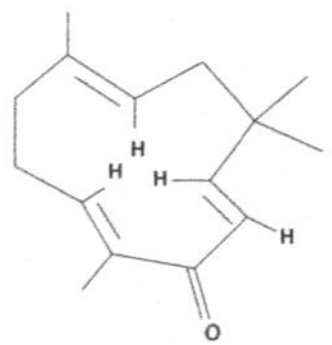 *Zingiber zerumbet*	Prostate cancer PC3 PC-3 and DU-145	Inhibition of growth of cancer cells Increases paclitaxel sensitivity Induces apoptosis and autophagy	Inhibition of JAK2/STAT3 pathway ER stress and mitochondria insult	[123, 286]
	Kidney cancer Xenograft mice model	Inhibits proliferation and induces apoptosis	Suppression of STAT 3 signalling cascade	[124]
	Gastric cancer SGC-7901	Induction of apoptosis	upregulated Bax levels, and triggered intrinsic pathway of apoptosis	[309]
	Cervical cancer SiHa	Induces cell cycle arrest and apoptosis	Production of ROS and a loss of the mitochondrial membrane potential and triggered intrinsic pathway of apoptosis	[215]
	Oral cancer ORL-48 and ORL-115	Reduces motility and proliferation	Down regulates PI3K/ mTOR signalling	[155]
	Non-small cell lung cancer A549	Suppresses cell invasion	Inhibition of FAK/AKT/ROCK pathway	[156, 330, 331]
	Liver cancer Hep G2 Sprague Dawley Rat model	Suppression of angiogenesis Inhibition of proliferation	Inhibition of MMP-9 activity Apoptosis Anti-proliferative and anti-angiogenic effect	[248, 249]

	Colorectal cancer cells and cancer stem cells (CSC) HCT-116 and SW-48 SW480 HCT-116 and SW480	Inhibits EMT Induces cell cycle arrest and has anti-migratory effect Supresses cancer invasion and metastasis	Inhibition of β-catenin pathway through micro-RNA Decreased mitochondrial membrane potential, and activated caspase cascade (caspase 3, caspase 8 and caspase 9) Modulation of FAk/PI3k/NF-κB-uPA Pathway	[210, 260, 323]
	Oesophageal cancer EC-109	Inhibits proliferation and apoptosis	Alteration of Bcl-2 levels	[310]
	Laryngeal cancer Hep-2	Inhibits proliferation and cell cycle arrest	Down-regulation of cyclin D1	[297]
	Breast cancer	Inhibition of cancer proliferation	Targeting β-catenin of the Wnt–β-catenin pathway	[261]
Kaempferol ($C_{15}H_{10}O_6$) 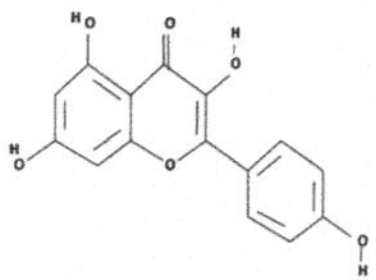 *Kaempferia galanga*	Prostate cancer LNCaP	Apoptosis	Induction of caspase cascade pathway	[283]
	Skin cancer A375	Cell cycle arrest Inhibition of cell migration Induction of apoptosis	Down-regulation of PI3K/AKT/mTOR signalling pathways	[158]
	Lung cancer A549 A549	Induction of EMT Inhibition of epithelial-mesenchymal transition.	Inhibition of Akt1-mediated phosphorylation of Smad 3 Modulating the expression of E-cadherin and vimentin	[330, 331]
	Kidney cancer 786-O	Inhibition of invasion and migration	Downregulation of EFGR pathway	[191]
	Ovarian cancer A2780 CP70	Promotes apoptosis	TRAIL pathway and death receptors	[171]
	MC3T3-E1 cells		Antioxidant upheaval and mitochondrial dysfunction	[216]

(continued)

Table 10.1 (continued)

Name of phytochemical and source plant	Type of cancer and cell line/ mouse model	Therapeutic effect	Mechanism	Reference
	Colorectal cancer cells	Apoptosis	ROS induction, p53 mediated action of p38 and caspase induction	[284]
	Pancreatic cancer	Cancer cell growth and proliferation inhibited	Blocking of EFGR related signal pathway	[186]
	Human fibrosarcoma HT1080	Inhibition of proliferation	Decrease in MMP-9	[316]
	Cervical cancer HeLa SiHa	Increase in apoptosis Induced apoptosis and inhibited proliferation	Down-regulation of PI3K/AKT and telomerase pathways Disruption of mitochondrial membrane potential and the disturbance of intracellular free Ca^{2+} concentration	[157, 284]
	Lymphoma Daudi cells	Anti-proliferative activity	–	[312]
	Breast cancer MDA-MB-231	Suppresses the migration and invasion	Down-regulation of the activities of RhoA and Rac1	[328]
Zingerone ($C_{11}H_{14}O_3$) 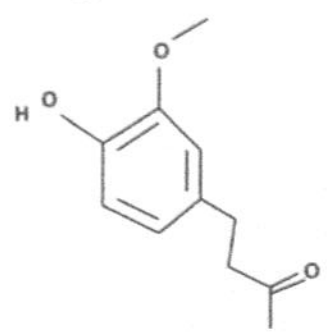 *Zingiber officinale*	Colon cancer HCT-116 Winster rats	Induction of apoptosis – Chemopreventive efficacy	Elevation of ROS Modulation of lipid peroxidation Elevation of ROS	[220–222]
	Breast cancer MCF-7 Experimental rat	Induction of apoptosis	Activation of caspase cascade	[277]
	Liver cancer SNU182 In rat model	Suppresses invasion and migration of cancer cells EMT phenomenon suppression Modulatory effect against cisplatin and radiation induced hepatotoxicity	Inhibition of TGF-β1 pathway Modulation of, p38, MAPK, JNK, ErK1/2 signal transduction cascades	[192, 332]
	Human neuroblastoma cells BALB/c mouse tumour model BE(2)-M17	Cell cycle arrest, inhibition of mitosis	Inhibition of activity of cyclin D1	[313]

	Mouse tumour Renca cells	Suppresses angiogenesis	By inhibition of MMPs	[252]
Zingiberene ($C_{15}H_{24}$) 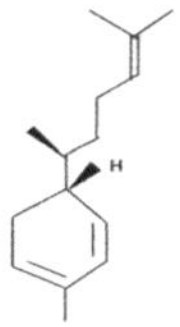 *Zingiber officinale*	Colon cancer HT-29	Induction of autophagy	Suppression of PI3K/AKT/mTOR pathway	[160]
Cardamonin ($C_{16}H_{14}O_4$) 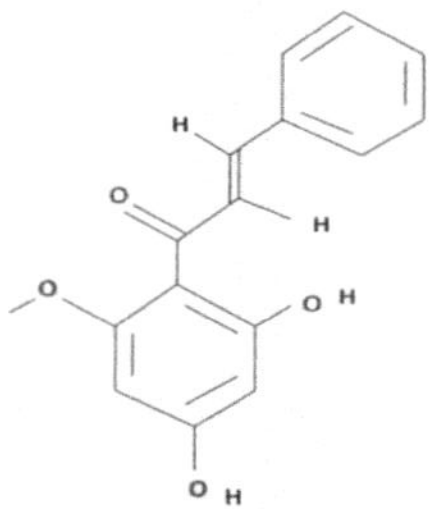 *Alpinia katsumadai*	Colon cancer and colorectal cancer HCT116 and LOVO cells SW480	Induction of autophagy and inhibition of cancer progression Cell cycle arrest at G2/M Anti-proliferative action	Formation of (LC3)-I–LC3-II, an incorporation of monodansylcadaverine (MDC), a marker for the acidic compartment of AVOs Degradation of β-catenin and thereby down-regulation of Wnt/β-catenin pathway	[264, 296]
	Breast cancer MCF-7, MDA-MB-231 and BT-549	Anti-invasive property Reversal of EMT	Down-regulation of Wnt/β-catenin signalling cascades	[265]
	Lung cancer A549 xenograft in mice A549 and HK1 cells	Cell cycle modulation Caspase dependent apoptosis	Suppression of NF-κB and decrease in expression of cyclin D1 Modulator of mTOR pathway	[105, 159]
	Prostate cancer (PC) PC androgen independent DU145 and androgen dependent LNCaP	Inhibits cancer proliferation and metastasis	Modulation of STAT3 pathway	[121]
	Gastric cancer AGS, MGC-803, BGC-823	Inhibition of cancer cell proliferation	Inhibition of STAT3 activity	[122]
	Skin cancer Human metastatic melanoma cell line A375 (CRL-1619)	Inhibition of cancer proliferation	Apoptosis of tumour cells	[219]

10.4 Generalized Mode of Anti-cancer Action of Some Important Secondary Metabolites of Zingiberaceae

The molecular basis of anti-carcinogenic and chemopreventive activities of curcumin is attributed to its effect on several targets including transcription factors, growth regulators, adhesion molecules, apoptotic genes, angiogenesis regulators and cellular signalling molecules. Excitingly, curcumin successfully acts at all three phases of cancer, namely initiation, advancement and development, by targeting critical processes involved in cancer development and progression [53]. The anti-carcinogenic nature of curcumin has been reported in preclinical models of lymphomas, multiple myeloma, leukaemia and brain, pancreatic, gastric and colorectal cancers [54]. Curcumin has been shown to down-regulate production of pro-inflammatory cytokines tumour necrosis factor-α (TNF-α), IL-1β and inhibit transcription factors like Nuclear Factor-κB (NF-κB), Signal Transducer and Activator of Transcription 3 (STAT3) and Activator Protein-1 (AP-1), which regulate signal cascades of genes involved in the pro-inflammatory pathways and protective antioxidant functions [55] that play key roles in cancer development and progression. It also inhibits Specificity Protein 1 (Sp-1) and its housekeeping genes to prevent cancer formation, migration and invasion [56]. Inhibition of downstream gene products like c-myc, Bcl-2, COX-2, NOS, Cyclin D1, TNF-α, interleukins (IL) and MMP-9 demonstrates the anti-proliferative property of curcumin. In addition, curcumin affects a variety of growth factor receptors and cell adhesion molecules involved in tumour growth, angiogenesis and metastasis [57]. Apart from action on STAT3 and NF-κB pathways, curcumin has been shown to inhibit cell proliferation, causing cell cycle arrest and stimulating apoptosis via modulation of other transcription factors, such as AP-1, Erg-1, p53, β-catenin, Notch-1, HIF-1 [58]. Curcumin asserts its anti-tumour activity in cancer cells by altering the deregulated cell cycle via cyclin-dependent, p53-dependent as well as p53-independent pathways. Curcumin has major positive influences on key signal transduction pathways of cell cycle and its effectiveness in animal model systems has made it eligible as a 'multiple edged sword' in combating the deadly disease—cancer [59]. Moreover recent studies show that curcumin can exert its anti-proliferative and pro-apoptotic action by modulating the expression of micro-RNA, like miR-33b, in case of gastric cancer lines BGC823 and SGC7901 [60] and down-regulation of Wt-1 protein through miR15a/16-1 in leukemic cells [61]. Figure 10.4 depicts the mode of action of curcumin.

Though many other metabolites from different members of Zingiberaceae are being investigated for their anti-cancer activities, their modes of action are not as thoroughly researched as curcumin. Induction of apoptosis by generation of oxidative stress is known to be induced by flavonoids. Galangin, 6-gingerol, 6-shogaol, zerumbone and kaempferol all reportedly induce ROS and arrest cell proliferation in different types of cancer. Another potent target of anti-cancer drugs is the mitochondria, as mitochondrial dysfunction brings on apoptosis. Galangin, apigenin, alpenitin, 6-gingerol, 6-shogaol, zerumbone and kaempferol are some of the prominent metabolites with proven capacity of inducing mitochondrial dysfunction in

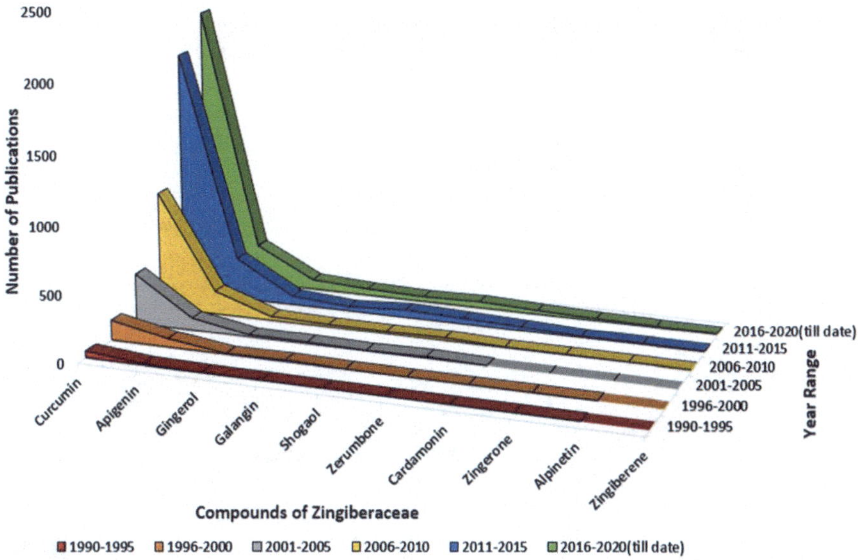

Fig. 10.3 Comparison of the number of research publications pertaining to the different anti-cancerous compounds of Zingiberaceae in the last three decades (data acquired from PubMed Central, accessed on 31 Jan 2020)

different cancer cell lines [46, 47]. Zerumbone and kaempferol are further implicated in the disturbance of the endoplasmic reticulum affecting a number of critical cellular processes that finally culminate in cell death [48, 51]. In some reports, galangin, apigenin, 6-shogaol and zerumbone are shown to interfere with autophagic cascades compromising cancer cell survival and potentiating apoptosis. Along similar lines, many of these compounds are known to modulate different cell cycle checkpoints and ushering in premature cell cycle. Probably the most interesting application of these zingiberaceous metabolites was in compromising cell adhesion and discouraging metastasis, the two most crucial requirements for proliferation of cancer [35].

In the next section, the authors have reviewed the cell signalling cascades and cellular pathways modulating different aspects of cancer induction and progression.

10.5 Cell Signalling Pathways Modulated by Zingiberaceous Metabolites Related to Their Anti-cancer Activities

10.5.1 NF-κB Pathway

Atypical activation of the NF-κB pathway has a significant role in cancer pathogenesis [62–64]. Curcumin can effectively disrupt many of integral activation steps of this pathway, thus promises to be of great use in cancer remediation. Curcumin

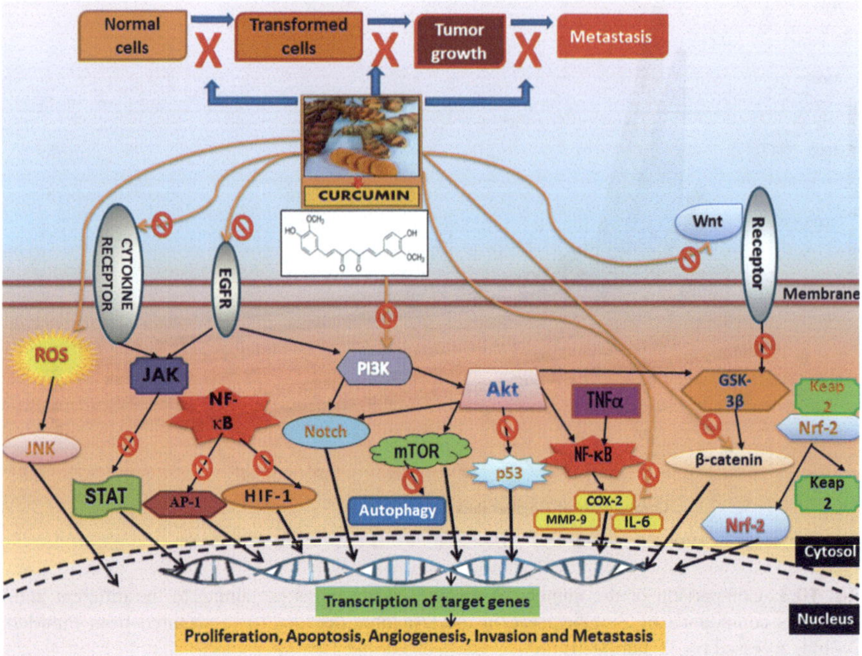

Fig. 10.4 Schematic depiction summarizing the mode of action of curcumin on cancer cells

reportedly blocked TNF-α-induced nuclear translocation of NF-κB, further compromising its DNA binding ability via inhibition of IκBα phosphorylation by down-regulating NF-κB-inducing kinase (NIK) and IκB kinase (IKK), which was followed by degeneration in the most common myeloid leukaemia cell line in human beings [65]. Similarly, curcumin is known to prevent cell proliferation, invasion, metastasis, angiogenesis as well as chemotherapy and radiotherapy resistance of numerous cancers by manifold manipulation of target molecules originally regulated through NF-κB pathway. In colon and ovarian cancer for example, curcumin induced apoptotic behaviour through inhibition of numerous genes namely survivin, BCL-2, specificity protein (Sp) transcription factors (Sp1, Sp3, and Sp4) and Sp-regulated genes, NF-κB (p65 and p50), hepatocyte growth factor receptor (c-MET), and cyclin D1, finally preventing tumorigenesis [53, 66]. On the other hand, in prostate cancer, curcumin promotes deactivation of androgen receptors and androgen receptor-related cofactors encompassing NF-κB as well [67]. Curcumin-induced down-regulation of NF-κB pathway also shows promise in treatment of lung cancer [68]. Hypopharyngeal cancer, one of the most hostile forms of head and neck malignancies with substandard prognosis [69], shows NF-κB playing a crucial link between neoplastic and inflammatory events in epithelial cells [70]. Vageli et al. [71] demonstrated in *in vitro* model that bile-related activation of NF-κB and its transcriptionally activated oncogenic factors can be inhibited by the usage of turmeric supplements (curcumin) in exposed normal human hypopharyngeal cells. Their discovery indicated that selective dietary supplements with anti-inflammatory and anti-apoptotic properties like curcumin can suppress acidic bile-induced oncogenic mRNA phenotype in human hypopharyngeal cells and may be useful for the prevention of extra-oesophageal reflux-related hypopharyngeal neoplasia. The fact that NF-κB inhibition could effectively decrease the expression of genes with cell proliferation or anti-apoptotic function in Head and Neck Squamous Cell Carcinoma (HNSCC) and the role of curcumin in potentiating it was also supported by various reports [72–75]. In breast cancer cells, the survival signalling molecules such as NF-κB play a pivotal role in cell proliferation [76]. It was reported that curcumin was able to inhibit NF-κB expression and toggled many downstream signalling pathways, silencing inflammatory cytokines (CXCL1 and CXCL2) and upregulating expression of matrix metalloproteinase 9 (MMP-9) in breast cancer cell lines [77], as well as repression of urokinase plasminogen activator (uPA), uPA receptor (uPAR), intercellular adhesion molecule 1 (ICAM-1) and chemokine receptor 4 (CXCR4) [78] ultimately leading to inhibition of colorectal cancer as well [79]. Reports reveal that effectiveness of curcumin and its cohorts in preventing breast cancer cell growth and invasion could partially be regulated through downregulation of NF-κB signalling pathways [80] and downregulation of the insulin-like growth factor 1 (IGF-1) signal cascades [81].

Preclinical studies have demonstrated that curcumin exerts anti-cancer effects against the deadly pancreatic cancer (PC) by modulating multiple molecular targets [82]. Curcumin can hinder survival of PC cells, under both in vitro and in vivo conditions, rendering activities of essential factors and modulators of different signalling cascades like COX-2, NF-κB, CD-31, VEGF- and IL-8 useless [83]. In vitro studies have shown potent cytotoxic effects of curcumin on different PC cell lines including MiaPaCa-2, Panc-1, AsPC-1 and BxPC-3. In addition, in vivo studies on PC models have shown that the anti-proliferative effects of curcumin are caused by inhibition of oxidative stress and angiogenesis through induction of apoptosis [84, 85]. Li et al. [83] demonstrated that curcumin down-regulated NF-κB and associated growth control molecules in human pancreatic cells in a time and dose-dependent manner. These effects were accompanied by marked growth inhibition and apoptosis, which was confirmed by other studies as well [86]. In hepatocellular carcinoma liver cancer cell lines SK-Hep-1, Huh7, etc. curcumin induced decreased NF-κB expression and lowering of COX-2 levels which led to inhibition of both migration and invasion of cancer cells [87]. Jutooru et al. [88] showed that curcumin inhibited NF-κB expression and cancer cell growth by down-regulation of the specificity protein Sp1. Tolfenamic acid and dietary spice curcumin co-treatment reportedly enhanced anti-proliferative effect in PC cells through Sp1 suppression, disruption of NF-κB translocation to the nucleus and cell cycle phase distribution [89]. Marquardt et al. [90] evaluated cancer cell depleting potential of NF-κB inhibition in liver cancer achieved by its inhibitor curcumin. Their work demonstrated that blocking NF-κB specifically target cancer stem cell populations and suggest a potential for combined inhibition of NF-κB and HDAC signalling for treatment of liver cancer patients with poor prognosis. This study along with others uncovered the potential of curcumin to diminish growth of difficult to treat hepatocellular cancer [91, 92].

The gamut of activities of other zingiberaceous metabolites echoes the action of curcumin on NF-κB. Clinical administration of extracted apigenin has shown to inhibit NF-κB activation in in vivo TRAMP mice models [93] and blocks IKKα activation as well as suppresses prostate cancer progression [94], however, apigenin did not show any inhibitory effect on NF-κB in A549-non-small cell lung cancer cell lines in human, but it did suppress translocation of NF-κB from to the nucleus [95], similar to curcumin's action as mentioned earlier. Galangin also inhibited activity of nuclear factor kappa B (NF-κB) and its binding activity to activator protein1 (AP-1) [96]. NF-κB inhibition and concomitant inhibition of MEK-ERK signalling occurred upon treatment of ovarian cancer cells with kaempferol [97, 98]. 6-Gingerol (6G), on the other hand, reportedly inhibited NF-κB activity and hence suppressed NF-κB pathway in liver cells [99, 100] and in cervical cancer cells as well [101]. Similar observations were reported in mouse skin where 6G application suppressed NF-κB DNA binding ability and activation of p53 MAPK [102, 103], as does 10-gingerol in cervical cancer according to reports [104]. Cardamonin is also known to modulate cell cycle through suppression of NF-κB and decrease in cyclin D1 expression in lung cancer A549 xenograft in mice [105]. 6-shogaol is reported to activate PPARc and suppress NF-κB expression in colorectal cancer lines HT29 and HCT116 inhibiting cancer proliferation [106].

10.5.2 STAT3 Pathway

STAT3, a pro-inflammatory transcription factor, is responsible for controlling and onset of various cancers [107]. Both chemo- and radioresistance in cancer cells are modulated by STAT3. Curcumin shows widespread success in inhibiting STAT3 activation pathway, as reported in multiple myeloma cells and both clinical trials and animal models [108, 109]. Curcumin either alone or in combination with 5-flourouracil is effective against gastric cancer and with cisplatin shows promising results in HNSCC cells by inhibiting STAT3 phosphorylation [110, 111] as well as in human non-small cell lung cancer (H460) cells [112]. In mice models, curcumin when administered intraperitoneally effectively inhibited STAT3 activation [113]. Similar results were documented in pancreatic cancer, multiple myeloma [114, 115] and dextran sulphate sodium (DSS)-induced colitis in mice model too [116]. Curcumin inhibited JAK-STAT3 phosphorylation in K562 chronic leukaemia cells through suppression of JAK2, cyclin D1 and v-src gene expression. In chronic lymphocytic leukaemia, curcumin down-regulated JAK-STAT3 pathway by inhibiting the kinase Jak 1 and influencing STAT3 phosphorylation, resulting in growth arrest and apoptosis [117].

Among the other metabolites, 6-shogaol reportedly down-regulated STAT3 pathway and suppressed breast cancer in a time-dependent manner [118]. Apigenin was also reported to suppress JAK/STAT pathway through decreased nuclear translocation of STAT3 [119]. In another study, ovarian cancer cells when treated with alpinetin showed variations in expression levels of STAT3, pSTAT3, c-myc and surviving cells, indicating decreased phosphorylation of STAT3 and suppression of the STAT3 pathway [120]. Modulation of STAT3 pathway and inhibition of cancer cell division as well as proliferation were reported in prostrate and gastric cancer cells treated with cardamonin [121, 122]. Suppression of the STAT3 and associated pathways upon application of zerumbone is reportedly responsible for the inhibition of prostate cancer line PC3 and kidney cancer xenograft mice model [123, 124].

10.5.3 PI3K/AKT/mTOR Pathway

Curcumin exhibited remarkable anti-apoptotic effects in different malignancies through modulation of the phosphatidylinositol 3 kinase/phosphatidylinositol 3-kinase and the mammalian target of rapamycin (PI3K/Akt/mTOR) signalling cascades [125]. Curcumin induced apoptosis in HNSCLC cell line by curbing PI3K/Akt and precluding miR-192-5p [126]. In breast cancer, curcumin degraded Akt protein, inducing autophagy and inhibiting ubiquitin–proteasome pathway depending on time and dose [127, 128], hence impeding metastasis. pAkt and MAPK pathways were also down-regulated in breast cancer by curcumin [129]. Moreover, synergistic action of curcumin with PI3K inhibitors caused apoptosis in MCF-7 breast cancer cells [130]. Curcumin showed cell cycle arrest in pancreatic cells too, by inducing FoxO1 expression which in turn deregulated PI3K/Akt signalling [131] by upregulating phosphatase and tensin homolog gene (PTEN) [132] and depletion of MMP1/7 and COX-2 proteins in thyroid cancer [133]. In LoVo cell line apoptosis was brought about by upregulation of caspase-3, cytochrome-c and Bax mRNA and inhibition of Akt phosphorylation by curcumin [134]. Similarly, in Burkitt's lymphoma expression of the PI3K/Akt was inhibited by curcumin [135]. Other than these curcumin also arrested cell cycle at G2/M and induced autophagy both in vitro and in vivo in melanoma cells [136] and in uterine leiomyosarcoma growth by suppressing mTOR and S6 phosphorylation [137]. In HNSCC cells, nicotine related Akt/mTOR regulation was thwarted by this molecule [138]. Synergistic role of curcumin with ECGC and imatinib effectively inhibited uterine leiomyosarcoma cell growth and down-regulated the Akt/mTOR pathway in many cell lines [139, 140].

ATP binding sites in PI3K were blocked by flavones like apigenin, directly leading to their inactivity and deactivation of Akt as well [141]. Apigenin also effectively down-regulated Akt phosphorylation and induced over-expression of FOXO3a target genes, p21WAF1/ CIP1 and p27KIP1, thus preventing PI3K/Akt/FOXO signalling midway in many cancers like breast, colon and hepatocellular carcinoma, culminating in cell cycle arrest, apoptosis and prevention of proliferation as well [142–144]. Induction of autophagy and programmed cell death by apigenin through mTOR/PI3K/Akt signalling pathway in cisplatin resistant colon cancer cells is also reported [145]. Inhibition of lung cancer progression and increase in sensitization of lung cancer lines were shown to be executed by alpinetin via repression of the PI3K/Akt pathway [146]. Galangin decreased phosphorylation of Akt and suppressed mTOR to ultimately down-regulate the PI3K/Akt/ mTOR pathway inducing autophagy or apoptosis in many cancer cells [147], including human nasopharyngeal carcinoma [148] and in laryngeal carcinoma TU 212 and M4e [149]. Modulation of the PI3K/Akt pathway leading to apoptosis and cell cycle arrest was shown in retinoblastoma cancer RB355 cell line upon 6-gingerol (6G) treatment [150]. In human oral and cervical cancer lines, OSCC, KB and SCC4, 6G reportedly down-regulated the PI3K/Akt/mTOR pathway [151]. 10-Gingerol is also reported to inhibit cancer by down-regulation of PI3K/Akt pathway in breast cancer cells [152] and cervical cancer HeLa cells [104, 153]. Inhibitory action of 6-shogaol against Akt/mTOR pathway was reported [154] in lung cancers. Zerumbone is also reported to inhibit the PI3K/Akt/ mTOR pathway [155]. In another study, zerumbone decreased lamellipodia formation in NSCLCs through inhibition of FAK/Akt/ROCK pathway and hence the downstream ROCK/LIMK/cofilin signalling in A549 cells [156]. Kaempferol was also found to inhibit PI3K/Akt pathway indicating a possible role in cancer therapy [157, 158]. Modulation of mTOR pathway and initiation of caspase dependant apoptosis have been reported in A549 and HK1 cells treated with cardamonin and its homologues [159]. In colon cancer HT-29 cell line, zingiberene is reported to induce autophagy through suppression of the PI3K/Akt/mTOR pathway [160].

10.5.4 Tumour Necrosis Factor Pathway

Cellular signalling pathway of TNF-related apoptosis-inducing ligand is known to be modulated by curcumin [161]. In addition to interrupting cell cycle, curcumin disrupts mitotic spindle structures, induces micronucleation and apoptosis and inhibits IL-2 gene expression, thus having immense anti-proliferative activity [162]. The multi-targeted action of curcumin is reported to be beneficial in early stages of chronic lymphocytic leukaemia. It prevents the progression of the disease, decreases CLL B-cell counts and also when administered together with conventional anti-cancer drugs, has synergistic actions in addition to lowering their dose and side effects [163]. Other studies have reported that curcumin enhanced the TNF-related apoptosis-inducing ligand (TRAIL)-induced apoptosis even in TRAIL-resistant breast cancer cells [164]. In Panc-1 cells, cell death was induced by reduced production of IAP (inhibitors of apoptosis) proteins using curcumin [165]. Further, curcumin inhibited cancer progression modulating epithelial-mesenchymal transition (EMT), Cyclooxygenase 2 (COX-2) and pro effector cytokines [166, 167].

It was shown that galangin sensitized TRAIL activity leading to human breast cancer cell apoptosis through TRAIL/Caspase-3/AMPK signalling pathway [168]. Ozbey et al. [169] reported the sensitization of human liver cancer cells to TRAIL pathway induced apoptosis after treatment with apigenin. That TRAIL pathway induced apoptosis of colon cancer cells upon treatment with zerumbone was also reported [170]. Promotion of apoptosis in ovarian cancer cell lines A2780 and CP70 through the TRAIL pathway was also shown on kaempferol application [171]. Kaempferol reportedly modulates telomerase pathways too [172]. Kaempferol decreased the expression level of human Telomerase Reverse Transcriptase (hTERT) (catalytic subunit of telomerase). As telomerase regulates ageing and hence apoptosis, decrease in expression of this gene resulted in apoptosis of cervical cancer HeLa cells [157, 173].

10.5.5 EGFR Pathway

EGFR is responsible for cellular proliferation, survival, migration, adhesion even differentiation down the cancer cascades [174], and curcumin was shown to interfere with this cascade in the cellular membrane microenvironment leading to inhibition of various enzymes in diffecent cancers [175]. In brain cancer LN229 cells, curcumin exerted anti-modulatory effects leading to cytotoxicity and inhibition of kinase inhibitors like AG494, AG1478 tyrphostins [176]. Additionally, in erlotinib-resistant NSCLC cells curcumin hindered proliferation led to apoptosis by downgrading EGFR, p-EGFR, survivin and other proteins in the pathway [177]. Further, curcumin suppressed COX-2, EGFR, ERK 1/2 activities ushering apoptosis in lung and pancreatic adenocarcinoma [178] and oral cancer as well [179]. In mice model too, it inhibited angiogenesis in cervical cancer cells, down-regulating the above-mentioned proteins [180] and curcumin-induced inactivation of EGFR and Cav-1 pathways controlled mouse hepatocellular carcinoma in a time and dose-dependent manner [181].

Curcumin induced breast cancer apoptosis by regulating expression of apoptosis-related genes. A group recently studied curcumin [182] treated triple-negative breast cancer cell lines (TNBC) and reported significant inhibition in phosphorylation levels of EGFR and downstream signalling molecules, such as ERK1/2. Another study, however, reported suppression of breast cancer cell growth due to attenuated levels of EGFR and Akt [183]. In addition, curcumin and paclitaxel synergistically inhibited growth and induced apoptosis in breast cancer cells by blocking EGFR signalling and modulating Bax/BCL-2 expression [184]. Curcumin also potentiated anti-tumour effect of gefitinib in NCSLC, both in vitro and in vivo, which was mediated through inhibition of proliferation and EGFR phosphorylation, and led to the induction of EGFR ubiquitination and apoptosis [185].

The EGFR and associated cascades are also potent targets of other zingiberaceous metabolites. While zerumbone is known to affect the MAPK/ERK cascade [186], ERK signalling pathway was also inhibited by apigenin through phosphorylation of focal adhesion kinase (FAK) and ERK, which reduced integrin protein levels and finally suppressed cancer cell migration and thereby inhibited metastasis [187]. Kwak et al. [188] proved that suppression of phosphorylation of threonine 179 residue in Smad3 linker region upon application of GA resulted in growth inhibition of human prostate cancer cells. GA strongly inhibited PKC activity and subsequently phosphorylated extracellular signal-regulated kinase 1/2 (ERK1/2) to reduce cancer [189]. In another study by Dang et al. [190], it was shown that kaempferol targeted MAPK pathway by modulation of c-Met activity, while Hung et al. reported down-regulation of the EFGR pathway in kidney cancer line 786-O which inhibited its invasion and migration [191]. Modulation of cisplatin or γ-radiation induced hepatotoxicity through the p38 MAPK/JNK/ErK1/2 signal pathway was shown in rat models treated with zingerone [192].

10.5.6 Nrf2 Pathway and ROS Induction

Nuclear factor 2-related factor (Nrf2) is a potent target for cancer chemoprevention because of its ability to regulate genes that are tangled in the electrophile and ROS detoxification cascades as well as in the repair or removal of damaged products [193]. Curcumin has shown in vivo potency to activate this pathway, renew p53 activity, thus controlling inflammatory signals [194]. Its role in prevention of metastasis in Nrf2 knockdown [195] and exerting chemoprevention in prostate cancer by epigenetic modification and activation of Nrf2-aided defence [196] is well documented. Curcumin affects down-regulation of Flap endonuclease 1 (Fen1) expression, whose over-expression promotes breast cancer, by interfering Nrf-2, eventually preventing breast cancer [197–199]. Curcumin induced apoptosis of breast cancer cells by ROS accumulation, finally leading to p53/p21- and p16/Rb-mediated breast cancer inhibition [76]. Further, in Head and Neck squamous carcinoma, tested in vivo and in vitro, curcumin improves the activity of cisplatin, reduces its chemosensitivity and helps in checking tumour proliferation by modulating pSTAT3 and Nrf2 [111] also in bladder carcinoma cells Keap1–Nrf2 pathway is hindered by curcumin along with cisplatin [200].

The generation of oxidative stress is a pivotal step for induction of apoptosis and researchers contend that flavonoids can increase ROS levels, thereby potentiating DNA damage of cancer cells. Galangin (GAL) treatment is known to induce ROS generation in cancer cells [201]. When selenium nanoparticles fused with galangin (Se@Ga) were used, ROS generation increased in liver cancer cells over controls [202]. Cancer inhibition by galangin was correlated with ROS generation in cervical cancer cells too [203]. Galangin reportedly mobilizes endogenous copper ions, which form a ternary complex with chromatin, generating ROS and inducing DNA cleavage [201]. Zhang et al. [204] showed that even low concentrations of apigenin increased ROS accumulation in human papillary thyroid carcinomas. The mechanism of generation of oxidative stress by treatment of cancer cells with gingerols is also well documented. 6-gingerol (6G) treatment of cancer cells inhibited fatty acid synthesis and subsequent malonyl-CoA accumulation, coupled to Carnitine Palmitoyltransferase-1 enzyme (CPT-1) inhibition, which triggered mitochondria mediated production of ROS in liver cancer cells [205]. Another study with 6G shows that ROS-induced oxidative stress resulted in the release of cathepsin D, a metastasis marker, into the cytosol by increasing permeabilization with consequent release of cytochrome-C [206]. 6G induced spike is ROS production was also associated with an enhanced p53 mediated G2/M cell cycle arrest in cervical cancer cells [207]. 6-shogaol induced intracellular ROS in breast cancer cells, ultimately triggering MAPK protein kinase and apoptosis [118, 208]; similarly MAPK activation in human colon cancer cells upon treatment with 10-gingerol

(10G) is reported, though the exact mechanism is not known [209]. Zerumbone (ZER) effectively increased ROS production in colorectal cancer cells in dose-dependent manner [210], and in melanoma cells as well, which in turn decreased the mitochondrial membrane potential, favouring apoptosis [211]. Zerumbone treated ROS-mediated apoptotic fate was reported in chronic myelogenous leukemic cells [212] and in cervical cancer cells [213]; though contrary reports indicate ROS-independent, thiol-dependent DNA damage and apoptosis of colorectal cancer cells [214]. Chiang et al. [215], however, reported insignificant increase in ROS production in human prostate cancer cells upon zerumbone treatment. Kaempferol (KMF) treatment significantly increased ROS production and triggered apoptosis in cancer cells. KMF reportedly inverted antimycin A (AMA)-induced toxicity by disruption of MMP and accumulation of intracellular calcium ions and ROS, via PI3K/Akt/CREB pathway [216, 217]. ROS accumulation via catalase inhibition upon kaempferol treatment is also reported [218]. ROS induction and cytotoxicity, leading to apoptosis, was shown upon cardamonin application in human tumour lines [219]. Elevation of ROS was also reported in human HCT-116 colon cancer cell lines and in Winster rats when treated with zingerone [220–222]. 8-shogaol reportedly induced cell death in leukaemia cells through ROS generation leading to glutathione depletion and caspase activation [223].

10.5.7 Notch 1 Pathway

Neurogenic locus notch homolog protein-1 (Notch 1) family members control cell fate by modulating cell differentiation, proliferation and apoptosis cascades, and are investigated as potent therapeutic targets for cancer therapy [224]. Curcumin controls Notch pathway effectively preventing cancer stem cells [225, 226]. Curcumin could effectively down-regulate this pathway in an array of cancers—colorectal, oesophageal, oral, decreasing γ-secretase complex proteins especially in oesophageal cancer [226]. Not only Notch-1 signalling but associated factors including early growth response-1 gene product (Egr-1), farnesyl-protein transferase (FPTase), telomerase, c-Myc, fibroblast growth factors (FGF) mediated cell signalling are also inhibited by curcumin [227, 228]. Curcumin reportedly blocked Notch-1 signalling pathways which play important roles in pancreatic tumour growth [229]. Only a few of the other metabolites of Zingiberaceae are known for their effects against Notch-1 signalling cascades. In a study, 6-shogaol was reported to inhibit breast cancer cells by modulation of notch signalling pathway [230]. Transfection experiments conducted with miR-211-5p and anti-miR-211-5p in oral squamous carcinomas showed that alpinetin upregulated microRNA levels and thereby suppressed Jagged-1 expression as well as Notch signalling pathways [231]. Alpinetin treated brain tumor (Glioma) of rat cells showed inhibition of Notch signaling cascade, where transcription of Notch target genes, such as HES and c-Myc, were both found to be suppressed in GSCs [232].

10.5.8 Activating Protein-1 Pathway

AP-1, a dimeric transcription factor, is involved in cellular proliferation, transformation and death [233]. Curcumin down-regulated androgen dependent and independent lines, halting transactivation of androgen receptor (AR)—AP-1, cAMP and NF-κB showcasing anti-tumour activities [234]. It also regulated pro-inflammation cytokines IL-1α, IL-1β of AP-1 and IL-6 in mice lymphoma model [235]. Curcumin reportedly induced apoptosis of monocytic leukaemia cells through AP-1 activation as well [236]. Galangin was also reported to interfere with the binding of NF-κB with AP-1 thus inhibiting cancer progression [237].

10.5.9 HIF-1 Pathway and Angiogenesis

Angiogenesis is an important physiological process promoting tumour growth and metastasis, as cancer growth depends upon establishment of new blood vessels [238]. Curcumin inhibits tumour generation via impeding Hypoxia-inducible (HIF-1α) protein [239]. HIF-1 is necessary for continual survival of cancer cells as it is crucial for glycolysis activation and also initiates angiogenesis [240]. It is known that, curcumin constrains pituitary adenoma by HIF-1α mRNA induction, while down grading aryl hydrocarbon receptors leading to inhibition [241]. Similarly, curcumin also remarkably upregulated in oral squamous cell carcinomas associated with areca quid chewing [242]. Further, curcumin and cisplatin or diamminedichloroplatinum (DDP) exerts positive additive effects in A549 cells, leading to apoptosis where HIF-1α is broken down but caspase-3 upregulated [243]. Among all the potent anti-cancer phytochemicals of Zingiberaceae, kaempferol is reported to be the most efficient in inhibiting angiogenesis. Kaempferol triggered apoptosis of Human Umbilical Vein Endothelial Cells (HUVECs) and inhibited angiogenesis [244, 245]. Kaempferol also inhibited expression of Vascular Endothelial Growth Factor (VEGF) that stimulated HUVECs and hence proved to be a key mediator of angiogenesis in ovarian cancer cells OVCAR-3 and A2780/CP70 [246]. Reports also indicate that inhibition of proliferation, migration and tubule formation in HUVEC can be achieved by zerumbone [212, 247, 248]. Similar anti-proliferative and anti-angiogenic effect of zerumbone was shown in Sprague Dawley mice model of hepatocellular carcinoma [249]. Anti-angiogenesis properties are also exhibited by galangin in ovarian cancer cells OVCAR-3 via regulation of VEGF expression [250]. Inhibition of angiogenesis and suppression of epithelial-mesenchymal transition in glioma cell lines after galangin application were reported recently [251]. Suppression of angiogenesis by inhibition of MMPs was reported in mouse tumour Renca cells after application of zingerone [252]. Angiogenesis and metastasis in orthotopic Ovarian Tumour Model through modulation of the AKT/P70S6K1/MMP-9 Pathway by apigenin is reported [253].

10.5.10 Wnt/β-Catenin Pathway

The Wnt/ β-catenin pathway with its immense importance in controlling apoptosis and cell survival [254] is known to be effectively controlled by curcumin. Migration of breast cancer stem cells was inhibited by curcumin, through reinstatement of E-cadherin expression, resulting in enhancement of E-cadherin–β-catenin complex formation [255].

Curcumin administration showed control in cell proliferation and cellular aggregation, which in turn is controlled by β-catenin transcription activity and β-catenins [256]. Hence, in Lymph Node Carcinoma of the Prostate (LNCaP) cell line, Wnt/β-catenin pathway was effectively controlled by curcumin as well [257]. Further, PLGA-CUR NPs [poly (lactic-co-glycolic acid) which are curcumin encapsulated nanoparticles, showed impressive result against prostate cancer by checking this pathway [258]. Curcumin also leads to cell cycle arrest at G0/G1-phase in Non-Small Cell Lung Carcinoma (NSCLC) cells by blocking this pathway [259].

Zerumbone reportedly inhibited Wnt/ Beta-catenin pathway to abolish cancer stem cells [260] and targeted the β-catenin disrupting the cascade in breast cancer [261], while apigenin also silences the Wnt/β-catenin signalling pathway [262, 263]. On the other hand, cardamonin is also found to exert anti-proliferative action on SW480 cells degradation of β-catenin and subsequent down-regulation of the Wnt/β-catenin signalling pathway [264]. In breast cancer lines MCF7, MDAMB231 and BT549 cardamonin showed reversal of epithelial-mesenchymal transition by down-regulation of the Wnt/β-catenin signalling pathway [265].

10.5.11 Induction of Dysfunction of Cellular Organelles

A reduction or disruption in mitochondrial function occurs as a result of loss of maintenance of the electrical and chemical transmembrane potential of the inner mitochondrial membrane, altering the electron transport chain, or reducing transport of critical metabolites into mitochondria [266], which culminates in a massive upheaval of cellular homeostasis and induces apoptosis. Anti-cancer drugs thus target mitochondria for ushering apoptosis of cancer cells. Curcumin is reported to prohibit apoptosis mitochondrial dysfunction ultimately leading to cell cycle arrest in human gastric cancer cell line MGC803 [267]. Galangin (GA) induced human colon cancer cell death through alteration of mitochondria membrane potential and dysfunction. GA exposure caused release of cytochrome-C (cyt C) and apoptosis-inducing factor (AIF) from the mitochondria to the cytoplasm, and translocation of pro- and anti-apoptotic proteins across the mitochondrial membranes, increasing the ratio of pro-apoptotic protein Bax and anti-apoptotic protein Bcl-2 in hepatocellular carcinoma [268]. Molecular mechanisms of galangin treatment suppressing tumour cell growth revealed that GA increased the expression of cleaved PARP and caspase-3 in human cancer cells [268]. Depletion of mitochondrial membrane potential and translocation of phosphatidyl serine was induced by selenium nanoparticle fused with GA triggering apoptosis [202]. Apigenin triggered both the intrinsic and the extrinsic pathways of apoptosis [269]. Apigenin caused increase in Bax/Bcl-2 ratio causing apoptosis in prostate cancer [270]; similar observations of apoptosis and a reduction in cell viability due to decreased Bcl-2 and Bcl-xL and increased active form of the Bax protein upon apigenin treatment in colon cancer are also known [271]. A dose-dependent suppression of the pro-survival member proteins, like XIAP, c-IAP1, c-IAP2, was also observed in their study. Apigenin application reduced outer membrane potential of mitochondria, leaking cytochrome-C and inducing procaspase-9 activation, through intrinsic pathway [272] and forming of Fas-associated death domain (FADD) [272]. Nonetheless, Bcl-2 and Bax activities were unaltered by apigenin [273]. Alpinetin is reported to promote Bax translocation and induction of mitochondrial pathway of apoptosis [274]. 6-gingerol too increased the ratio of Bax to Bcl2 at the messenger RNA (mRNA) level in both 143B and MG63 osteosarcoma cell lines in a dose-dependent manner [275], and also elevated levels of cleaved caspase-3, caspase-8 and caspase-9. 6G treatment of Human cervical adenocarcinoma cell (HeLa) showed a gradual decline in membrane potential of mitochondria [101]. On a similar note, caspase3 and PARP were altered to induce apoptosis with 6-gingerol. In MDAMB231TNBC, a breast cancer line, 10-gingerol caused mitochondrial dysfunction leading to cytochrome-c and caspase cascade initiation [276]. Gan et al. reported activation of caspase cascade and induction of apoptosis in breast cancer line MCF-7 after zingerone application [277]. 6-shogaol activated caspase-8, caspase-9 and caspase-3 and caused PARP cleavage in MDA-MB-231 (human breast cancer), DU145 (human prostate cancer), SCC4 (human squamous cell carcinoma), HepG2 (human hepatocellular carcinoma), A549 (human lung adenocarcinoma) and also suppressed expression of Bcl-2, Bcl-xL and Survivin in tumour tissues [278, 279]. Its effect on colorectal cancer line inducing apoptosis through mitochondrial membrane damage, cytochrome-c leakage and caspase cascade activation is also documented [106]. Zerumbone decreased mitochondrial membrane potential and pushed cancer cells towards apoptosis [280]. Mitochondrial dysfunction via elevation of ROS was induced in a variety of cancer cells upon kaempferol treatment. A study with melanoma cells showed that zerumbone efficiently inhibited mitochondrial biogenesis by suppressing activity of a mitochondrial biogenesis factor—TFAM [211]. Apoptosis of MCF-7 breast cancer cells accompanied with nuclear condensation and mitochondria dysfunction [281] and mitochondrial dysfunction via increase in ROS in colorectal cancer cells [282] were observed after treatment with kaempferol. Induction of caspase cascade was reported in prostate LNCaP cells by kaempferol culminated in apoptosis [283]. Tu et al. reported disruption of mitochondrial membrane potential and perturbation in intracellular free Ca^{+} concentration in SiHa cells treated with kaempferol inhibiting their proliferation and inducing apoptosis [284].

Endoplasmic reticulum (ER) is the principal organelle responsible for multiple cellular functions including macromolecular trafficking involving protein folding and protein translocation and maintenance of cellular homeostasis. Disturbance in the ER (a pool of free calcium ions) environment by external or internal stimuli causes calcium depletion, altered glycosylation and oxidative stress. Experiments have revealed that galangin altered signal transduction pathways, which in turn induced ER stress and inhibited calcium channels, resulting in a significant increase of Ca^{2+} concentration in the cytoplasm and mitochondria, leading to apoptosis of hepatocellular carcinomas [285]. Zerumbone also induced a significant increase of intracellular Ca^{2+} concentration in prostate cancer cell lines, PC-3 and DU-145. Further, calpain I (calcium dependent protease) was induced upon zerumbone treatment facilitating apoptosis in those cell lines [286]. Kaempferol induced liver cancer cell death via ER stress and CHOP-autophagy signalling pathway [287].

10.5.12 Induction of Autophagy

Autophagy marks the transport and compartmentalization of cellular (cytoplasmic) material in vacuoles for degradation by lysosomal enzymes [288]. Reports show that galangin (GA) induced autophagy and apoptosis concomitantly. Beclin1, autophagy-related gene (ATG) 6, reportedly reacted to increased levels Bcl-XL and decreased Bcl-2 levels upon treatment with GA. Subsequent signal transduction cascades ultimately culminate in the formation of autophagosomes [289]. GA also induced autophagy via deacetylation of LC3 by SIRT1 in HepG2 liver cancer cells [290]. That galangin induced autophagy through upregulation of p53 in HepG2 liver cancer cells was also reported [147].

That apigenin induced autophagy was evident from presence of acidic vesicular organelles (AVOs) as well as the Atg5/Atg7 dependent autophagy marker, LC3-II [291]; reports also indicated concomitant autophagy and apoptosis upon apigenin application [292]. In macrophages, apigenin treatment upregulated Beclin 1, Atg5 and Atg7, to usher autophagy and the appearance of LC3-II in human colon carcinoma HCT-116 cells confirmed the same [293]. Vital staining with acridine orange showed accumulation of AVO (acidic vacuoles) in cytoplasm of HeLa cervical cancer cells exposed to 6G-gingerol [101]. Onset of apoptosis and mitochondrial damage was reported in U-118MG glioblastoma cells upon treatment with 6-G as well [294]. SSi6 (a 6-gingerol analogue) markedly blocked the autophagic flux and subsequently increased levels of LC3B-II, which contributed to cell death in triple-negative breast cancer cell line MDA-MB-231 [295].

6-shogaol (6-SG) induced a large number of cytoplasmic vacuoles in MCF-7 breast cancer cells. Localization of LC3 to autophagosomes in these cells was also noted in the same study [230]. AVO formation in non-small lung cancer cell line A549 cells was observed after exposure to 6-SG. Moreover, 6-shogaol mediated autophagy was blocked by 3-MA, an autophagy inhibitor, confirming that 6 SG induced autophagy in cancer cells [154]. Instances of zerumbone-induced increase of LC3-II formation indicating autophagy in human hormone-refractory prostate cancers were also reported [286]. Similarly, cardamonin is known to induce autophagy and inhibit cell cycle progression in HCT116 and LOVO cells where formation of AVOs and LC3 were also reported [296].

10.5.13 Modulation of Cell Cycle

Uncontrolled and rapid cell division is another hallmark of cancer. As evidenced, anti-cancer drugs inhibit cancer cell proliferation by modulating the cell cycle and blocking it at the G2/M or G0/G1 or the S-phase checkpoints. Reports pertaining to the cell cycle modulation by novel anti-cancer phytochemicals like apigenin, alpenitin, zerumbone and kaempferol show inhibition of the activities of some key players, namely cyclin D1, D3, E and A and cyclin-dependent kinase (CDK) 4 and CDK 6 [274, 281, 297, 298].

Inhibition of proliferation of pancreatic cancer cell lines like BxPC-3 using curcumin showed its therapeutic prospect. One of the pathways involved initiates cell cycle arrest at G2/M by preventing expression of cyclin B1/ Cyclin-dependent kinase 1 (Cdk1). Activation of ataxia telangiectasia mutated (ATM)/Checkpoint kinase 1(Chk1)/Cell Division Cycle 25C (Cdc25C) showed similar data [299]. Cell cycle arrest at different checkpoints using Aurora-A and kinase activity using Curcumin in breast cancer have also been projected [53]. Curcumol, a novel anti-cancer metabolite from *Curcuma longa*, too has been shown to be effective in arresting cell cycle and inducing apoptosis in a number of cancer lines through modulation of Cdk and p53 signalling cascades [300].

Cell cycle arrest at G2/M stage as well as G0/G1 checkpoint was shown by the treatment of apigenin on human colorectal carcinoma, prostate cancer cells, breast cancer line MDA-MB-231 and various other cancer cells [301, 302]. Similarly, G2/M transition in cell cycle and p53 reactivation via proteasome inhibition and upregulation of p21-p53/p21 by treatment of cervical cancer cells with 6-gingerol (6G) was also reported [207, 303]. Apoptotic nuclear shrinkage and membrane blebbing were shown in oral and cervical cancer cell lines upon 6G treatment [100]. Cell cycle arrest and apoptosis through modulation of cyclins by 6G have also been shown in hepatocarcinoma and colorectal carcinoma lines [304–306]. Cyclin A and CDK expression decreased via down-regulation of Rb phosphorylation and upregulation of p21 in pancreatic cancer cells upon 6G treatment [306]. Application of 10-gingerol (10G) resulted in down-regulation of the cell cycle regulatory proteins like CDK2, CDK4, cyclin D and cyclin E [152] in breast cancer cells. 10G also induced cell cycle arrest at G2/M phase in ovarian cancer HEY, OVCAR3 and SKOV3 lines due to decrease in cyclin B1 and D3 [307]. Cell cycle arrest in the G0/G1 phase by degradation of β-catenin, decreased c-myc expression and inhibition of activity of cyclins and CDKs upon treatment of cancer cells with galangin were reported [188]. Zerumbone, besides inducing cell cycle arrest of cervical carcinoma at G1 [213] and at G2/M stage [210] of cell cycle, also potentially inhibited ATM phosphorylation, and hence ATM activation, thereby sensitizing the cervical cancer cells and prostate cancer cells towards radiation [48, 215]. A combination of zerumbone and cisplatin was found to be effective in inducing cell cycle arrest in cervical intraepithelial neoplasia in female BALB/c mice [308]. Upregulation of Bax levels and alteration of Bcl levels were reported to induce apoptosis in gastric cancer SGC-7901 and oesophageal cancer EC-09 by zerumbone [309, 310]. Evidences pertaining to cell cycle arrest by kaempferol indicate that G1 and G2/M cell cycle arrest of cancer cells take place by inhibition of cyclin A, cyclin B1, cyclin E, cyclins D1 protein expressions, as well as inhibition of CDK2 and CDK4 activities, reduction in phosphorylation of retinoblastoma (Rb) protein and lowering the expression levels of Cdc2, Cdc25C`[311]. Several cell cycle related genes like CHK1, CHK2 and p21waf1/Cip1 were found to be upregulated and p35 and cyclinB1 genes were down-regulated upon treatment with kaempferol [191]. Anti-proliferative property of kaempferol was shown in lymphoma Daudi cells by Parmar et al (2016) [312]. Zingerone is implicated in cell cycle arrest and inhibition of mitosis through suppression cyclin D1 in human neuroblastoma BALB/c mouse tumour model BE(2)-M17 [313]. Alpinetin treatment of gastric cancer lines AGS and N87 reportedly induce cell cycle arrest at G2/M phase and apoptosis by translocation of Bax and triggering of the mitochondrial apoptotic cascade [274, 298]. Moreover arrest of colorectal cancer HT-29 cell cycle at G0/G1 and S-phases by alpinetin was reported with p53 mediated cell cycle arrest and uridine-cytidine kinase 2 inhibition [314].

10.5.14 Inhibition of Cancer Cell Adhesion and Metastasis

Cell adhesion is defined as the binding capability of one cell to another cell or to the extracellular matrix (ECM), and is a critical process through which cancer cells establish new tumours in the body. Metastasis involves the over-expression of the proteolytic enzymes, such as matrix metalloproteinases (MMPs). Studies have shown that MMP-2 and MMP-9 and TPA (12-O-tetradecanoylphorbol-13-acetate) are the main players of tumour metastasis. Several reports of curcumin inhibiting multiple metastatic steps including invasion and migration exist, which include its effect on thyroid carcinoma BCPAP cell line through the TGF/Smad2/3 pathway as well [315]. Galangin effectively inhibited adhesion of TPA-treated HepG2 cells in a dose-dependent manner, further RT-PCR studies with galangin treated liver cancer cells showed alteration in F-actin pattern, inhibition of transcription of MMP 2 and MMP 9 mRNAs [237] thereby inhibiting the cell adhesion and cancer cell metastasis. Inhibition of proliferation of human fibrosarcoma HT-1080 cell lines was attributed to galangin induced decrease in MMP-9 expression as well [316]. Another investigation revealed that the anti-metastatic ability of galangin resulted from repression of ADAM9 expression in the Glioma cells [189]. It was also reported by different authors that galangin administration suppressed cancer cell migration and metastasis of renal carcinoma cell lines through increased ROS levels and downregulation of the PI3K/Akt/mTOR signalling pathway [317, 318].

Apigenin strongly inhibited tumour cell invasion and migration in a dose-dependent manner in prostate cancer cells [270]. Cell migration and invasion of human and murine melanoma B16F10 cells in mice was prevented by down-regulating STAT3 phosphorylation and its target genes MMP-2, MMP-9, VEGF and Twist1 upon apigenin treatment [319]. It has been reported that in human ovarian cancer in vitro, cellular migration and onslaught of invasion by prohibiting FAK expression thus stopping metastasis as seen in mice model when treated with apigenin [320]. In an orthotopic colorectal cancer model, apigenin was shown to upregulated transgelin (active protein in actin cross-linking) and downregulate MMP-9 expression [321].

Alpinetin was reported to suppress proliferation and sensitize liver cancer cells towards chemotherapeutic agents by activation of mitogen-activated protein kinase kinase-7 (MKK-7) [322]. With respect to metastasis, alpinetin reportedly decreased migratory capacity of ovarian cancer cells; moreover, MMP2 and MMP9 protein expression levels were significantly decreased in alpinetin-treated cells in comparison to control cells. Conversely, tissue inhibitor of metalloproteinase TIMP1 and TIMP2 expression levels were increased in the alpinetin treated ovarian cancer cells [120]. Inhibition of invasion of glioma cancer cells by treatment with alpinetin was also reported [232]. It was found that 6G-treated liver cancer cells showed reduced metastatic burden and necrotic areas; moreover, 6G reduced the levels of MSE and improved the tumour microenvironment to inhibit metastasis. In the same study, Immunohistochemical (IHC) staining demonstrated a reduction in the expression of HIF1α, MMP2 and MMP9 in the 6G-treated cancer cells when compared to the untreated cancer cells. 6G treatment of osteosarcoma cells 143B and MG63 underwent apoptosis after inhibition of cell proliferation through activation of AMPK, caspase cascades and alteration of BCl2 levels [275].

Sithara et al. [210] studied the anti-migratory effect of zerumbone, and claimed that, colorectal cancer cells grew only in the internal space of the wound, rather spreading in form of dense cell mass with reduction in density due to zerumbone treatment. Hosseini et al. showed that zerumbone can induce suppression of cancer invasion and metastasis can be attributed to modulation of the FAk/PI3k/NF-KB-uPA pathway in human colorectal cancer lines HCT-116 and SW-48 [323]. Other studies with liver cancer cells [324], ovarian cancer cells [325] and breast cancer cells [326] also revealed anti-migratory properties of zerumbone. Kaempferol effectively blocked the development of metastatic cancer by inhibiting MMP-3 activity in highly invasive breast cancer cell line MDA-MB-231 [327] or through down-regulation of the RhoA and Rac1 cascades [328]. Migratory activity of pancreatic cancer cells was also reported to be inhibited by low doses of kaempferol, without having any cytotoxic effects on normal cells [329]. Inhibition of epithelial-mesenchymal transition through repression of multiple pathways was reported upon kaempferol treatment in A549 lung cancer cell line and human non-small cell lung cancer [330, 331]. Invasion and migration of liver cancer cells SNU182 were suppressed by inhibition of TGF-β-1 signalling pathway upon zingerone application [332].

Some important phytochemicals like apigenin [333] and 6-gingerol [334] are reported to improve the tumour microenvironment thus helping in prevention of metastasis. In prostate cancer cells, apigenin treatment induced downregulation of SPOCK1, improved tumour microenvironment bringing in reduced expressions of mesenchymal markers and significant depletion of the invasive abilities of metastatic cells [333]. 6G is known to prevent metastasis and improve tumour microenvironment through modulation of the pVEGFR2/VEcadherin/β-catenin/actin pathway [334].

10.6 Novel Anti-cancer Targets

10.6.1 Inhibition of Leukotriene Activity

The leukotrienes belong to a class of hormones which are formed by leukocytes, macrophages and other tissues in response to immunological and nonimmunological stimuli, and are found at high levels in most inflammatory lesions, probably playing an important role in cancer development as well. 6-gingerol readily suppressed Leukotriene A4 hydrolase (LTA4H) activity in human colorectal cancer cells HCT116 as well as in in vivo in nude mice under in vivo conditions [335]. As LTA4H has a positive role in carcinogenesis and is known to be responsible for anchorage-independent growth of cancer cells, this aspect of target of cancer cells might be a good strategy for inhibition of proliferation of cancers.

10.6.2 Effect on Cancer Stem Cell (CSC) and Glioma Stem Cells (GSC)

Cancer stem cells (CSCs) are chemo-resistant, self-renewing, tumorigenic sub-population of cells which are present in a very small percentage in the total tumour. CSCs reside in specified niches of the total tumour [336]. CSCs play a very important role in cancer development and progression. Hence, chemotherapeutic as well as phytochemical drugs targeting the main tumour along with their CSCs might be considered as a useful strategy to inhibit human cancers.

So far as the phytochemicals from members of Zingiberaceae are considered, only apigenin (AP), alpinetin (ALP) and 6-shogaol (6-SG) can reportedly target cancer stem cells. In ovarian cancer (SKOV3) apigenin compromised self-renewal ability of SFCs [337] and cervical cancer (HeLa) cells by downregulation of Casein Kinase 2 α (CK2α) expression [338], alpinetin suppressed proliferation and invasiveness of GSCs through suppression of Notch signalling [339]. 6-shogaol on the other hand was reported to be effective in inducing apoptosis of both breast cancer monolayer as well as the interior spheroid cells (SFCs) like cancer stem cells [230]. Apigenin can inhibit stem cell like phenotype of glioblastoma lines U87MG and U373MG by suppression of c-met signalling [232].

10.6.3 Cancer Immunotherapy

Cancer immunotherapy is the choicest of all ways to treat cancerous growth these days and is executed through upregulating the inherent immunity of a patient. A specific protein-programmed cell death 1 protein (PD1), commonly found in metastatic immune cells—T and B cells, monocytes and natural killer cells, is a means to disintegrate such immunity, with PD1/PD-L1 proteins being used for such surveillance [340]. T-regulatory cells (Tregs) are also used in an alternate strategy whereby effector T cells are prevented in the background of a hostile environment [341].

With respect to cancer immunotherapy, apigenin seems to be very promising. In a murine pancreatic cancer model, apigenin treatment enhanced CD4+CD8+ T cells and decreased the percentage of Tregs and ultimately showed prolonged mouse survival time [342]. Zerumbone is also reported to modulate CD1d expression and lipid antigen presentation pathway in breast cancer cells [280]. Recently it was shown that apigenin can inhibit transcription of interleukin-6, a potent pro-inflammatory cytokine that has a prominent presence in human oesophagus cancer patients. Repression of IL-6 transcription in human oesophagus cancer Eca-109 and Kyse-30 cells by apigenin has opened a novel dimension in cancer therapy [343]. The following table represents a summary of information known about different compounds from members of Zingiberaceae with known anti-cancer activities. Since the same compound may be present in multiple members, the major source of the particular metabolite is mentioned as the source plant in Table 10.1.

10.7 Cumulative Mechanism of Phytochemicals Towards Cancer Therapy

The intricate modulation of signal transduction pathways with a detailed study of the multi-step processes leading to cancer origin reveals that over-expression of the ligands and proteins related to respective signal transduction pathways, results in overall gain of function of distinct oncogenes and subsequently, loss of function of tumour suppressor genes, result in uncontrolled cell division, resulting in cancer. Table 10.1 summarizes the effect of the various compounds extracted from Zingiberaceae family used for cancer prevention and treatment. Elucidation of the exact modes of action and improvements in analytical techniques have also propelled a look in on many lesser known compounds like zingerone, zingiberine, cardamonin, which show promising results.

The anti-cancer phytochemicals discussed in detail here target multiple signal transduction pathways like NF-κB-IκB, PI3K/AKT/mTOR, TGF-β1/ Smad pathway, GSK3β-PKC, STAT3 and cause downregulation of the pathways. Down-regulation of NF-κB-IκB results in the decrease in activities of mitochondrial metalloproteinases (MMPs) (MMP-2, MMP-9). Down-regulation of PI3K/AKT/mTOR pathway results in elevation in levels of p53, this subsequently results in elevation of ROS levels in tumour cells (Figs. 10.4 and 10.5). The suppression of MAPK/ERK signalling pathway leads to the alteration in MMPs on the one hand, and on the other hand results in decrease of FAK activity. Inhibition of Smad pathway results in elevation in levels of Beclin 1 protein, leading towards autophagy. The suppression of the concerned signal transduction processes leads to either total inhibition or decrease in expression of genes, responsible for cell division, like c-myc, cyclins (in most cases cyclin D1). This results in cell cycle arrest of cancer cells at G0/G1 phase or G2/M phase (Figs. 10.4 and 10.5). Intracellular ROS hires the caspases and direct the cancer cells towards either extrinsic pathway for apoptosis or result in release of cytochrome-c, followed by formation of apaptosome

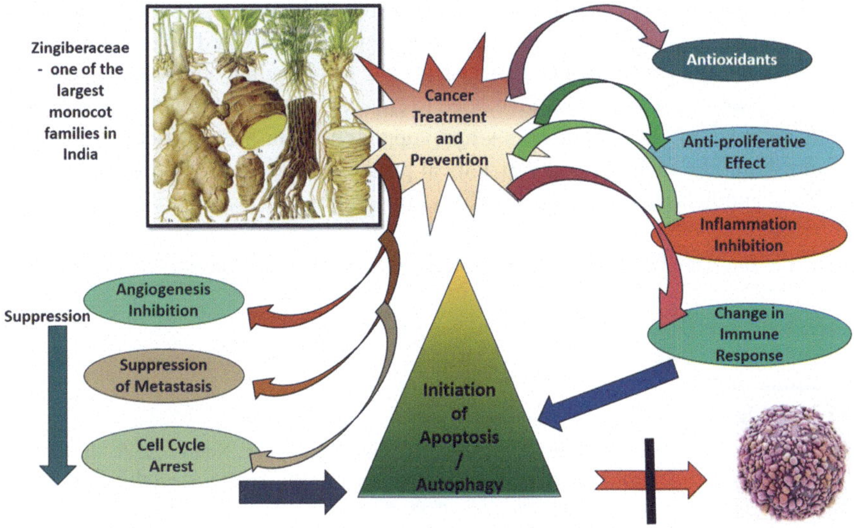

Fig. 10.5 Generalized diagram of effectiveness of the above-mentioned compounds in treating cancerous growth

and triggers the intrinsic pathway of apoptosis via mitochondrial dysfunction. On the other hand, the decrease in the MMPs and decrease in FAK activity ultimately lead towards decreased expression of F-actin and integrin protein levels, which ultimately inhibits cancer cell progression and migration (Fig. 10.5) and hence metastasis, thus the malignancy of cancer is inhibited.

10.8 Conclusion

The uses of plant-based nutraceuticals-pharmaceuticals have manifold in the recent years. People worldwide, searching for plausible solution against deadly cancers, are taking keen interest in such products as well. The age-old novelties and knowledge passed on for thousands of years are being rebranded presently to meet market trends.

The members of Zingiberaceae as discussed in the chapter have been used by the Asians for hundreds of years against a variety of diseases. The option to use easy to find, cheap herbs and rhizomes as alternatives and supplementary aids to traditional cancer medications and therapies is an exciting field. However, as discussed, bioavailability has still remained the major issue when it comes to human consumption and dose determination. Curcumin, for example, has lost some of its glory because of issues relating to bioavailability being insoluble in water. Preclinical trials with many of the compounds from the family have shown impressive result in treating or preventing angiogenesis, metastasis or tumour formation as a whole. However, clinical trials under strict control conditions should be ensured to meet the end objective of providing safe alternative to presently available drugs with multiple side effects and toxic aftermaths. Plants for a long time have provided better alternatives; these clinical trials should ensure such standards from leading authorities like FDA. Marketing strategies should be developed to ensure that maximum number of people could get hold of such medications and supplements, in a cost effective manner. Efforts should be made to popularize local products to reach global population and share the goodness and richness of plants universally.

Conflict of Interest The authors declare that they have no conflict of interest.

References

1. Kooti W, Servatyari K, Behzadifar M, Asadi-Samani M, Sadeghi F, Nouri B, Zare Marzouni H (2017) Effective medicinal plant in cancer treatment, part 2: review study. Evid Based Complemenary Alternate 22(4):982–995. https://doi.org/10.1177/2156587217696927
2. Kuruppu AI, Paranagama P, Goonasekara C (2019) Medicinal plants commonly used against cancer in traditional medicine formulae in Sri Lanka. Saudi Pharm J. https://doi.org/10.1016/j.jsps.2019.02.004
3. Wang H, Oo Khor T, Shu L, Su ZY, Fuentes F, Lee JH, Tony Kong AN (2012) Plants vs. cancer: a review on natural phytochemicals in preventing and treating cancers and their druggability. Anti-cancer Agent Med 12(10):1281–1305. https://doi.org/10.2174/187152012803833026
4. El-Shami K, Oeffinger KC, Erb NL, Willis A, Bretsch JK, Pratt-Chapman ML, Stein KD (2015) American Cancer Society colorectal cancer survivorship care guidelines. CA Cancer J Clin 65(6):427–455. https://doi.org/10.3322/caac.21286-*/
5. Jamwal R (2018) Bioavailable curcumin formulations: a review of pharmacokinetic studies in healthy volunteers. J Integr Med 16(6):367–374. https://doi.org/10.1016/j.joim.2018.07.001
6. Nurgali K, Jagoe RT, Abalo R (2018) Editorial: adverse effects of cancer chemotherapy: anything new to improve tolerance and reduce sequelae? Front Pharm 9:245. https://doi.org/10.3389/fphar.2018.00245
7. Greenwell M, Rahman PKSM (2015) Medicinal plants: their use in anticancer treatment. Int J Pharm Sci 6(10):4103. https://doi.org/10.13040/IJPSR.0975-8232.6(10).4103-12
8. Patwardhan B, Mutalik G, Tillu G (2015) Integrative approaches for health: biomedical research, aayurveda and yoga. Academic, New York
9. Akhtar MA, Swamy MK (2018) Anticancer plants: natural products and biotechnological implements, vol 2. Springer, Singapore. https://doi.org/10.1007/978-981-10-8064-7
10. Kocyigit A, Guler EM, Dikilitas M (2017) Role of antioxidant phytochemicals in prevention, formation and treatment of cancer. Reac Oxy Sps (ROS) Liv Cells. https://doi.org/10.5772/intechopen.72217
11. Scarpa ES, Ninfali P (2015) Phytochemicals as innovative therapeutic tools against cancer stem cells. Int J Mol 16(7):15727–15742. https://doi.org/10.3390/ijms160715727
12. Kaur V, Kumar M, Kumar A, Kaur K, Dhillon VS, Kaur S (2018) Pharmacotherapeutic potential of phytochemicals: Implications in cancer chemoprevention and future perspectives. Biomed Pharmacol Ther 97:564–586. https://doi.org/10.1016/j.biopha.2017.10.124
13. Amararathna M, Johnston MR, Rupasinghe HP (2016) Plant polyphenols as chemopreventive agents for lung cancer. Int J Mol 17(8):1352. https://doi.org/10.3390/ijms17081352
14. Dutt R, Garg V, Khatri N, Madan AK (2019) Phytochemicals in anticancer drug development. Anti-cancer Agent Med 19(2):172–183

15. Kingston DG (1993) Taxol, an exciting anticancer drug from Taxus brevifolia: an overview. Human medicinal agents from plants. ACS Symp Ser 534(10):138–148. https://doi.org/10.1021/bk-1993-0534.ch010
16. Nakamura H, Jun F, Maeda H (2015) Development of next-generation macromolecular drugs based on the EPR effect: challenges and pitfalls. Expert Opin Drug Deliv 12(1):53–64. https://doi.org/10.1517/17425247.2014.955011
17. Lee KW, Ching SM, Hoo FK, Ramachandran V, Swamy MK (2018) Traditional medicinal plants and their therapeutic potential against major cancer types. In: Anticancer plants: natural products and biotechnological implements. Springer, Singapore, pp 383–410. https://doi.org/10.1007/978-981-10-8064-7_16
18. Dai L, Wang G, Pan W (2017) Andrographolide inhibits proliferation and metastasis of SGC7901 gastric cancer cells. Biomed Res Int 2017:10
19. Smilkov K, Ackova DG, Cvetkovski A, Ruskovska T, Vidovic B, Atalay M (2019) Piperine: old spice and new nutraceutical? Curr Phar 25(15):1729–1739
20. Honari M, Shafabakhsh R, Reiter RJ, Mirzaei H, Asemi Z (2019) Resveratrol is a promising agent for colorectal cancer prevention and treatment: focus on molecular mechanisms. Cancer Cell Int 19(1):180
21. Jamshidi-Kia F, Lorigooini Z, Amini-Khoei H (2018) Medicinal plants: past history and future perspective. J Herbmed Pharmacol 7(1):1–7
22. Lichota A, Gwozdzinski K (2018) Anticancer activity of natural compounds from plant and marine environment. Int J Mol 19(11):3533. https://doi.org/10.3390/ijms19113533
23. Kress WJ (1990) The phylogeny and classification of the Zingiberales. Ann Missouri Bot 77:698–721. https://doi.org/10.2307/2399669
24. Kress WJ, Prince LM, Williams KJ (2002) The phylogeny and a new classification of the gingers (Zingiberaceae): evidence from molecular data. Am J Bot 89(10):1682–1696. https://doi.org/10.3732/ajb.89.10.1682
25. Sabu M (2006) Zingiberaceae and Costaceae of South India. Indian Association of Angiosperm Taxonomy, Kerala
26. Barbosa GB, Jayasinghe NS, Natera SH, Inutan ED, Peteros NP, Roessner U (2017) From common to rare zingiberaceae plants-a metabolomics study using GC-MS. Phytochemistry 140:141–150
27. Ghasemzadeh A, Jaafar HZ, Rahmat A (2011) Effects of solvent type on phenolics and flavonoids content and antioxidant activities in two varieties of young ginger (Zingiber officinale Roscoe) extracts. J Med Plant Res 5(7):1147–1154
28. Quideau S, Deffieux D, Douat-Casassus C, Pouysegu L (2011) Plant polyphenols: chemical properties, biological activities, and synthesis. Angew Chem Int Ed 50(3):586–621
29. Russo GL, Tedesco I, Spagnuolo C, Russo M (2017) Antioxidant polyphenols in cancer treatment: friend, foe or foil? In: Semin Cancer Biol, vol 46. Academic, New York, pp 1–13
30. Yoon JH, Baek SJ (2005) Molecular targets of dietary polyphenols with anti-inflammatory properties. Yonsei Med J 46(5):585–596
31. Hasimun P, Adnyana IK (2019) Zingiberaceae family effects on alpha-glucosidase activity: implication for diabetes. In: Bioactive food as dietary interventions for diabetes. Academic, New York, pp 387–393
32. Wal P, Saraswat N, Pal RS, Wal A, Dubey S (2019) A detailed review on traditionally used and potent sources showing anti-pyretic action. RJPT 12(10):5107–5112
33. Rodríguez-Pérez C, Segura-Carretero A, Del Mar CM (2019) Phenolic compounds as natural and multifunctional anti-obesity agents: a review. Crit Rev Food Sci Nutr 59(8):1212–1229
34. Sunilson JAJ, Suraj R, Rejitha G, Anandarajagopal K, Kumari AVAG, Promwichit P (2009) In vitro antimicrobial evaluation of Zingiber officinale, Curcuma longa and Alpinia galanga extracts as natural food preservatives. Am J Food Technol 4(5):192–200
35. Danciu C, Vlaia L, Fetea F, Hancianu M, Coricovac DE, Ciurlea SA, Trandafirescu C (2015) Evaluation of phenolic profile, antioxidant and anticancer potential of two main representants of zingiberaceae family against B164A5 murine melanoma cells. Biol Res 48(1):1–9

36. Basri AM, Taha H, Ahmad N (2017) A review on the pharmacological activities and phytochemicals of Alpinia officinarum (Galangal) extracts derived from bioassay-guided fractionation and isolation. Phcog Rev 11(21):43–56
37. Chanda S, Ramachandra TV (2019) Phytochemical and pharmacological importance of turmeric (Curcuma longa): a review. RRJoP 9(1):16–23
38. Sharifi-Rad M, Varoni EM, Salehi B, Sharifi-Rad J, Matthews KR, Ayatollahi SA, Sharifi-Rad M, Rigano D (2017) Plants of the Genus Zingiber as source of antimicrobial agents: from tradition to pharmacy. Molecules 22(2145):1–20
39. Maiti K, Mukherjee K, Gantait A, Saha BP, Mukherjee PK (2007) Curcumin–phospholipid complex: preparation, therapeutic evaluation and pharmacokinetic study in rats. Int J Pharm 330(1-2):155–163. https://doi.org/10.1016/j.ijpharm.2006.09.025
40. Priyadarsini K (2014) The chemistry of curcumin: from extraction to therapeutic agent. Molecules 19(12):20091–20112. https://doi.org/10.3390/molecules191220091
41. Chempakam B, Parthasarathy VA (2006) Turmeric. In: Parthasarathy VA, Chempakam B, Zachariah TJ (eds) Chemistry of spice. CABI, Cambridge, pp 97–123
42. Jayaprakasha GK, Rao LJ, Sakariah KK (2006) Antioxidant activities of curcumin, demethoxycurcumin and bisdemethoxycurcumin. Food Chem 98(4):720–724. https://doi.org/10.1016/j.foodchem.2005.06.037
43. Dairam A, Limson JL, Watkins GM, Antunes E, Daya S (2007) Curcuminoids, curcumin, and demethoxycurcumin reduce lead-induced memory deficits in male Wistar rats. J Agric Food Chem 55(3):1039–1044. https://doi.org/10.1021/jf063446t
44. Pfeiffer E, Höhle S, Solyom AM, Metzler M (2003) Studies on the stability of turmeric constituents. J Food Eng 56(2-3):257–259. https://doi.org/10.1016/S0260-8774(02)00264-9
45. Benguedouar L, Lahouel M, Gangloff SC, Durlach A, Grange F, Bernard P, Antonicelli F (2016) Ethanolic extract of algerian propolis and galangin decreased murine melanoma tumor progression in mice. Anti Cancer Agents Med Chem 16(9):1172–1183. https://doi.org/10.2174/1871520616666160211124459
46. He X, Wei Z, Wang J, Kou J, Liu W, Fu Y, Yang Z (2016) Alpinetin attenuates inflammatory responses by suppressing TLR4 and NLRP3 signaling pathways in DSS-induced acute colitis. Sci Rep 6:28370. https://doi.org/10.1038/srep28370
47. Dugasani S, Pichika MR, Nadarajah VD, Balijepalli MK, Tandra S, Korlakunta JN (2010) Comparative antioxidant and anti-inflammatory effects of [6]-gingerol,[8]-gingerol, [10]-gingerol and [6]-shogaol. J Ethnopharmacol 127(2):515–520
48. Haque MA, Jantan I, Arshad L, Bukhari SNA (2017) Exploring the immunomodulatory and anticancer properties of zerumbone. Food Funct 8(10):3410–3431. https://doi.org/10.1039/C7FO00595D
49. Antonious GF, Kochhar TS (2003) Zingiberene and curcumene in wild tomato. J Environ Sci Health B 38(4):489–500
50. Swindell WR, Bojanowski K, Chaudhuri RK (2019) A zingerone analog, acetyl zingerone, bolsters matrisome synthesis, inhibits matrix metallopeptidases, and represses IL-17A target gene expression. JID. https://doi.org/10.1016/j.jid.2019.07.715
51. Chen AY, Chen YC (2013) A review of the dietary flavonoid, kaempferol on human health and cancer chemoprevention. Food Chem 138(4):2099–2107. https://doi.org/10.1016/j.foodchem.2012.11.139
52. Gonçalves LM, Valente IM, Rodrigues JA (2014) An overview on cardamonin. J Med Food 17(6):633–640. https://doi.org/10.1089/jmf.2013.0061
53. Ke CS, Liu HS, Yen CH, Huang GC, Cheng HC, Huang CYF, Su CL (2014) Curcumin-induced Aurora-A suppression not only causes mitotic defect and cell cycle arrest but also alters chemosensitivity to anticancer drugs. J Nutr Biochem 25(5):526–539. https://doi.org/10.1016/j.jnutbio.2014.01.003
54. Aggarwal BB, Kumar A, Bharti AC (2003) Anticancer potential of curcumin: preclinical and clinical studies. Anticancer Res 23(1/A):363–398

55. Nagaraju GP, Aliya S, Zafar SF, Basha R, Diaz R, El-Rayes BF (2012) The impact of curcumin on breast cancer. Integr Biol 4:996–1007. https://doi.org/10.1039/c2ib20088k
56. Surh YJ, Han SS, Keum YS, Seo HJ, Lee SS (2000) Inhibitory effects of curcumin and capsaicin on phorbol ester-induced activation of eukaryotic transcription factors, NF-kappaB and AP-1. Biofactors 12(1–4):107–112
57. Vallianou NG, Evangelopoulos A, Schizas N, Kazazis C (2015) Potential anticancer properties and mechanisms of action of curcumin. Anticancer Res 35(2):645–651
58. Wilken R, Veena MS, Wang MB, Srivatsan ES (2011) Curcumin: a review of anti-cancer properties and therapeutic activity in head and neck squamous cell carcinoma. Mol Cancer 10 (1):12
59. Das L, Vinayak M (2015) Long term effect of curcumin in restoration of tumour suppressor p53 and phase-II antioxidant enzymes via activation of Nrf2 signalling and modulation of inflammation in prevention of cancer. PLoS ONE 10:e0124000. https://doi.org/10.1371/journal.pone.0124000
60. Sun SQ, Zhang W, Guo Y, Li Z, Chen X, Wang Y, Zhao G (2016) Curcumin inhibits cell growth and induces cell apoptosis through upregulation of miR-33b in gastric cancer. Tumour Biol 37(10):13177–13184. https://doi.org/10.1007/s13277-016-5221-9
61. Gao SM, Yang JJ, Chen CQ, Chen JJ, Ye LP, Wang LY (2012) Pure curcumin decreases the expression of WT1 by upregulation of miR-15a and miR-16-1 in leukemic cells. J Exp Clin Cancer Res 31(1):27. https://doi.org/10.1186/1756-9966-31-27
62. Sa G, Das T (2008) Anti cancer effects of curcumin: cycle of life and death. Cell Div 3(1):14
63. Sovak MA, Bellas RE, Kim DW, Zanieski GJ, Rogers AE, Traish AM et al (1997) Aberrant nuclear factor-kappaB/Rel expression and the pathogenesis of breast cancer. J Clin Invest 100:2952–2960. https://doi.org/10.1172/JCI119848
64. Sun Y, Ai X, Shen S, Lu S (2015) NF-κB-mediated miR-124 suppresses metastasis of non-small-cell lung cancer by targeting MYO10. Oncotarget 6(10):8244–8254. https://doi.org/10.18632/oncotarget.3135
65. Singh S, Aggarwal BB (1995) Activation of transcription factor NF-kappa B is suppressed by curcumin (diferuloylmethane). J Biol Chem 270:24995–25000
66. Lin YG, Kunnumakkara AB, Nair A, Merritt WM, Han LY, Armaiz-Pena GN (2007) Curcumin inhibits tumor growth and angiogenesis in ovarian carcinoma by targeting the nuclear factor-kappaB pathway. Clin Cancer Res 13:3423–3430. https://doi.org/10.1158/1078-0432.CCR-06-3072
67. Nakamura K, Yasunaga Y, Segawa T, Ko D, Moul JW, Srivastava S (2002) Curcumin down-regulates AR gene expression and activation in prostate cancer cell lines. Int J Oncol 21:825–830. https://doi.org/10.3892/ijo.21.4.825
68. Zhang BY, Shi YQ, Chen X, Dai J, Jiang ZF, Li N et al (2013) Protective effect of curcumin against formaldehyde-induced genotoxicity in A549 Cell Lines. J Appl Toxicol 33:1468–1473. https://doi.org/10.1002/jat.2814
69. Kuo P, Chen MM, Decker RH, Yarbrough WG, Judson BL (2014) Hypopharyngeal cancer incidence, treatment, and survival: temporal trends in the United States. Laryngoscope 124 (9):2064–2069. https://doi.org/10.1002/lary.24651
70. Aggarwal BB, Harikumar KB (2009) Potential therapeutic effects of curcumin, the anti-inflammatory agent, against neurodegenerative, cardiovascular, pulmonary, metabolic, auto-immune and neoplastic diseases. Int J Biochem Cell Biol 41(1):40–59. https://doi.org/10.1016/j.biocel.2008.06.010
71. Vageli DP, Doukas SG, Spock T, Sasaki CT (2018) Curcumin prevents the bile reflux-induced NF-κB-related mRNA oncogenic phenotype, in human hypopharyngeal cells. J Cell Mol Med 22:4209–4220. https://doi.org/10.1111/jcmm.13701
72. LoTempio MM, Veena MS, Steele HL, Ramamurthy B, Ramalingam TS, Cohen AN, Wang MB (2005) Curcumin suppresses growth of head and neck squamous cell carcinoma. Clin Cancer Res 11(19):6994–7002. https://doi.org/10.1158/1078-0432.CCR-05-0301

73. Vander Broek R, Snow GE, Chen Z, Van Waes C (2014) Chemoprevention of head and neck squamous cell carcinoma through inhibition of NF-κB signaling. Oral Oncol 50(10):930–941. https://doi.org/10.1016/j.oraloncology.2013.10.005
74. Karin M (2006) Nuclear factor-κB in cancer development and progression. Nature 441 (7092):431. https://doi.org/10.1038/nature04870
75. Xi Y, Gao H, Callaghan MU, Fribley AM, Garshott DM, Xu ZX (2015) Induction of BCL2-interacting killer, BIK, is mediated for anti-cancer activity of curcumin in human head and neck squamous cell carcinoma cells. J Cancer 6(4):327. https://doi.org/10.7150/jca.11185
76. Calaf GM, Echiburú-Chau C, Roy D, Chai Y, Wen G, Balajee AS (2011) Protective role of curcumin in oxidative stress of breast cells. Oncol Rep 26(4):1029–1035. https://doi.org/10.3892/or.2011.1386
77. Liu Q, Loo WT, Sze SCW, Tong Y (2009) Curcumin inhibits cell proliferation of MDA-MB-231 and BT-483 breast cancer cells mediated by down-regulation of NFκB, cyclinD and MMP-1 transcription. Phytomedicine 16(10):916–922. https://doi.org/10.1016/j.phymed.2009.04.008
78. Bachmeier B, Nerlich A, Iancu C, Cilli M, Schleicher E, Vené R, Pfeffer U (2007) The chemopreventive polyphenol curcumin prevents hematogenous breast cancer metastases in immunodeficient mice. Cell Physiol Biochem 19(1-4):137–152. https://doi.org/10.1159/000099202
79. Kunnumakkara AB, Diagaradjane P, Anand P, Kuzhuvelil HB, Deorukhkar A, Gelovani J (2009) Curcumin sensitizes human colorectal cancer to capecitabine by modulation of cyclin D1, COX-2, MMP-9, VEGF and CXCR4 expression in an orthotopic mouse model. Int J Cancer 125(9):2187–2197. https://doi.org/10.1002/ijc.24593
80. Yodkeeree S, Ampasavate C, Sung B, Aggarwal BB, Limtrakul P (2010) Demethoxycurcumin suppresses migration and invasion of MDA-MB-231 human breast cancer cell line. Eur J Pharmacol 627(1-3):8–15. https://doi.org/10.1016/j.ejphar.2009.09.052
81. Xia Y, Jin L, Zhang B, Xue H, Li Q, Xu Y (2007) The potentiation of curcumin on insulin-like growth factor-1 action in MCF-7 human breast carcinoma cells. Life Sci 80(23):2161–2169. https://doi.org/10.1016/j.lfs.2007.04.008
82. Sahebkar A (2016) Curcumin: a natural multitarget treatment for pancreatic cancer. Integr Cancer Ther 15(3):333–334
83. Li L, Aggarwal BB, Shishodia S, Abbruzzese J, Kurzrock R (2004) Nuclear factor-kappaB and ikappaB kinase are constitutively active in human pancreatic cells, and their down-regulation by curcumin (diferuloylmethane) is associated with the suppression of proliferation and the induction of apoptosis. Cancer 101:2351–2362
84. Bimonte S, Barbieri A, Palma G, Luciano A, Rea D, Arra C (2013). Curcumin inhibits tumor growth and angiogenesis in an orthotopic mouse model of human pancreatic cancer. Biomed Res. https://doi.org/10.1155/2013/810423
85. Zhao Z, Li C, Xi H, Gao Y, Xu D (2015) Curcumin induces apoptosis in pancreatic cancer cells through the induction of forkhead box O1 and inhibition of the PI3K/Akt pathway. Mol Med Rep 12:5415–5422. https://doi.org/10.3892/mmr.2015.4060
86. Hosseini M, Hassanian SM, Mohammadzadeh E, ShahidSalesm S, Maftouh M, Fayazbakhsh H, Avan A (2017) Therapeutic potential of curcumin in treatment of pancreatic cancer: current status and future perspectives. J Cell Biochem 118(7):1634–1638. https://doi.org/10.1002/jcb.25897
87. Darvesh SA, Aggarwal BB, Bishayee A (2012) Curcumin and liver cancer: a review. Curr Pharm Biotechnol 13(1):218–228. https://doi.org/10.2174/138920112798868791
88. Jutooru I, Chadalapaka G, Lei P, Safe S (2010) Inhibition of NfkappaB and pancreatic cancer cell and tumor growth by curcumin is dependent on specificity protein down-regulation. J Biol Chem 285:25332–25344. https://doi.org/10.1074/jbc.M109.095240
89. Basha R, Connelly SF, Sankpal UT, Nagaraju GP, Patel H, Vishwanatha JK, Shelake S, Tabor-Simecka L, Shoji M, Simecka JW, El-Rayes B (2016) Small molecule tolfenamic acid and dietary spice curcumin treatment enhances antiproliferative effect in pancreatic cancer

cells via suppressing Sp1, disrupting NF-kB translocation to nucleus and cell cycle phase distribution. J Nutr Biochem 31:77–87. https://doi.org/10.1016/j.jnutbio.2016.01.003

90. Marquardt JU, Gomez-Quiroz L, Camacho LOA, Pinna F, Lee YH, Kitade M (2015) Curcumin effectively inhibits oncogenic NF-κB signaling and restrains stemness features in liver cancer. J Hepatol 63(3):661–669. https://doi.org/10.1016/j.jhep.2015.04.018
91. Zeng Y, Shen Z, Gu W, Mianhua W (2018) Inhibition of hepatocellular carcinoma tumorigenesis by curcumin may be associated with CDKN1A and CTGF. Gene 651:183–193. https://doi.org/10.1016/j.gene.2018.01.083
92. Pan Z, Zhuang J, Ji C, Cai Z, Liao W, Huang Z (2018) Curcumin inhibits hepatocellular carcinoma growth by targeting VEGF expression. Oncol Lett 15:4821–4826. https://doi.org/10.3892/ol.2018.7988
93. Shukla S, Bhaskaran N, Babcook MA, Fu P, MacLennan GT, Gupta S (2013) Apigenin inhibits prostate cancer progression in TRAMP mice via targeting PI3K/Akt/FoxO pathway. Carcinogenesis 35(2):452–460
94. Shukla S, Kanwal R, Shankar E, Datt M, Chance MR, Fu P, Gupta S (2015) Apigenin blocks IKKα activation and suppresses prostate cancer progression. Oncotarget 6(31):31216
95. Chen M, Wang X, Zha D, Cai F, Zhang W, He Y, Huang Q, Zhuang H, Hua ZC (2016) Apigenin potentiates TRAIL therapy of non-small cell lung cancer via upregulating DR4/DR5 expression in a p53-dependent manner. Sci Rep 6(35468):1–17
96. Choi M (2017) Galangin suppresses pro-inflammatory gene expression in polyinosinic-polycytidylic acid-stimulated microglial cells. Biomol Ther 25(6):641–647. https://doi.org/10.4062/biomolther.2017.173
97. Luo H, Rankin GO, Li Z, DePriest L, Chen YC (2011) Kaempferol induces apoptosis in ovarian cancer cells through activating p53 in the intrinsic pathway. Food Chem 128(2):513–519
98. Yang S, Si L, Jia Y, Jian W, Yu Q, Wang M, Lin R (2019) Kaempferol exerts anti-proliferative effects on human ovarian cancer cells by inducing apoptosis, G0/G1 cell cycle arrest and modulation of MEK/ERK and STAT3 pathways. J BUON 24:975–981
99. Oyagbemi AA, Saba AB, Azeez OI (2010) Molecular targets of [6]-gingerol: Its potential roles in cancer chemoprevention. Biofactors 36(3):169–178
100. Weng CJ, Chou CP, Ho CT, Yen GC (2012) Molecular mechanism inhibiting human hepatocarcinoma cell invasion by 6-shogaol and 6-gingerol. Mol Nutr Food Res 56(8):1304–1314. https://doi.org/10.1002/mnfr.201200173
101. Chakraborty D, Bishayee K, Ghosh S, Biswas R, Mandal SK, Khuda-Bukhsh AR (2012) [6]-Gingerol induces caspase 3 dependent apoptosis and autophagy in cancer cells Drug–DNA interaction and expression of certain signal genes in HeLa cells. Eur J Pharmacol 694(1-3):20–29
102. Kim SO, Chun KS, Kundu JK, Surh YJ (2004) Inhibitory effects of [6]-gingerol on PMA-induced COX-2 expression and activation of NF-κB and p38 MAPK in mouse skin. Biofactors 21(1-4):27–31
103. Park KK, Chun KS, Lee JM, Lee SS, Surh YJ (1998) Inhibitory effects of [6]-gingerol, a major pungent principle of ginger, on phorbol ester-induced inflammation, epidermal ornithine decarboxylase activity and skin tumor promotion in ICR mice. Cancer Lett 129(2):139–144. https://doi.org/10.1016/S0304-3835(98)00081-0
104. Zhang F, Thakur K, Hu F, Zhang JG, Wei ZJ (2017) 10-Gingerol, a phytochemical derivative from "tongling white ginger", inhibits cervical cancer: insights into the molecular mechanism and inhibitory targets. J Agric Food Chem 65(10):2089–2099. https://doi.org/10.1021/acs.jafc.7b00095
105. He W, Jiang Y, Zhang X, Zhang Y, Ji H, Zhang N (2014) Anticancer cardamonin analogs suppress the activation of NF-kappaB pathway in lung cancer cells. Mol Cell Biochem 389(1-2):25–33. https://doi.org/10.1007/s11010-013-1923-0
106. Pan MH, Hsieh MC, Kuo JM, Lai CS, Wu H, Sang S, Ho CT (2008) 6-Shogaol induces apoptosis in human colorectal carcinoma cells via ROS production, caspase activation, and

GADD 153 expression. Mol Nutr Food Res 52(5):527–537. https://doi.org/10.1002/mnfr.200700157
107. Aggarwal BB, Sethi G, Ahn KS, Sandur SK, Pandey MK, Kunnumakkara AB (2006) Targeting signal-transducer-and-activator-of-transcription-3 for prevention and therapy of cancer: modern target but ancient solution. Ann N Y Acad Sci 1091:151–169. https://doi.org/10.1196/annals.1378.063
108. Gupta SC, Kannappan R, Reuter S, Kim JH, Aggarwal BB (2011) Chemosensitization of tumors by resveratrol. Ann N Y Acad Sci 1215:150–160. https://doi.org/10.1111/j.1749-6632.2010.05852.x
109. Bharti AC, Donato N, Aggarwal BB (2003) Curcumin (diferuloylmethane) inhibits constitutive and IL-6-inducible STAT3 phosphorylation inhuman multiple myeloma cells. J Immunol 171:3863–3871. https://doi.org/10.4049/jimmunol.171.7.3863
110. Pandey A, Vishnoi K, Mahata S, Tripathi SC, Misra SP, Misra V (2015) Berberine and curcumin target survivin and STAT3 in gastric cancer cells and synergize actions of standard chemotherapeutic 5-Fluorouracil. Nutr Cancer 67:1293–1304. https://doi.org/10.1080/01635581.2015.1085581
111. Fetoni AR, Paciello F, Mezzogori D, Rolesi R, Eramo SL, Paludetti G (2015) Molecular targets for anticancer redox chemotherapy and cisplatin-induced ototoxicity: the role of curcumin on pSTAT3 and Nrf-2 signalling. Br J Cancer 113:1434–1444. https://doi.org/10.1038/bjc.2015.359
112. Wu SH, Hang LW, Yang JS, Chen HY, Lin HY, Chiang JH, Chung JG (2010) Curcumin induces apoptosis in human non-small cell lung cancer NCI-H460 cells through ER stress and caspase cascade-and mitochondria-dependent pathways. Anticancer Res 30(6):2125–2133
113. Alexandrow MG, Song LJ, Altiok S, Gray J, Haura EB, Kumar NB (2012) Curcumin: a novel Stat3 pathway inhibitor for chemoprevention of lung cancer. Eur J Cancer Prev 21:407–412. https://doi.org/10.1097/CEJ.0b013e32834ef194
114. Dhillon N, Aggarwal BB, Newman RA, Wolff RA, Kunnumakkara AB, Abbruzzese JL et al (2008) Phase II trial of curcumin in patients with advanced pancreatic cancer. Clin Cancer Res 14:4491–4499. https://doi.org/10.1158/1078-0432.CCR-08-0024
115. Vadhan-Raj S, Weber DM, Wang M, Giralt SA, Thomas SK, Alexanian R (2007) Curcumin downregulates NF-kB and related genes in patients with multiple myeloma: results of a phase I/II study. Blood 11:1177
116. Yang JY, Zhong X, Yum HW, Lee HJ, Kundu JK, Na HK (2013) Curcumin inhibits STAT3 signaling in the colon of dextran sulphate sodium-treated mice. J Cancer Prev 18:186–191. https://doi.org/10.15430/jcp.2013.18.2.18
117. Kelkel M, Jacob C, Dicato M, Diederich M (2010) Potential of the dietary antioxidants resveratrol and curcumin in prevention and treatment of hematologic malignancies. Molecules 15(10):7035–7074. https://doi.org/10.3390/molecules15107035
118. Kim SM, Kim C, Bae H, Lee JH, Baek SH, Nam D, Sethi G (2015) 6-Shogaol exerts anti-proliferative and pro-apoptotic effects through the modulation of STAT3 and MAPKs signaling pathways. Mol Carcinog 54(10):1132–1146
119. Seo HS, Ku JM, Choi HS, Woo JK, Jang BH, Shin YC, Ko SG (2014) Induction of caspase-dependent apoptosis by apigenin by inhibiting STAT3 signaling in HER2-overexpressing MDA-MB-453 breast cancer cells. Anticancer Res 34:2869–2882
120. Zhao X, Guo X, Shen J, Hua D (2018) Alpinetin inhibits proliferation and migration of ovarian cancer cells via suppression of STAT3 signaling. Mol Med Rep 18:4030–4036
121. Zhang J, Sikka S, Siveen KS, Lee JH, Um JY, Kumar AP, Ahn KS (2017) Cardamonin represses proliferation, invasion, and causes apoptosis through the modulation of signal transducer and activator of transcription 3 pathway in prostate cancer. Apoptosis 22(1):158–168. https://doi.org/10.1007/s10495-016-1313-7
122. Wang Z, Tang X, Wu X, Yang M, Wang W, Wang L, Wang D (2019) Cardamonin exerts anti-gastric cancer activity via inhibiting LncRNA-PVT1-STAT3 axis. Biosci Rep 39(5):1–9. https://doi.org/10.1042/BSR20190357

123. Jorvig JE, Chakraborty A (2015) Zerumbone inhibits growth of hormone refractory prostate cancer cells by inhibiting JAK2/STAT3 pathway and increases paclitaxel sensitivity. Anti-Cancer Drugs 26(2):160–166. https://doi.org/10.1097/CAD.0000000000000171
124. Shanmugam MK, Rajendran P, Li F, Kim C, Sikka S, Siveen KS, Sethi G (2015) Abrogation of STAT3 signaling cascade by zerumbone inhibits proliferation and induces apoptosis in renal cell carcinoma xenograft mouse model. Mol Carcinog 54(10):971–985. https://doi.org/10.1002/mc.22166
125. Kizhakkayil J, Thayyullathil F, Chathoth S, Hago A, Patel M, Galadari S (2010) Modulation of curcumin-induced Akt phosphorylation and apoptosis by PI3K inhibitor in MCF-7 cells. Biochem Biophys Res Commun 394:476–481. https://doi.org/10.1016/j.bbrc.2010.01.132
126. Jin H, Qiao F, Wang Y, Xu Y, Shang Y (2015) Curcumin inhibits cell proliferation and induces apoptosis of human non-small cell lung cancer cells through the upregulation of miR-192-5p and suppression of PI3K/Akt signaling pathway. Oncol Rep 34:2782–2789. https://doi.org/10.3892/or.2015.4258
127. Guan F, Ding Y, Zhang Y, Zhou Y, Li M, Wang C (2016) Curcumin suppresses proliferation and migration of MDA-MB-231 breast cancer cells through autophagy-dependent Akt degradation. PLoS ONE 11:e0146553. https://doi.org/10.1371/journal.pone.0146553
128. Akkoc Y, Berrak O, Arsan ED, Obakan P, Coker-Gurkan A, Palavan-Unsal N (2015) Inhibition of PI3K signaling triggered apoptotic potential of curcumin which is hindered by Bcl2 through activation of autophagy in MCF-7 cells. Biomed Pharmacother 71:161–171. https://doi.org/10.1016/j.biopha.2015.02.029
129. Lai HW, Chien SY, Kuo SJ, Tseng LM, Lin HY, Chi CW (2012) The potential utility of curcumin in the treatment of HER-2-overexpressedbreast cancer: an in vitro and in vivo comparison study with herceptin. Evid Based Complement Alternat Med 2012:486568. https://doi.org/10.1155/2012/486568
130. Tan BL, Norhaizan ME (2019) Curcumin combination chemotherapy: the implication and efficacy in cancer. Molecules 24(14):2527. https://doi.org/10.3390/molecules24142527
131. Hao F, Kang J, Cao Y (2015) Curcumin attenuates palmitate-induced apoptosis in MIN6 pancreatic β-cells through PI3K/Akt/FoxO1 and mitochondrial survival pathways. Apoptosis 20:1420–1432. https://doi.org/10.1007/s10495-015-1150-0
132. Li XJ, Li Y, Jin CT, Fan J, Li HJ (2015) Curcumin induces apoptosis by PTEN/PI3K/AKT pathway in EC109 cells. Chin J Appl Physiol 31(2):174–177
133. Xu X, Qin J, Liu W (2014) Curcumin inhibits the invasion of thyroid cancer cells via down-regulation of PI3K/Akt signaling pathway. Gene 546:226–232. https://doi.org/10.1016/j.gene.2014.06.006
134. Jiang QG, Li TY, Liu DN, Zhang HT (2014) PI3K/Akt pathway involving into apoptosis and invasion in human colon cancer cells LoVo. Mol Biol Rep 41:3359–3367
135. Qiao Q, Jiang Y, Li G (2013) Inhibition of the PI3K/AKT-NF-κB pathway with curcumin enhanced radiation-induced apoptosis in human Burkitt's lymphoma. J Pharmacol Sci 121:247–256. https://doi.org/10.1254/jphs.12149FP
136. Zhao G, Han X, Zheng S, Li Z, Sha Y, Ni J (2016) Curcumin induces autophagy, inhibits proliferation and invasion by downregulating AKT/mTOR signaling pathway in human melanoma cells. Oncol Rep 35:1065–1074. https://doi.org/10.3892/or.2015.4413
137. Li B, Takeda T, Tsuiji K, Wong TF, Tadakawa M, Kondo A (2013) Curcumin induces cross-regulation between autophagy and apoptosis in uterine leiomyosarcoma cells. Int J Gynecol Cancer 23:803–808. https://doi.org/10.1097/IGC.0b013e31828c9581
138. Clark CA, McEachern MD, Shah SH, Rong Y, Rong X, Smelley CL (2010) Curcumin inhibits carcinogen and nicotine-induced mammalian target of rapamycin pathway activation in head and neck squamous cell carcinoma. Cancer Prev Res 3:1586–1595. https://doi.org/10.1158/1940-6207.CAPR-09-0244
139. Kondo A, Takeda T, Li B, Tsuiji K, Kitamura M, Wong TF et al (2012) Epigallocatechin-3-gallate potentiates curcumin's ability to suppress uterine leiomyosarcoma cell growth and induce apoptosis. Int J Clin Oncol 18:380–388. https://doi.org/10.1007/s10147-012-0387-7

140. Guo Y, Li Y, Shan Q, He G, Lin J, Gong Y (2015) Curcumin potentiates the anti-leukemia effects of imatinib by downregulation of the AKT/mTOR pathway and BCR/ABL gene expression in Ph+ acute lymphoblastic leukemia. Int J Biochem Cell Biol 65:1–11. https://doi.org/10.1016/j.biocel.2015.05.00
141. Lin CH, Chang CY, Lee KR, Lin HJ, Chen TH, Wan L (2015) Flavones inhibit breast cancer proliferation through the Akt/FOXO3a signaling pathway. BMC Cancer 15:958
142. Tong X, Pelling JC (2013) Targeting the PI3K/Akt/mTOR axis by apigenin for cancer prevention. Anti Cancer Agents Med Chem 13:971–978
143. Tong J, Shen Y, Zhang Z, Hu Y, Zhang X, Han L (2019) Apigenin inhibits epithelial-mesenchymal transition of human colon cancer cells through NF-κB/Snail signaling pathway. Biosci Rep 39(5). https://doi.org/10.1042/BSR20190452
144. Yang J, Pi C, Wang G (2018) Inhibition of PI3K/Akt/mTOR pathway by apigenin induces apoptosis and autophagy in hepatocellular carcinoma cells. Biomed Pharmacother 103:699–707
145. Chen X, Xu H, Yu X, Wang X, Zhu X, Xu X (2019) Apigenin inhibits in vitro and in vivo tumorigenesis in cisplatin-resistant colon cancer cells by inducing autophagy, programmed cell death and targeting mTOR/PI3K/Akt signalling pathway. J BUON 24:488–493
146. Wu L, Yang W, Zhang SN, Lu JB (2015) Alpinetin inhibits lung cancer progression and elevates sensitization drug-resistant lung cancer cells to cis-diammined dichloridoplatium. Drug Des Devel Ther 9:6119–6127. https://doi.org/10.2147/DDDT.S92702
147. Wen M, Wu J, Luo H, Zhang H (2012) Galangin induces autophagy through upregulation of p53 in HepG2 cells. Pharmacology 89:247–255
148. Lee CC, Lin ML, Meng M, Chen SS (2018) Galangin induces p53-independent S-phase arrest and apoptosis in human nasopharyngeal carcinoma cells through inhibiting PI3K–AKT signaling pathway. Anticancer Res 38(3):1377–1389
149. Wang HX, Tang C (2017) Galangin suppresses human laryngeal carcinoma via modulation of caspase-3 and AKT signaling pathways. Oncol Rep 38(2):703–714
150. Meng B, Ii H, Qu W, Yuan H (2018) Anticancer effects of 6-gingerol in retinoblastoma cancer cells (RB355 Cell Line) are mediated via apoptosis induction, cell cycle arrest and upregulation of PI3K/Akt signaling pathway. Med Sci Monit 24:1980. https://doi.org/10.12659/msm.905450
151. Kapoor V, Aggarwal S, Das SN (2016) 6-Gingerol mediates its anti tumor activities in human oral and cervical cancer cell lines through apoptosis and cell cycle arrest. Phytother Res 30(4):588–595. https://doi.org/10.1002/ptr.5561
152. Joo JH, Hong SS, Cho YR, Seo DW (2016) 10-Gingerol inhibits proliferation and invasion of MDA-MB-231 breast cancer cells through suppression of Akt and p38MAPK activity. Oncol Rep 35(2):779–784
153. Loung CY, Rasmussen AN, Hoskin DW (2019) The phenolic gingerols and gingerol-derived shogaols: features and properties related to the prevention and treatment of cancer and chronic inflammation. In: Polyphenols in plant. Academic, New York, pp 395–405. https://doi.org/10.1016/B978-0-12-813768-0.00024
154. Hung JY, Hsu YL, Li CT, Ko YC, Ni WC, Huang MS, Kuo PL (2009) 6-Shogaol, an active constituent of dietary ginger, induces autophagy by inhibiting the AKT/mTOR pathway in human non-small cell lung cancer A549 cells. J Agric Food Chem 57(20):9809–9816. https://doi.org/10.1021/jf902315e
155. Zainal NS, Gan CP, Lau BF, San Yee P, Tiong KH, Rahman ZAA, Cheong SC (2018) Zerumbone targets the CXCR4-RhoA and PI3K-mTOR signaling axis to reduce motility and proliferation of oral cancer cells. Phytomedicine 39:33–41
156. Kang CG, Lee HJ, Kim SH, Lee EO (2015) Zerumbone suppresses osteopontin-induced cell invasion through inhibiting the FAK/AKT/ROCK pathway in human non-small cell lung cancer A549 cells. J Nat Prod 79(1):156–160

157. Kashafi E, Moradzadeh M, Mohamadkhani A, Erfanian S (2017) Kaempferol increases apoptosis in human cervical cancer HeLa cells via PI3K/AKT and telomerase pathways. Biomed Pharmacother 89:573–577
158. Yang J, Xiao P, Sun J, Guo L (2018) Anticancer effects of kaempferol in A375 human malignant melanoma cells are mediated via induction of apoptosis, cell cycle arrest, inhibition of cell migration and downregulation of m-TOR/PI3K/AKT pathway. J BUON 23:218–223
159. Break MKB, Hossan MS, Khoo Y, Qazzaz ME, Al-Hayali MZ, Chow SC, Khoo TJ (2018) Discovery of a highly active anticancer analogue of cardamonin that acts as an inducer of caspase-dependent apoptosis and modulator of the mTOR pathway. Fitoterapia 125:161–173. https://doi.org/10.1016/j.fitote.2018.01.006
160. Chen H, Tang X, Liu T, Jing L, Wu J (2019) Zingiberene inhibits in vitro and in vivo human colon cancer cell growth via autophagy induction, suppression of PI3K/AKT/mTOR pathway and caspase 2 deactivation. J BUON 24(4):1470–1475
161. Nagy LI, Fehér LZ, Szebeni GJ, Gyuris M, Sipos P, Alföldi R (2015) Curcumin and its analogue induce apoptosis in leukemia cells and have additive effects with bortezomib in cellular and xenograft models. BioMed Res. https://doi.org/10.1155/2015/968981
162. Jagetia GC, Aggarwal BB (2007) "Spicing up" of the immune system by curcumin. Clin Immunol 27(1):19–35. https://doi.org/10.1007/s10875-006-9066-7
163. Shanbhag VKL (2017) Curcumin in chronic lymphocytic leukemia–A review. Asian Pac J Trop Biomed 7(6):505–512. https://doi.org/10.1016/j.apjtb.2017.05.003
164. Park S, Cho DH, Andera L, Suh N, Kim I (2013) Curcumin enhances TRAIL-induced apoptosis of breast cancer cells by regulating apoptosis-related proteins. Mol Cell Biochem 383(1-2):39–48. https://doi.org/10.1007/s11010-013-1752-1
165. Díaz Osterman CJ, Gonda A, Stiff T, Sigaran U, Valenzuela MM, Ferguson Bennit HR et al (2016) Curcumin induces pancreatic adenocarcinoma cell death via reduction of the inhibitors of apoptosis. Pancreas 45(1):101–109. https://doi.org/10.1097/MPA.0000000000000411
166. Sarkar S, Dubaybo H, Ali S, Goncalves P, Kollepara SL, Sethi S (2013) Down-regulation of miR-221 inhibits proliferation of pancreatic cancer cells through up-regulation of PTEN, p27kip1, p57kip2, and PUMA. Am J Cancer Res 3(5):465
167. Sun XD, Liu XE, Huang DS (2013) Curcumin reverses the epithelial-mesenchymal transition of pancreatic cancer cells by inhibiting the Hedgehog signaling pathway. Oncol Rep 29 (6):2401–2407. https://doi.org/10.3892/or.2013.2385
168. Song W, Yan CY, Zhou QQ, Zhen LL (2017) Galangin potentiates human breast cancer to apoptosis induced by TRAIL through activating AMPK. Biomed Pharmacother 89:845–856
169. Ozbey U, Attar R, Romero MA, Alhewairini SS, Afshar B, Sabitaliyevich UY, Farooqi AA (2019) Apigenin as an effective anticancer natural product: spotlight on TRAIL, WNT/β-catenin, JAK-STAT pathways, and microRNAs. J Cell Biochem 120(2):1060–1067
170. Yodkeeree S, Sung B, Limtrakul P, Aggarwal BB (2009) Zerumbone enhances TRAIL-induced apoptosis through the induction of death receptors in human colon cancer cells: Evidence for an essential role of reactive oxygen species. Cancer Res 69(16):6581–6589
171. Gao Y, Yin J, Rankin G, Chen Y (2018) Kaempferol induces G2/M cell cycle arrest via checkpoint kinase 2 and promotes apoptosis via death receptors in human ovarian carcinoma A2780/CP70 Cells. Molecules 23(5):1095. https://doi.org/10.3390/molecules23051095
172. Kashyap D, Sharma A, Tuli HS, Sak K, Punia S, Mukherjee TK (2017) Kaempferol–A dietary anticancer molecule with multiple mechanisms of action: Recent trends and advancements. J Funct Foods 30:203–219
173. Imran M, Salehi B, Sharifi-Rad J, Aslam Gondal T, Saeed F, Imran A, Shahbaz M, Tsouh Fokou PV, Umair Arshad M, Khan H, Guerreiro SG, Martins N, Estevinho LM (2019) Kaempferol: a key emphasis to its anticancer potential. Molecules 24(12):2277. https://doi.org/10.3390/molecules24122277
174. Yarden Y (2001) The EGFR family and its ligands in human cancer: signalling mechanisms and therapeutic opportunities. Eur J Cancer 37:S3–S8. https://doi.org/10.1016/S0959-8049 (01)00230-1

175. Starok M, Preira P, Vayssade M, Haupt K, Salome L, Rossi C (2015) EGFR inhibition by curcumin in cancer cells: a dual mode of action. Biomacromolecules 16:1634–1642. https://doi.org/10.1021/acs.biomac.5b00229
176. Bojko A, Cierniak A, Adamczyk A, Ligeza J (2015) Modulatory effects of curcumin and tyrphostins (AG494 and AG1478) on growth regulation and viability of LN229 human brain cancer cells. Nutr Cancer 67:1170–1182. https://doi.org/10.1080/01635581.2015.1073764
177. Li S, Liu Z, Zhu F, Fan X, Wu X, Zhao H (2013) Curcumin lowers erlotinib resistance in non-small cell lung carcinoma cells with mutated EGF receptor. Oncol Res 21:137–144. https://doi.org/10.3727/096504013X13832473330032
178. Lev-Ari S, Starr A, Vexler A, Karaush V, Loew V, Greif J (2006) Inhibition of pancreatic and lung adenocarcinoma cell survival by curcumin is associated with increased apoptosis, down-regulation of COX-2 and EGFR and inhibition of Erk1/2 activity. Anticancer Res 26:4423–4430
179. Chen JW, Tang YL, Liu H, Zhu ZY, Lü D, Geng N, Chen Y (2011) Anti-proliferative and anti-metastatic effects of curcumin on oral cancer cells. Hua xi kou qiang yi xue za zhi Huaxi kouqiang yixue zazhi. West Chin J Stomatol 29(1):83–86
180. Yoysungnoen-Chintana P, Bhattarakosol P, Patumraj S (2014) Antitumor and antiangiogenic activities of curcumin in cervical cancer xenografts in nude mice. Biomed Res Int 2014:817972. https://doi.org/10.1155/2014/817972
181. Wang S, Yu S, Shi W, Ge L, Yu X, Fan J (2011) Curcumin inhibits the migration and invasion of mouse hepatoma Hca-F cells through down-regulating caveolin-1 expression and epidermal growth factor receptor signaling. IUBMB Life 63:775–782. https://doi.org/10.1002/iub.507
182. Sun XD, Liu XE, Huang DS (2012) Curcumin induces apoptosis of triple-negative breast cancer cells by inhibition of EGFR expression. Mol Med Rep 6(6):1267–1270. https://doi.org/10.3892/mmr.2012.1103
183. Somers-Edgar TJ, Scandlyn MJ, Stuart EC, Le Nedelec MJ, Valentine SP, Rosengren RJ (2008) The combination of epigallocatechin gallate and curcumin suppresses ER alpha-breast cancer cell growth in vitro and in vivo. Int J Cancer 122:1966–1971. https://doi.org/10.1002/ijc.23328
184. Zhan Y, Chen Y, Liu R, Zhang H, Zhang Y (2014) Potentiation of paclitaxel activity by curcumin in human breast cancer cell by modulating apoptosis and inhibiting EGFR signaling. Arch Pharm Res 37:1086–1095. https://doi.org/10.1007/s12272-013-0311
185. Lee JY, Lee YM, Chang GC, Yu SL, Hsieh WY, Chen JJ (2011) Curcumin induces EGFR degradation in lung adenocarcinoma and modulates p38 activation in intestine: the versatile adjuvant for gefitinib therapy. PLoS ONE 6:e23756. https://doi.org/10.1371/journal.pone.0023756
186. Lv T, Zhang W, Han X (2018) Zerumbone suppresses the potential of growth and metastasis in hepatoma HepG2 cells via the MAPK signaling pathway. Oncol Lett 15(5):7603–7610
187. Pham H, Chen M, Takahashi H, King J, Reber HA, Hines OJ, Pandol S, Eibl G (2012) Apigenin inhibits NNK-induced focal adhesion kinase activation in pancreatic cancer cells. Pancreas 41:1306–1315
188. Kwak MK, Yang KM, Park J, Lee S, Park Y, Hong E, Lee J (2017) Galangin enhances TGF-β1-mediated growth inhibition by suppressing phosphorylation of threonine 179 residue in Smad3 linker region. Biochem Biophys Res 494(3-4):706–713
189. Lei D, Zhang F, Yao D, Xiong N, Jiang X, Zhao H (2018) Galangin increases ERK1/2 phosphorylation to decrease ADAM9 expression and prevents invasion in A172 glioma cells. Mol Med Rep 17:667–673. https://doi.org/10.3892/mmr.2017.7920
190. Dang Q, Song W, Xu D, Ma Y, Li F, Zeng J, Li L (2015) Kaempferol suppresses bladder cancer tumor growth by inhibiting cell proliferation and inducing apoptosis. Mol Carcinog 54 (9):831–840
191. Hung TW, Chen PN, Wu HC, Wu SW, Tsai PY, Hsieh YS, Chang HR (2017) Kaempferol inhibits the invasion and migration of renal cancer cells through the downregulation of AKT and FAK pathways. Int J Res Med Sci 14(10):984. https://doi.org/10.7150/ijms.20336

192. Mohamed HE, Badawy MM (2019) Modulatory effect of zingerone against cisplatin or γ-irradiation induced hepatotoxicity by molecular targeting regulation. Appl Radiat Isot 154:108891. https://doi.org/10.1016/j.apradiso.2019.10889
193. Kwak MK, Kensler TW (2010) Targeting NRF2 signaling for cancer chemoprevention. Toxicol Appl Pharmacol 244:66–76. https://doi.org/10.1016/j.taap.2009.08.028
194. Ashrafizadeh M, Ahmadi Z, Mohamamdinejad R, Farkhondeh T, Samarghandian S (2020) Curcumin activates the Nrf2 pathway and induces cellular protection against oxidative injury. Curr Mol Med 20(2):116–133. https://doi.org/10.2174/1566524019666191016150757
195. Boyanapalli SS, Paredes-Gonzalez X, Fuentes F, Zhang C, Guo Y, Pung D (2014) Nrf2 knockout attenuates the anti-inflammatory effects of phenethyl isothiocyanate and curcumin. Chem Res Toxicol 27:2036–2043. https://doi.org/10.1021/tx500234h
196. Khor TO, Huang Y, Wu TY, Shu L, Lee J, Kong AN (2011) Pharmacodynamics of curcumin as DNA hypomethylation agent in restoring the expression of Nrf2 via promoter CpGs demethylation. Biochem Pharmacol 82:1073–1078. https://doi.org/10.1016/j.bcp.2011.07.065
197. Avtanski D, Poretsky L (2018) Phyto-polyphenols as potential inhibitors of breast cancer metastasis. Mol Med 24(1):29
198. Chen J, Wang FL, Chen WD (2014) Modulation of apoptosis-related cell signaling pathways by curcumin as a strategy to inhibit tumor progression. Mol Biol Rep 41:4583–4594. https://doi.org/10.1007/s11033-014-3329-
199. DeNicola GM, Karreth FA, Humpton TJ, Gopinathan A, Wei C, Frese K (2011) Oncogene-induced Nrf2 transcription promotes ROS detoxification and tumorigenesis. Nature 475 (7354):106. https://doi.org/10.1038/nature10189
200. Zhang SN, Yong Q, Wu XL, Liu XP (2014) Synergism inhibition of curcumin combined with cisplatin on T24 bladder carcinoma cells and its related mechanism. ZhongYaoCai 37:2043–2046
201. Fang D, Xiong Z, Xu J, Yin J, Luo R (2019) Chemopreventive mechanisms of galangin against hepatocellular carcinoma: a review. Biomed Pharmacother 109:2054–2061
202. Li Y, Guo M, Lin Z, Zhao M, Xia Y, Wang C, Zhu B (2018) Multifunctional selenium nanoparticles with Galangin-induced HepG2 cell apoptosis through p38 and AKT signalling pathway. R Soc Open Sci 5(11):180509. https://doi.org/10.1098/rsos.180509
203. Kumar R, Tiku AB (2018) Galangin induces cell death by modulating the expression of glyoxalase-1 and Nrf-2 in HeLa cells. Chem Biol Interact 279:1–9. https://doi.org/10.1016/j.cbi.2017.11.001
204. Zhang L, Cheng X, Gao Y, Zheng J, Xu Q, Sun Y, Sun Z (2015) Apigenin induces autophagic cell death in human papillary thyroid carcinoma BCPAP cells. Food Funct 6(11):3464–3472
205. Impheng H, Richert L, Pekthong D, Scholfield CN, Pongcharoen S, Pungpetchara I, Srisawang P (2015) [6]-Gingerol inhibits de novo fatty acid synthesis and carnitine palmitoyltransferase-1 activity which triggers apoptosis in HepG2. Am J Cancer Res 5(4):1319
206. Yang G, Wang S, Zhong L, Dong X, Zhang W, Jiang L, Ma Y (2012) 6-Gingerol induces apoptosis through lysosomal-mitochondrial axis in human hepatoma G2 cells. Phytother Res 26(11):1667–1673
207. Rastogi N, Duggal S, Singh SK, Porwal K, Srivastava VK, Maurya R, Mishra DP (2015) Proteasome inhibition mediates p53 reactivation and anti-cancer activity of 6-gingerol in cervical cancer cells. Oncotarget 6(41):43310. https://doi.org/10.18632/oncotarget.6383
208. Liang T, He Y, Chang Y, Liu X (2019) 6-shogaol a active component from ginger inhibits cell proliferation and induces apoptosis through inhibition of STAT-3 translocation in ovarian cancer cell lines (A2780). Biotechnol Bioproc E 24(3):560–567
209. Ryu MJ, Chung HS (2015) [10]-Gingerol induces mitochondrial apoptosis through activation of MAPK pathway in HCT116 human colon cancer cells. In Vitro Cell Dev Biol Anim 51 (1):92–101. https://doi.org/10.1007/s11626-014-9806-6
210. Sithara T, Dhanya BP, Arun KB, Sini S, Dan M, Kokkuvayil Vasu R, Nisha P (2018) Zerumbone, a cyclic sesquiterpene from Zingiber zerumbet induces apoptosis, cell cycle

arrest, and antimigratory effects in SW480 colorectal cancer cells. J Agric Food Chem 66 (3):602–612. https://doi.org/10.1021/acs.jafc.7b04472

211. Yan H, Ren MY, Wang ZX, Feng SJ, Li S, Cheng Y, Zhang GQ (2017) Zerumbone inhibits melanoma cell proliferation and migration by altering mitochondrial functions. Oncol Lett 13 (4):2397–2402. https://doi.org/10.3892/ol.2017.5742
212. Girisa S, Shabnam B, Monisha J, Fan L, Halim CE, Arfuso F, Kunnumakkara AB (2019) Potential of zerumbone as an anti-cancer agent. Molecules 24(4):734. https://doi.org/10.3390/molecules24040734
213. Zhou YZ, Tu WW, Shu CZ, Gu XF, Huang Y, Shen XJ, Gu HF (2017) Zerumbone induces G1 cell cycle arrest and apoptosis in cervical carcinoma cells. Int J Clin Exp Med 10 (4):6640–6647
214. Deorukhkar A, Ahuja N, Mercado AL, Diagaradjane P, Raju U (2015) Zerumbone increases oxidative stress in a thiol-dependent ROS-independent manner to increase DNA damage and sensitize colorectal cancer cells to radiation. Cancer Med 4(2):278–292
215. Chiang PK, Tsai WK, Chen M, Lin WR, Chow YC, Lee CC, Chen YJ (2018) Zerumbone regulates DNA repair responding to ionizing radiation and enhances radiosensitivity of human prostatic cancer cells. Integr Cancer Ther 17(2):292–298
216. Choi EM (2011) Kaempferol protects MC3T3-E1 cells through antioxidant effect and regulation of mitochondrial function. Food Chem Toxicol 49(8):1800–1805
217. Kalyani C, Narasu ML, Devi YP (2017) Synergistic growth inhibitory effect of flavonol–kaempferol and conventional chemotherapeutic drugs on cancer cells. Int J Pharm Pharm Sci 9:123–127
218. Imran M, Rauf A, Shah ZA, Saeed F, Imran A, Arshad MU, Mubarak MS (2019) Chemopreventive and therapeutic effect of the dietary flavonoid kaempferol: a comprehensive review. Phytother Res 33(2):263–275
219. Berning L, Scharf L, Aplak E, Stucki D, von Montfort C, Reichert AS, Brenneisen P (2019) In vitro selective cytotoxicity of the dietary chalcone cardamonin (CD) on melanoma compared to healthy cells is mediated by apoptosis. PLoS One 14(9). https://doi.org/10.1371/journal.pone.0222267
220. Su P, Veeraraghavan VP, Mohan SK, Lu W (2019) A ginger derivative, zingerone—a phenolic compound—induces ROS-mediated apoptosis in colon cancer cells (HCT-116). J Biochem Mol Toxicol 2019:e22403. https://doi.org/10.1002/jbt.22403
221. Vinothkumar R, Sudha M, Nalini N (2014) Chemopreventive effect of zingerone against colon carcinogenesis induced by 1,2-dimethylhydrazine in rats. Eur J Cancer Prev 23(5):361–371. https://doi.org/10.1097/CEJ.0b013e32836473ac
222. Ganaie MA, Al Saeedan A, Madhkali H, Jan BL, Khatlani T, Sheikh IA, Wani K (2019) Chemopreventive efficacy zingerone (4-[4-hydroxy-3-methylphenyl] butan-2-one) in experimental colon carcinogenesis in Wistar rats. Environ Toxicol 34(5):610–625. https://doi.org/10.1002/tox.22727
223. Shieh PC, Chen YO, Kuo DH, Chen FA, Tsai ML, Chang IS, Pan MH (2010) Induction of apoptosis by [8]-shogaol via reactive oxygen species generation, glutathione depletion, and caspase activation in human leukemia cells. J Agric Food Chem 58(6):3847–3854
224. Subramaniam D, Ramalingam S, Houchen CW, Anant S (2010) Cancer stem cells: a novel paradigm for cancer prevention and treatment. Mini Rev Med Chem 10:359–371. https://doi.org/10.2174/138955710791330954
225. Banik U, Parasuraman S, Adhikary AK, Othman NH (2017) Curcumin: the spicy modulator of breast carcinogenesis. J Exp Clin Cancer Res 36(1):98
226. Subramaniam D, Ponnurangam S, Ramamoorthy P, Standing D, Battafarano RJ, Anant S (2012) Curcumin induces cell death in esophageal cancer cells through modulating Notch signaling. PLoS One 7:e30590. https://doi.org/10.1371/journal.pone.0030590
227. Kunnumakkara AB, Anand P, Aggarwal BB (2008) Curcumin inhibits proliferation, invasion, angiogenesis and metastasis of different cancers through interaction with multiple cell signaling proteins. Cancer Lett 269(2):199–225. https://doi.org/10.1016/j.canlet.2008.03.009

228. Ghosh AK, Kay NE, Secreto CR, Shanafelt TD (2009) Curcumin inhibits prosurvival pathways in chronic lymphocytic leukemia B cells and may overcome their stromal protection in combination with EGCG. Clin Cancer Res 15(4):1250–1258. https://doi.org/10.1158/1078-0432.CCR-08-1511
229. Bimonte S, Barbieri A, Leongito M, Piccirillo M, Giudice A, Pivonello C, Izzo F (2016) Curcumin anticancer studies in pancreatic cancer. Nutrients 8(7):433
230. Ray A, Vasudevan S, Sengupta S (2015) 6-Shogaol inhibits breast cancer cells and stem cell-like spheroids by modulation of Notch signaling pathway and induction of autophagic cell death. PLoS One 10(9):e0137614
231. Guo Y, Chen Y, Liu H, Yan W (2019) Alpinetin inhibits oral squamous cell carcinoma proliferation via miR-211-5p upregulation and notch pathway deactivation. Nutr Cancer 72 (5):757–767
232. Wang J, Yan Z, Liu X, Che S, Wang C (2016) Alpinetin targets glioma stem cells by suppressing notch pathway. Tumor Biol 37:9243–9248. https://doi.org/10.1007/s13277-016-4827-2
233. Shaulian E, Karin M (2002) AP-1 as a regulator of cell life and death. Nat Cell Biol 4(5):E131. https://doi.org/10.1038/ncb0502-e131
234. Liu S, Wang Z, Hu Z, Zeng X, Li Y, Su Y (2011) Anti-tumor activity of curcumin against androgen-independent prostate cancer cells via inhibition of NF-κB and AP-1 pathway in vitro. J Huazhong Univ Sci Technolog Med Sci 31:530–534. https://doi.org/10.1007/s11596-011-0485-1
235. Das L, Vinayak M (2014) Curcumin attenuates carcinogenesis by down regulating proinflammatory cytokine interleukin-1 (IL-1α and IL-1β) via modulation of AP-1 and NF-IL6 in lymphoma bearing mice. Int Immunopharmacol 20:141–147. https://doi.org/10.1016/j.intimp.2014.02.024
236. Yang CW, Chang CL, Lee HC, Chi CW, Pan JP, Yang WC (2012) Curcumin induces the apoptosis of human monocytic leukemia THP-1cells via the activation of JNK/ERK pathways. BMC Complement Altern Med 12:22. https://doi.org/10.1186/1472-6882-12-22
237. Chien ST, Shi MD, Lee YC, Te CC, Shih YW (2015) Galangin, a novel dietary flavonoid, attenuates metastatic feature via PKC/ERK signaling pathway in TPA-treated liver cancer HepG2 cells. Cancer Cell Int 15(1):15
238. Bertl E, Bartsch H, Gerhäuser C (2006) Inhibition of angiogenesis and endothelial cell functions are novel sulforaphane-mediated mechanisms in chemoprevention. Mol Cancer Ther 5:575–585
239. Bae MK, Kim SH, Jeong JW, Lee YM, Kim HS, Kim SR (2006) Curcumin inhibits hypoxia-induced angiogenesis via down-regulation of HIF-1. Oncol Rep 15:1557–1562. https://doi.org/10.3892/or.15.6.1557
240. Upadhyay J, Kesharwani RK, Misra K (2009) Comparative study of antioxidants as cancer preventives through inhibition of HIF-1 alpha activity. Bioinformation 4(6):233–236. https://doi.org/10.6026/97320630004233
241. Shan B, Schaaf C, Schmidt A, Lucia K, Buchfelder M, Losa M (2012) Curcumin suppresses HIF1A synthesis and VEGFA release in pituitary adenomas. J Endocrinol 214:389–398
242. Lee SS, Tsai CH, Yang SF, Ho YC, Chang YC (2010) Hypoxia inducible factor-1α expression in areca quid chewing-associated oral squamous cell carcinomas. Oral Dis 16:696–701. https://doi.org/10.1111/j.1601-0825.2010.01680.x
243. Ye MX, Zhao YL, Li Y, Miao Q, Li ZK, Ren XL (2012) Curcumin reverses cisplatin resistance and promotes human lung adenocarcinoma A549/DDP cell apoptosis through HIF-1α and caspase-3 mechanisms. Phytomedicine 19:779–787. https://doi.org/10.1016/j.phymed.2012.03.005
244. Lee C, Yang J, Tsai F, Chiang N, Lu C, Huang Y, Chen F (2016) Kaempferol induces ATM/p53-mediated death receptor and mitochondrial apoptosis in human umbilical vein endothelial cells. Int J Oncol 48:2007–2014. https://doi.org/10.3892/ijo.2016.3420

245. Chin HK, Horng CT, Liu YS, Lu CC, Su CY, Chen PS, Yang JS (2018) Kaempferol inhibits angiogenic ability by targeting VEGF receptor-2 and downregulating the PI3K/AKT, MEK and ERK pathways in VEGF-stimulated human umbilical vein endothelial cells. Oncol Rep 39 (5):2351–2357. https://doi.org/10.3892/or.2018.6312
246. Luo H, Rankin GO, Liu L, Daddysman MK, Jiang B-H, Chen YC (2009) Kaempferol inhibits angiogenesis and VEGF expression through both HIF dependent and independent pathways in human ovarian cancer cells. Nutr Cancer 61(4):554–563. https://doi.org/10.1080/01635580802666281
247. Park JH, Park GM, Kim JK (2015) Zerumbone, sesquiterpene photochemical from ginger, inhibits angiogenesis. Korean J Physiol Pharm 19(4):335–340
248. Samad NA, Abdul AB, Rahman HS, Rasedee A, Ibrahim TAT, Keon YS (2017) Zerumbone suppresses angiogenesis in HepG2 cells through inhibition of matrix Metalloproteinase-9, vascular endothelial growth factor, and vascular endothelial growth factor receptor expressions. Pharmacogn Mag 13(Suppl 4):S731
249. Samad NA, Abdul AB, Rahman HS, Abdullah R, Ibrahim TAT, Othman HH (2019) Antiproliferative and antiangiogenic effects of zerumbone from Zingiber zerumbet L. Smith in sprague dawley rat model of hepatocellular carcinoma. Pharmacogn Mag 15(61):277. https://doi.org/10.4103/pm.pm_118_18
250. Huang H, Chen AY, Rojanasakul Y, Ye X, Rankin GO, Chen YC (2015) Dietary compounds galangin and myricetin suppress ovarian cancer cell angiogenesis. J Funct Foods 15:464–475. https://doi.org/10.1016/j.jff.2015.03.051
251. Chen D, Li D, Xu XB, Qiu S, Luo S, Qiu E, Zheng D (2019) Galangin inhibits epithelial-mesenchymal transition and angiogenesis by downregulating CD44 in glioma. J Cancer 10 (19):4499–4508. https://doi.org/10.7150/jca.31487
252. Bae WY, Choi JS, Kim JE, Park C, Jeong JW (2016) Zingerone suppresses angiogenesis via inhibition of matrix metalloproteinases during tumor development. Oncotarget 7(30):47232. https://doi.org/10.18632/oncotarget.10030
253. He J, Xu Q, Wang M, Qian X, Shi Z, Jiang BH (2012) Oral administration of apigenin inhibits metastasis through AKT/P70S6K1/MMP-9 pathway in orthotopic ovarian tumor model. Int J Mol 13(6):7271–7282. https://doi.org/10.3390/ijms13067271
254. Pai SG, Carneiro BA, Mota JM, Costa R, Leite CA, Barroso-Sousa R, Giles FJ (2017) Wnt/beta-catenin pathway: modulating anticancer immune response. J Hematol Oncol 10:101. https://doi.org/10.1186/s13045-017-0471-6
255. Mukherjee S, Mazumdar M, Chakraborty S, Manna A, Saha S, Khan P (2014) Curcumin inhibits breast cancer stem cell migration by amplifying the E-cadherin/β-catenin negative feedback loop. Stem Cell Res Ther 5:116. https://doi.org/10.1186/scrt506
256. Sundram V, Chauhan SC, Ebeling M, Jaggi M (2012) Curcumin attenuates β-catenin signaling in prostate cancer cells through activation of protein kinase D1. PLoS ONE 7:e35368. https://doi.org/10.1371/journal.pone.0035368
257. Choi HY, Lim JE, Hong JH (2010) Curcumin interrupts the interaction between the androgen receptor and Wnt/β-catenin signaling pathway in LNCaP prostate cancer cells. Prostate Cancer Prostatic Dis 13:343–349. https://doi.org/10.1038/pcan.2010.26
258. Yallapu MM, Khan S, Maher DM, Ebeling MC, Sundram V, Chauhan N (2014) Anti-cancer activity of curcumin loaded nanoparticles in prostate cancer. Biomaterials 35:8635–8648. https://doi.org/10.1016/j.biomaterials.2014.06.04
259. Lu Y, Wei C, Xi Z (2014) Curcumin suppresses proliferation and invasion in non-small cell lung cancer by modulation of MTA1-mediated Wnt/β-catenin pathway. In Vitro Cell Dev Biol Anim 50:840–850. https://doi.org/10.1007/s11626-014-9779-5
260. Dermani FK, Amini R, Saidijam M, Pourjafar M, Saki S, Najafi R (2018) Zerumbone inhibits epithelial-mesenchymal transition and cancer stem cells properties by inhibiting the β-catenin pathway through miR-200c. J Cell Physiol 233(12):9538–9547. https://doi.org/10.1002/jcp.26874

261. Fatima A, Abdul ABH, Abdullah R, Karjiban RA, Lee VS (2018) Docking studies reveal zerumbone targets β-catenin of the Wnt–β-catenin pathway in breast cancer. J Serb Chem Soc 83(5):575–591. https://doi.org/10.2298/JSC170313108F
262. Liu X, Li L, Lv L, Chen D, Shen L, Xie Z (2015) Apigenin inhibits the proliferation and invasion of osteosarcoma cells by suppressing the Wnt/beta-catenin signaling pathway. Oncol Rep 34:1035–1041. https://doi.org/10.3892/or.2015.4022
263. Xu M, Wang S, Song YU, Yao J, Huang K, Zhu X (2016) Apigenin suppresses colorectal cancer cell proliferation, migration and invasion via inhibition of the Wnt/β-catenin signaling pathway. Oncol Lett 11(5):3075–3080
264. Park S, Gwak J, Han SJ, Oh S (2013) Cardamonin suppresses the proliferation of colon cancer cells by promoting β-catenin degradation. Biol Pharm Bull 13:00158. https://doi.org/10.1248/bpb.b13-00158
265. Shrivastava S, Jeengar MK, Thummuri D, Koval A, Katanaev VL, Marepally S, Naidu VGM (2017) Cardamonin, a chalcone, inhibits human triple negative breast cancer cell invasiveness by downregulation of Wnt/β-catenin signaling cascades and reversal of epithelial–mesenchymal transition. Biofactors 43(2):152–169. https://doi.org/10.1002/biof.1315
266. Nicolson GL (2014) Mitochondrial dysfunction and chronic disease: treatment with natural supplements. J Integr Med 13(4):35–43
267. Cai XZ, Wang J, Xiao-Dong L, Wang GL, Liu FN, Cheng MS, Li F (2009) Curcumin suppresses proliferation and invasion in human gastric cancer cells by downregulation of PAK1 activity and cyclin D1 expression. Cancer Biol Ther 8(14):1360–1368. https://doi.org/10.4161/cbt.8.14.8720
268. Zhang HT, Luo H, Wu J, Lan LB, Fan DH, Zhu KD, Liu HM (2010) Galangin induces apoptosis of hepatocellular carcinoma cells via the mitochondrial pathway. World J Gastroenterol 16(27):3377. https://doi.org/10.3748/wjg.v16.i27.3377
269. Şirin N, Elmas L, Seçme M, Dodurga Y (2020) Investigation of possible effects of apigenin, sorafenib and combined applications on apoptosis and cell cycle in hepatocellular cancer cells. Gene 737:144428. https://doi.org/10.1016/j.gene.2020.144428
270. Gupta S, Afaq F, Mukhtar H (2002) Involvement of nuclear factor-kappa B, Bax and Bcl-2 in induction of cell cycle arrest and apoptosis by apigenin in human prostate carcinoma cells. Oncogene 21(23):3727–3738
271. Maeda Y, Takahashi H, Nakai N, Yanagita T, Ando N, Okubo T, Saito K, Shiga K, Hirokawa T, Hara M, Ishiguro H, Matsuo Y, Takiguchi S (2018) Apigenin induces apoptosis by suppressing Bcl-xl and Mcl-1 simultaneously via signal transducer and activator of transcription 3 signaling in colon cancer. Int J Oncol 52(5):1661–1673. https://doi.org/10.3892/ijo.2018.4308
272. Wang W, Heideman L, Chung CS, Pelling JC, Koehler KJ, Birt DF (2000) Cell-cycle arrest at G2/M and growth inhibition by apigenin in human colon carcinoma cell lines. Mol Carcinog 28(2):102–110
273. Seo HS, Jo JK, Ku JM, Choi HS, Choi YK, Woo JK, Shin YC (2015) Induction of caspase-dependent extrinsic apoptosis by apigenin through inhibition of signal transducer and activator of transcription 3 (STAT3) signalling in HER2-overexpressing BT-474 breast cancer cells. Biosci Rep 35(6):1–14. https://doi.org/10.1042/BSR20150165
274. Wang Z, Lu W, Li Y, Tang B (2013) Alpinetin promotes Bax translocation, induces apoptosis through the mitochondrial pathway and arrests human gastric cancer cells at the G2/M phase. Mol Med Rep 7:915–920. https://doi.org/10.3892/mmr.2012.1243
275. Fan J, Yang X, Bi Z (2015) 6-Gingerol inhibits osteosarcoma cell proliferation through apoptosis and AMPK activation. Tumor Biol 36(2):1135–1141
276. Fuzer AM, Martin AC, Becceneri AB, da Silva JA, Vieira PC, Cominetti MR (2019) [10]-Gingerol affects multiple metastatic processes and induces apoptosis in MDAMB-231 breast tumor cells. Anti-cancer Agent Med 19(5):645–654
277. Gan H, Zhang Y, Zhou Q, Zheng L, Xie X, Veeraraghavan VP, Mohan SK (2019) Zingerone induced caspase-dependent apoptosis in MCF-7 cells and prevents 7, 12-dimethylbenz

(a) anthracene-induced mammary carcinogenesis in experimental rats. J Biochem Mol Toxicol 33(10):e22387. https://doi.org/10.1002/jbt.22387

278. Hu R, Zhou P, Peng YB, Xu X, Ma J, Liu Q, Li P (2012) 6-Shogaol induces apoptosis in human hepatocellular carcinoma cells and exhibits anti-tumor activity in vivo through endoplasmic reticulum stress. PLoS One 7(6):e39664. https://doi.org/10.1371/journal.pone.0039664
279. Tan BS, Kang O, Mai CW, Tiong KH, Khoo ASB, Pichika MR, Leong CO (2013) 6-Shogaol inhibits breast and colon cancer cell proliferation through activation of peroxisomal proliferator activated receptor γ (PPARγ). Cancer Lett 336(1):127–139. https://doi.org/10.1016/j.canlet.2013.04.014
280. Shyanti RK, Sehrawat A, Singh SV, Mishra JPN, Singh RP (2017) Zerumbone modulates CD1d expression and lipid antigen presentation pathway in breast cancer cells. Toxicol In Vitro 44:74–84
281. Liao W, Chen L, Ma X, Jiao R, Li X, Wang Y (2016) Protective effects of kaempferol against reactive oxygen species-induced hemolysis and its antiproliferative activity on human cancer cells. Eur J Med Chem 114:24–32
282. Choi JB, Kim JH, Lee H, Pak JN, Shim BS, Kim SH (2018) Reactive oxygen species and p53 mediated activation of p38 and caspases is critically involved in kaempferol induced apoptosis in colorectal cancer cells. J Agric Food Chem 66(38):9960–9967
283. Halimah E, Diantini A, Destiani DP, Pradipta IS, Sastramihardja HS, Lestari K, Koyama H (2015) Induction of caspase cascade pathway by kaempferol-3-O-rhamnoside in LNCaP prostate cancer cell lines. Biomed Rep 3(1):115–117
284. Tu LY, Bai HH, Cai JY, Deng SP (2016) The mechanism of kaempferol induced apoptosis and inhibited proliferation in human cervical cancer SiHa cell: From macro to nano. Scanning 38 (6):644–653. https://doi.org/10.1002/sca.21312
285. Su L, Chen X, Wu J, Lin B, Zhang H, Lan L, Luo H (2013) Galangin inhibits proliferation of hepatocellular carcinoma cells by inducing endoplasmic reticulum stress. Food Chem Toxicol 62:810–816. https://doi.org/10.1016/j.fct.2013.10.019
286. Chan ML, Liang JW, Hsu LC, Chang WL, Lee SS, Guh JH (2015) Zerumbone, a ginger sesquiterpene, induces apoptosis and autophagy in human hormone-refractory prostate cancers through tubulin binding and crosstalk between endoplasmic reticulum stress and mitochondrial insult. Naunyn Schmiedeberg's Arch Pharmacol 388(11):1223–1236
287. Guo H, Lin W, Zhang X, Zhang X, Hu Z, Li L, Ren F (2017) Kaempferol induces hepatocellular carcinoma cell death via endoplasmic reticulum stress-CHOP-autophagy signaling pathway. Oncotarget 8(47):82207
288. Yue Z, Jin S, Yang C, Levine AJ, Heintz N (2003) Beclin 1, an autophagy gene essential for early embryonic development, is a haploinsufficient tumor suppressor. Proc Natl Acad Sci U S A 100:15077–15082
289. Ren K, Zhang W, Wu G, Ren J, Lu H, Li Z, Han X (2016) Synergistic anti-cancer effects of galangin and berberine through apoptosis induction and proliferation inhibition in oesophageal carcinoma cells. Biomed Pharmacother 84:1748–1759
290. Li X, Wang Y, Xiong Y, Wu J, Ding H, Chen X, Zhang H (2016) Galangin induces autophagy via deacetylation of LC3 by SIRT1 in HepG2 Cells. Sci Rep 6:30496. https://doi.org/10.1038/srep30496
291. Cao X, Liu B, Cao W, Zhang W (2013) Autophagy inhibition enhances apigenin-induced apoptosis in human breast cancer cells. Chin J Cancer Res 25(2):212–222. https://doi.org/10.3978/j.issn.1000-9604.2013.04.01
292. Lee Y, Sung B, Kang YJ, Kim DH, Jang J, Hwang SY, Kim ND (2014) Apigenin-induced apoptosis is enhanced by inhibition of autophagy formation in HCT116 human colon cancer cells. Int J Oncol 44:1599–1606. https://doi.org/10.3892/ijo.2014.2339
293. Wang B, Zhao XH (2017) Apigenin induces both intrinsic and extrinsic pathways of apoptosis in human colon carcinoma HCT-116 cells. Oncol Rep 37(2):1132–1140

294. Czarnik-Kwaśniak J, Kwaśniak K, Kwasek P, Świerzowska E, Strojewska A, Tabarkiewicz J (2020) The influence of lycopene,[6]-gingerol, and silymarin on the apoptosis on U-118MG glioblastoma cells in vitro model. Nutrients 12(1):96
295. Luna-Dulcey L, Tomasin R, Naves MA, da Silva JA, Cominetti MR (2018) Autophagy-dependent apoptosis is triggered by a semi-synthetic [6]-gingerol analogue in triple negative breast cancer cells. Oncotarget 9(56):30787
296. Ko H, Kim YJ, Amor EC, Lee JW, Kim HC, Kim HJ, Yang HO (2011) Induction of autophagy by dimethyl cardamonin is associated with proliferative arrest in human colorectal carcinoma HCT116 and LOVO cells. J Cell Biochem 112(9):2471–2479. https://doi.org/10.1002/jcb.23171
297. Jegannathan SD, Arul S, Dayalan H (2016) Zerumbone, a sesquiterpene, controls proliferation and induces cell cycle arrest in human laryngeal carcinoma cell line Hep-2. Nutr Cancer 68 (5):865–872. https://doi.org/10.1080/01635581.2016.1159701
298. Du J, Tang B, Wang J, Sui H, Jin X (2012) Antiproliferative effect of alpinetin in BxPC-3 pancreatic cancer cells. Int J Mol Med 29:607–612. https://doi.org/10.3892/ijmm.2012.884
299. Sahu RP, Batra S, Srivastava SK (2009) Activation of ATM/Chk1 by curcumin causes cell cycle arrest and apoptosis in human pancreatic cancer cells. Br J Cancer 100(9):1425. https://doi.org/10.1038/sj.bjc.6605039
300. Wei W, Azhar Rasul AS, Sarfraz I, Hussain G, Nageen B, Liu X, Li J (2019) Curcumol: from plant roots to cancer roots. Int J Biol Sci 15(8):1600–1609. https://doi.org/10.7150/ijbs.34716
301. Yan X, Qi M, Li P, Zhan Y, Shao H (2017) Apigenin in cancer therapy: anti-cancer effects and mechanisms of action. Cell Biosci 7(1):50
302. Tseng TH, Chien MH, Lin WL, Wen YC, Chow JM, Chen CK, Lee WJ (2017) Inhibition of MDA-MB-231 breast cancer cell proliferation and tumor growth by apigenin through induction of G2/M arrest and histone H3 acetylation-mediated p21WAF1/CIP1 expression. Environ Toxicol 32(2):434–444. https://doi.org/10.1002/tox.2224
303. Lee SH, Cekanova M, Baek SJ (2008) Multiple mechanisms are involved in 6-gingerol-induced cell growth arrest and apoptosis in human colorectal cancer cells. Mol Carcinog 47 (3):197–208. https://doi.org/10.1002/mc.20374
304. Yagihashi S, Miura Y, Yagasaki K (2008) Inhibitory effect of gingerol on the proliferation and invasion of hepatoma cells in culture. Cytotechnology 57(2):129–136. https://doi.org/10.1007/s10616-008-9121-8
305. Lin CB, Lin CC, Tsay GJ (2012) 6-Gingerol inhibits growth of colon cancer cell LoVo via induction of G2/M arrest. Evid Based Complementary Alternate Med 2012:1–7. https://doi.org/10.1155/2012/326096
306. Park YJ, Wen J, Bang S, Park SW, Song SY (2006) [6]-Gingerol induces cell cycle arrest and cell death of mutant p53-expressing pancreatic cancer cells. Yonsei Med J 47(5):688–697
307. Rasmussen A, Murphy K, Hoskin DW (2019) 10-Gingerol inhibits ovarian cancer cell growth by inducing G2 arrest. Adv Pharm Bull. https://doi.org/10.15171/apb.2019.080
308. Abdul AB, Abdelwahab SI, Jalinas JB, Al-Zubairi AS, Taha MME (2009) Combination of zerumbone and cisplatin to treat cervical intraepithelial neoplasia in female BALB/c mice. Int J Gynecol Cancer 19(6):1004–1010
309. Wang D, Li Y, Cui P, Zhao Q, Tan BB, Zhang ZD, Jia N (2016) Zerumbone induces gastric cancer cells apoptosis: Involving cyclophilin A. Biomed Pharmacother 83:740–745. https://doi.org/10.1016/j.biopha.2016.07.034
310. Ma S, Lei Y, Zhang L, Wang J (2018) Effects of zerumbone on proliferation and apoptosis of esophageal cancer cells and on P53 and Bcl-2 expression levels. Oncol Lett 16(4):4379–4383
311. Kannan M, Jayamohan S, Mohanakumar AK, Moorthy RK, Purushothama KM, Arockiam AJV (2019) Bcl-2/BCL2L12 mediated apoptosis and cell cycle arrest induced by Kaempferol through the suppression of PI3K/AKT signaling pathway in Hepatocellular carcinoma. J Adv Appl Sci Res 2(1):2454–3225

312. Parmar F, Patel C, Highland H, Pandya H, George LB (2016) Antiproliferative efficacy of kaempferol on cultured Daudi cells: an in silico and in vitro study. Adv Biol 2016:1–10. https://doi.org/10.1155/2016/9521756
313. Choi JS, Ryu J, Bae WY, Park A, Nam S, Kim JE, Jeong JW (2018) Zingerone suppresses tumor development through decreasing cyclin D1 expression and inducing mitotic arrest. Int J Mol Sci 19(9):2832. https://doi.org/10.3390/ijms19092832
314. Malami I, Abdul AB, Abdullah R, Kassim NK, Rosli R (2017) Crude extracts, flavokawain B and alpinetin compounds from the Rhizome of Alpinia mutica induce cell death via UCK2 enzyme inhibition and in turn reduce 18S rRNA biosynthesis in HT-29 cells. PLoS One 12: e0170233. https://doi.org/10.1371/journal.pone.0170233
315. Zhang L, Cheng X, Gao Y, Zhang C, Bao J, Guan (2016) Curcumin inhibits metastasis in human papillary thyroid carcinoma BCPAP cells via down-regulation of the TGF-β/Smad2/3 signaling pathway. Exp Cell Res 341(2):157–165. https://doi.org/10.1016/j.yexcr.2016.01.00
316. Choi YJ, Lee YH, Lee ST (2015) Galangin and kaempferol suppress phorbol-12-myristate-13-acetate-induced matrix metalloproteinase-9 expression in human fibrosarcoma HT-1080 cells. Mol Cells 38(2):151–155
317. Cao J, Wang H, Chen F, Fang J, Xu A, Xi W, Wang Z (2016) Galangin inhibits cell invasion by suppressing the epithelial-mesenchymal transition and inducing apoptosis in renal cell carcinoma. Mol Med Rep 13(5):4238–4244
318. Zhu Y, Rao Q, Zhang X, Zhou X (2018) Galangin induced antitumor effects in human kidney tumor cells mediated via mitochondrial mediated apoptosis, inhibition of cell migration and invasion and targeting PI3K/AKT/mTOR signalling pathway. J BUON 23(3):795–799
319. Cao HH, Chu JH, Kwan HY, Su T, Yu H, Cheng CY, Chou GX (2016) Inhibition of the STAT3 signaling pathway contributes to apigenin-mediated anti-metastatic effect in melanoma. Sci Rep 6:21731
320. Hu XW, Meng D, Fang J (2008) Apigenin inhibited migration and invasion of human ovarian cancer A2780 cells through focal adhesion kinase. Carcinogenesis 29(12):2369–2376
321. Chunhua L, Donglan L, Xiuqiong F, Lihua Z, Qin F, Yawei L, Liang Z, Ge W, Linlin J, Ping Z, Kun L, Xuegang S (2013) Apigenin up-regulates transgelin and inhibits invasion and migration of colorectal cancer through decreased phosphorylation of AKT. J Nutr Biochem 24:1766–1775
322. Tang B, Du J, Wang J, Tan G, Gao Z (2012) Alpinetin suppresses proliferation of human hepatoma cells by the activation of MKK7 and elevates sensitization to cis-diammined dichloridoplatium. Oncol Rep 27:1090–1096. https://doi.org/10.3892/or.2011.1580
323. Hosseini N, Khoshnazar A, Saidijam M, Azizi Jalilian F, Najafi R, Mahdavinezhad A, Amini R (2019) Zerumbone suppresses human colorectal cancer invasion and metastasis via modulation of FAk/PI3k/NFκB-uPA pathway. Nutr Cancer 71(1):159–171. https://doi.org/10.1080/01635581.2018.1540719
324. Wani NA, Zhang B, Teng KY, Barajas JM, Motiwala T, Hu P, Jacob ST (2018) Reprograming of glucose metabolism by zerumbone suppresses hepatocarcinogenesis. Mol Cancer Res 16 (2):256–268
325. Abdelwahab SI, Abdul AB, Zain ZNM, Hadi AHA (2012) Zerumbone inhibits interleukin-6 and induces apoptosis and cell cycle arrest in ovarian and cervical cancer cells. Int Immunopharmacol 12(4):594–602
326. Kim S, Lee J, Jeon M, Lee JE, Nam SJ (2016) Zerumbone suppresses the motility and tumorigenicity of triple negative breast cancer cells via the inhibition of TGF-β1 signaling pathway. Oncotarget 7(2):1544–1558
327. Phromnoi K, Yodkeeree S, Anuchapreeda S, Limtrakul P (2009) Inhibition of MMP-3 activity and invasion of the MDA-MB-231 human invasive breast carcinoma cell line by bioflavonoids. Acta Pharmacol Sin 30(8):1169–1176
328. Li S, Yan T, Deng R, Jiang X, Xiong H, Wang Y, Zhu Y (2017) Low dose of kaempferol suppresses the migration and invasion of triple-negative breast cancer cells by downregulating

the activities of RhoA and Rac1. Oncotargets Ther 10:4809. https://doi.org/10.2147/OTT.S140886
329. Kim JH (2016) Kaempferol inhibits pancreatic cancer cell growth and migration through the blockade of EGFR-related pathway in vitro. PLoS One 11(5):e0155264
330. Jo E, Park SJ, Choi YS, Jeon WK, Kim BC (2015) Kaempferol suppresses transforming growth factor-β1–induced epithelial-to-mesenchymal transition and migration of A549 lung cancer cells by inhibiting Akt1-mediated phosphorylation of Smad3 at threonine-179. Neoplasia 17(7):525–537. https://doi.org/10.1016/j.neo.2015.06.004
331. Hang M, Zhao F, Chen SB, Sun Q, Zhang CX (2015) Kaempferol modulates the metastasis of human non-small cell lung cancer cells by inhibiting epithelial-mesenchymal transition. Bangladesh J Pharmacol 10(2):267–270. https://doi.org/10.3329/bjp.v10i2.21739
332. Kim YJ, Jeon Y, Kim T, Lim WC, Ham J, Park YN, Ko H (2017) Combined treatment with zingerone and its novel derivative synergistically inhibits TGF-β1 induced epithelial-mesenchymal transition, migration and invasion of human hepatocellular carcinoma cells. Bioorg Med Chem 27(4):1081–1088. https://doi.org/10.1016/j.bmcl.2016.12.04
333. Chien MH, Lin YW, Wen YC, Yang YC, Hsiao M, Chang JL, Lee WJ (2019) Targeting the SPOCK1-snail/slug axis-mediated epithelial-to-mesenchymal transition by apigenin contributes to repression of prostate cancer metastasis. J Exp Clin Cancer Res 38(1):246. https://doi.org/10.1186/s13046-019-1247-3
334. Zhong W, Yang W, Qin Y, Gu W, Xue Y, Tang Y, Sun B (2019) 6-Gingerol stabilized the p-VEGFR2/VE-cadherin/β-catenin/actin complex promotes microvessel normalization and suppresses tumor progression. J Exp Clin Cancer Res 38(1):285
335. Jeong CH, Bode AM, Pugliese A, Cho YY, Kim HG, Shim JH, Dong Z (2009) [6]-Gingerol suppresses colon cancer growth by targeting leukotriene A4 hydrolase. Cancer Res 69 (13):5584–5591
336. Krause M, Dubrovska A, Linge A, Baumann M (2017) Cancer stem cells: radioresistance, prediction of radiotherapy outcome and specific targets for combined treatments. Adv Drug Deliv 109:63–73
337. Tang AQ, Cao XC, Tian L, He L, Liu F (2015) Apigenin inhibits the self-renewal capacity of human ovarian cancer SKOV3derived sphere-forming cells. Mol Med Rep 11:2221–2136
338. Liu J, Cao XC, Xiao Q, Quan MF (2015) Apigenin inhibits HeLa sphere forming cells through inactivation of casein kinase 2α. Mol Med Rep 11(1):665–669
339. Das PK, Zahan T, Abdur MR, Khanam JA, Pillai S, Islam F (2019) Natural compounds targeting cancer stem cells: a promising resource for chemotherapy. Anti Cancer Agents Med Chem 19(15):1796–1808
340. Mahoney KM, Rennert PD, Freeman GJ (2015) Combination cancer immunotherapy and new immunomodulatory targets. Nat Rev Drug Discov 14:561–584
341. Mishra AK, Kadoishi T, Wang X, Driver E, Chen Z, Wang XJ, Wang JH (2016) Squamous cell carcinomas escape immune surveillance via inducing chronic activation and exhaustion of CD8+ T Cells co-expressing PD-1 and LAG-3 inhibitory receptors. Oncotarget 7:81341–81356
342. Nelson N, Szekeres K, Iclozan C, Rivera IO, McGill A, Johnson G, Ghansah T (2017) Apigenin: selective CK2 inhibitor increases Ikaros expression and improves T cell homeostasis and function in murine pancreatic cancer. PLoS One 12(2):e0170197. https://doi.org/10.1371/journal.pone.0170197
343. Jiang BH, Qiu JG, Wang L, Liu WJ, Wang JF, Zhao EJ, Wang W (2019) Apigenin inhibits IL-6 transcription and suppresses esophageal carcinogenesis. Front Pharmacol 10:1002. https://doi.org/10.3389/fphar.2019.01002

3 Dimensional Cell Culture Techniques in Cancer Research

11

K. C. Dhanya and Aditya Menon

Abstract

In vivo animal models are vital tools for cancer research for identification of carcinogens, to provide valuable insights regarding the molecular aspects of tumorigenesis, growth, metastasis, etc., and it aids in the screening of anticancer molecules. Though in vivo models are the primary models and prerequisite for the preclinical trials, they are actually considered as poor models for most of the human ailments. Genetic, molecular, immunological, and cellular variances are there between most of the animal models and humans and this makes such models, up to an extent, unreliable. To better reproduce the biological systems under in vitro conditions in a lab set-up, 2D cell culture models were developed and even though 2D or 2 dimensional cell cultures provide inexpensive and homogenous materials, they fail to accurately represent real cell microenvironment. This further lead to 3D or 3 dimensional cell culturing techniques. The present chapter deals with details and types of various 3D cell culture approaches used in cancer studies. The various methods by which cells could be cultured in 3D have led to better reproduction of cellular physiology and mechanics and have enabled to in creating reliable models for tumour micro environment, cell migration, and testing drug delivery. The chapter proposes to discuss the importance and applications of 3D cell culture techniques and their utility in cancer related research.

Keywords

Cell culture · Cancer · Tumour microenvironment · Screening

K. C. Dhanya (✉)
Department of Microbiology, St. Mary's College, Thrissur, Kerala, India

A. Menon
Origin Diagnostics and Research, Karunagappally, Kerala, India

M. Kumar et al. (eds.), *Pharmacotherapeutic Botanicals for Cancer Chemoprevention*, https://doi.org/10.1007/978-981-15-5999-0_11

11.1 Introduction

Plants, marine flora and fauna, bacteria, and fungi are major sources of bioactive organic molecules. These molecules are of interest for pharmaceutical industry and are being screened for new drugs and chemicals [1]. Majority of anticancer drugs have been derived from these bioactive molecules. The screening for molecules with anticancer properties is a laborious process requiring years of research and huge amount of financial investments. In spite of these, only a small number of potential anticancer molecules get approval to be used for treatment. This situation is primarily due to the inability to test these drugs on a clinically relevant tumour model [2].

Primarily, preclinical testing of molecules was performed on in vivo models. The lack of reliability of such models in mimicking the intricacies of human cancers has rendered them not very useful in screening molecules. Ethical concerns and cost of maintaining such models have also curtailed their wide usage. The development in the field of culturing of cancer cells under in vitro conditions has augmented the in vivo models of drug screening. Initially cells cultured in 2D monolayer helped in faster identification of cytotoxic drugs. Co-culturing of 2 different cells helped in understanding the interaction between them and also in identifying the immunological mechanisms of cancers. The constraints of 2D cell culturing in mimicking the cancer morphology and physiology have prevented it from being used as an alternative for in vivo modes [3].

The technology of culturing cells in a 3D substrate has enabled in reproducing various aspects of a tumour in an in vitro condition. This has enabled in screening potential anticancer molecules with much more precision and reliability. The rest of the chapter describes the advances in tumour model systems and use of 3 dimensional culturing techniques in cancer research.

11.2 In Vivo and 2D Cell Culture Cancer Models

The basis of several of the currently available pharmaceuticals is botanicals or phytochemicals which are the primary or secondary metabolite molecules obtained from plants. They were used either as a medicine or as an ingredient of medicines from historic times and continue to be so. The biosynthetic capacity of plants for the production of such metabolites is being utilized as a valuable therapeutics source [1]. These phytochemicals or compounds derived from these molecules are also being recognized as potential anticancer agents. Screening the vast pool of these phytochemicals for their therapeutic potential, dosage, effective combinations, etc., requires enormous screening procedures and that necessitates the availability of potential test models. Most of the currently used anticancer agents have undergone such vigorous screening procedures. The discovery, development, and optimization of anticancer phytopharmaceuticals are carried out by several pharmaceutical companies, government organizations, and other non-government organizations [4]. They employ various in vivo and/or in vitro models for such screening.

There are four stages in a drug discovery and drug development process; they are (1) Discovery of the novel drug; (2) Preclinical development; (3) Clinical development and trials, and (4) Regulatory approval of the drug. There are various phases in each of these stages. The drug discovery and further development is a very extensive and expensive process. Most drug candidates that manage to reach the clinical trials fail at phase 2 and phase 3 and this failure is generally attributed to their poor efficacy and safety concerns [2]. The cost estimated to take a drug from the lab to market is approximated to be around $800 million to $2 billion and the procedure may take up to 10–15 years.

The knowledge in the field of cancer biology is ever increasing which have revealed the more and more complex nature of the malady and it has resulted in offering different prospects for novel diagnostic as well as therapeutic approaches. Various cancer cell complexities that present challenges during the trials of drug development include tumour microenvironment intricacies, intra-tumour and inter-tumour heterogeneity at molecular and physiologic levels, systemic and local tumour immune response, heterogenic metabolic retorts, etc., and the most important one being the acquiring of drug-resistant stem-like cancer cells which repopulate cancer after treatment [5].

Vital tools in cancer research for identifying the molecular mechanisms of cancer cell origin, progression, development, and metastasis, for carcinogen identification, etc., in vivo models (animal based) and in vitro models (cell culture techniques) are of significant importance.

The most advanced preclinical in vivo approach to study human cancers are the murine cancer models and they allow traversing of diverse mechanisms for therapeutic development rationale [6]. A few approaches of murine cancer models are standard cell line derived xenograft or CDX models, genetically engineered mouse or GEM models, patient-derived xenograft or PDX models, GEM-derived allograft or GDA models, etc. [5].

CDX or cell line derived xenograft models are obtained by transplanting murine tumour cells into host mice which are immunocompetent and this was the earliest in vivo cancer models approach and it has enabled to identify a large number of effective and potential cytotoxic drugs, for example, procarbazine and vincristine [7]. Here immunocompromised mice develop tumour mass after subcutaneous transplantation of in vitro maintained human cancer cells. The NCI60 cell line panel [7] acts as a valuable resource to generate CDX and helps to identify cytotoxic therapeutic drugs. The major drawback of CDX identified drugs was that they failed to be effective during clinical therapies and therefore received no or low approval rate from FDA, about 5–7% only for targeted therapeutic molecules with an approximate span of 12 years between the drug discovery point to clinical practice stage, and the average estimated cost come up to $0.50–2.00 billion [8].

PDX model is derived in immunocompromised mice by transplanting them subcutaneously with clinical tumour tissue samples. This model better mimics human scenario since the tumour is developed from surgically derived tumour tissue, that is, a piece of tumour tissue is transplanted directly in to recipient mice to become an in vivo model. Generally immunocompromised mice such as SCID, nude, and

NOD/SCID mice strains are used to overcome the problems of immunological rejection of the transplanted tumour tissue.

Another in vivo model is the genetically engineered mouse models or GEM model. This model is the most accurate representation of cancer with all stages of carcinogenesis, coevolve with intrinsic stroma, and immune system and this model allows better evaluation of therapeutic response, drug delivery modes, and biomarker expression patterns.

The advantages of GEM models and PDXs are combined in GDA models, GEM-derived allograft models [9]. In this model, immunocompetent animals are transplanted with tissue fragments obtained from GEM tumours. Here too, similar to GEM model, immune systems will be fully functional maintaining the interactions among tumour cells and intrinsic microenvironment and allow the assessment of metastatic system. GEMs and GDA models offer only preclinical platform for immunomodulatory therapy evaluations and optimizations.

Since PDX and GEM models express maximum similarity to human cancer biology, they are the best employed models for preclinical cancer research.

Zebrafish models of cancer allows visualization of cellular and molecular mechanisms of cancer progression, vascularization, clonal evolution, tumour heterogeneity, invasion, etc. [10, 11] and are utilized as paediatric cancer research models due to their genetically tractable nature, potential to be directly visualized, shorter generation time, etc., and they are cost-effective when compared to murine models.

Even though in vivo murine models are primarily used for the preclinical phase trials, they poorly represent human diseases due to the immunological, genetic, cellular, and molecular differences between mice and humans. The reliability of murine models is further impeded by the facts that even the environment where the animals are housed influences the results of the study, the high cost of maintenance, ethical clearance requirement, ethical considerations involved in large-scale screening, etc. Thus it is always advantageous to consider drug screening protocols under in vitro. So nowadays, cellular based or target based and high throughput in vitro screening assays usually precede the in vivo evaluation of the prospective anticancer drug. The drug has to be tested under in vivo animal model only if the in vitro screening results show a promising anticancer agent. In such an effective combined screening, the in vitro and in vivo models should be in sync with each other for the better translation of the in vitro conditions to in vivo systems. The in vitro cell culture characteristics should mimic the in vivo conditions such as cellular microenvironment and cellular complexities. Mammalian cell culturing techniques offer a platform to analyse cell and/or tissue physiologies and patho-physiologies outside the organism. The conventional in vitro models used in cancer research are 2D in vitro cancer cell lines (CCLs) cultures. Here, cells are grown on flat dishes with a culture condition optimized to enhance maximum cell attachment and cell growth and are used to test response of cells to drug candidates. An example is the NCI60 anticancer drug screening protocol which is followed for initial in vitro screening of drugs where the drug molecule is screened using a panel of sixty human tumour cell lines [12].

Two-dimensional or 2D cell culture helps to study cell biology, tissue morphology, mechanisms of disease origin and progression, mechanism of actions of drugs, protein synthesis, and the development of tissue engineering [13]. The appropriate cell culture method allows better understanding of tumour cell biology, and thus helps to optimize regimens of radiotherapy and/or chemotherapy, or to design and test new treatment protocols [14]. In cell culture the cells are dispersed in an artificial nutritional environment having proper nutrients, optimum levels of temperature, humidity, and gaseous levels in an appropriate surface [15]. Generally, the cells are grown in a sterile humidified incubator at a temperature of 37 °C with 5% CO_2 for an incubation period, days or weeks, until adequate number of cell population is achieved. Now the cell culture is ready to be used to study the cellular response to different agonists/antagonists or physiological stimulants or pathogens or potential drugs.

2D culture conditions generally vary for different cell types. The most important is the appropriate cell culture media that allow the growth of cells. The cell culture media may either be prepared in laboratory or it may be commercially obtained. The cell culture media is usually supplemented with serum and required antibiotics. MEM supplemented with serum and antibiotics, RPMI medium with serum, DMEM supplemented with serum and antibiotics, etc., are examples of cell culture media [3].

The culturing can be done either using adherent conditions where the cells are grown attached onto a glass or plastic surface or may be grown in a suspension, which, for example, in the case of lymphocyte culture resembles the natural environment more [16]. In adherent 2 dimensional cultures, cells grow as a monolayer attached to the surface in a culture bottle [culture flask or petri dish] [17]. The cells need to sub-cultured or passaged in fresh sterile medium. To sub-culture adherent cells, the cells are to be detached from the plate or flask by trypsin-EDTA treatment or by using a cell scraper to physically scrap off the cells with utmost care so as to avoid any mechanical and chemical damage to the cell structure. After this, fresh sterile medium is added so that the cell suspension is diluted or the trypsin-EDTA activity is inhibited and the cell suspension can be inoculated into fresh culture media. This is followed by incubation. For sub culturing suspended cell culture, cells from the culture is inoculated into a sterile dish or flask containing sterile growth media followed by incubation at 37 °C with of 5% CO_2 atmosphere.

The 2D cultures possess several advantages such that they are simple, involve only low-cost for maintenance and they perform well with various functional tests and are extensively used in cell-based screening assays for anticancer agents [4]. Most of the malignant tumour cells are available as immortal or continuous cell lines commercially, the reproducibility of which in a culture is highly subjective to culture conditions and thus by modifying the culture conditions, the cellular metabolisms may be modulated and used for screening. 2D cell culture is used for several in vitro assays to determine drug absorption, metabolism, distribution, toxicity, and excretion. For example, the human colon carcinoma cells or Caco-2 cells are used in drug absorption studies as they form a monolayer that mimics intestinal epithelium and express proteins involved in drug transport

[18]. Madin-Darby canine kidney or MDCK-MDR1 cell line is used to study drug absorption and hepatocytes to study drug metabolism [19].

Studies using 2D cell culture have increased our understanding of mechanisms of action of various drugs, but still the 2D cell culture has several limitations. The major limitation is due to the monolayer nature which results in atypical growth kinetics and cellular attachments. 2D system does not represent the natural microenvironments of the cells and/or tissue [20]. A few disadvantages are listed below

- 2D cultured cells do not mimic natural tissue structures or tumour architecture. Various interactions between cells and with their environment that are required for the cellular differentiation, cell proliferation, gene expression, metabolic response towards drugs, cell viability and vitality, several other metabolic and physiological functions, etc., will not be properly represented in the culture method, as they would be in the tumour tissue [21, 22].
- During the isolation of cells from tissue, their transfer to 2D conditions and culturing, the morphology as well as cell division is altered. The loss of phenotype diversity and morphological alterations of the cells further affect their function, organization of cellular organelles, cellular secretion, and cell signalling pathways [23–25].
- Cells lose their polarity and there may be changes in the response of cells, such as apoptosis in response to changes in external environment interactions [26, 27].
- 2D culture offers an extremely different growth conditions to the cells by providing unlimited access to nutrients, oxygen, signal molecules, and metabolites while the cancer cells under in vivo conditions lack such a uniform availability of oxygen, nutrients, etc., due to the 3D architecture of the tumour mass [28].
- 2D culture changes the patterns of gene expression, RNA splicing, and overall biochemistry of the cell [29, 30].
- 2D cultures are typically monocultures and could be used to study a single cell type. The tumour microenvironment, or niche, which is present in the in vivo condition is absent in the 2D system [31, 32].

These limitations of 2D culture system has made it a failure to correctly identify only those drug candidates which can perform successfully during the later preclinical and/or clinical stages of drug testing. This is due to the incapability of 2D culture system to represent the tumour complexity, to distinguish tumour cells from normal cells, or to measure cell viability rather than cell proliferation.

To overcome these problems researchers tried to improve in vitro cell culture systems to make them resemble more the in vivo conditions and thus came the three-dimensional cell cultures and co-cultures which better mimic in vivo tissue physiology [28].

11.3 3D Cell Culture Techniques

The central emphasis of drug discovery field over the last decade is on the development of more appropriate in vitro systems mimicking the in vivo systems. This have led to developing more and more systems of culturing cells in laboratory and there are thousands of research articles on the application level and well as basic research leading to more understanding regarding to cell physiology and developmental biology and improvements in cancer drug discovery. Generally, a 2 dimensional polystyrene or glass plate is used as the substrate for the growth of cells. The intricate yet complicated mechanisms and activities of normal cells, such as its division, differentiation, proliferation, motility, migration, and processes of apoptosis are precisely controlled and are mainly relied on the chronological and spatial organization of cells. There have been lots of studies which all proved that the 2 dimensional cell culture and its monolayer nature fails to represent the biochemistry, physiology, geometrical organization, etc., of cells in a living tissue [3]. This shortcoming of 2 dimensional cell culture adversely affected the field of drug discovery to the most extent as evident from the fact that only a few out of 900 drug candidates came out of the in vitro 2D culture based studies received approval under clinical trials for cancer therapy [33].

To overcome 2 dimensional cell culture system limitations and also to better mimic the tissue and organ specific cellular architecture, cells could be cultured in a 3 dimensional or 3D matrix (Fig. 11.1). As per studies there is difference in the morphology of cells growing in 2D culture and 3D culture systems [34]. 3D culture systems were initiated in view to act as bridge between 2D culture and in vivo animal models so as to make the gap minimum, between the in vitro and the in vivo. The 3D culture could be achieved only by optimizing several settings such as the culture conditions, choice of matrix, type of cells, origin of cells, etc. All the efforts to

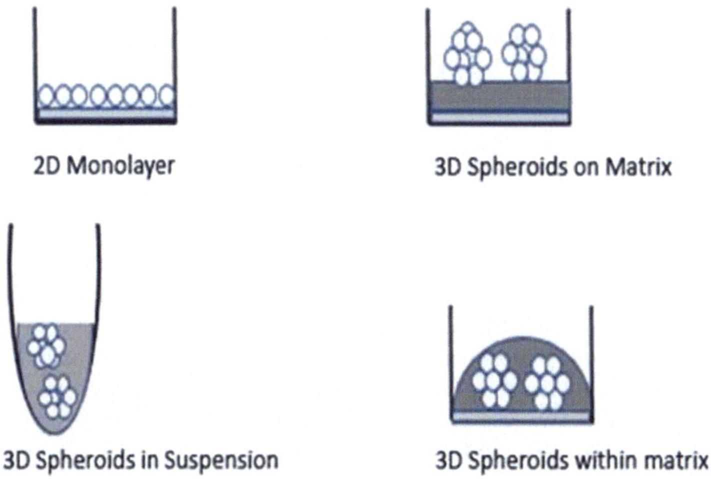

Fig. 11.1 Schematic diagrams of different culture systems

Table 11.1 Differences in the cellular and metabolic processes in 2D and 3D cell cultures

Cellular characteristics	2D cell cultures	3D cell cultures	References
Morphology of cells	Flat, stretched	Similar to natural shape in spheroid or aggregate structures	[21, 35]
Proliferation of cells	Generally proliferate at faster rate than in vivo	Proliferation rate may vary with type of cells and the culture model used	[36]
Exposure to media/ growth factors/drug molecules	All the cells equally exposed	Exposure in gradient manner due to 3D structure	[35, 37]
Stages of cell cycle	Almost all the cells in same stage of growth	Cells at diverse stages of cell cycle	[35, 38]
Gene expression	Different from that of in vivo	Similar to that in vivo tissues	[36, 39]
Drug sensitivity	Cells less resistant and drugs are very effective	Cells are more resistant to drugs	[36, 40]

explore and refine these factors significantly have improved the cell culturing techniques. It could be rightly hoped from the progress rate at which the 3D cell culture systems are getting improved that one day it will be equivalent enough to replace the in vivo biological research models.

The differences in various cellular and metabolic processes in 2D and 3D cell cultures are summarized in Table 11.1.

11.3.1 Types of 3D Cell Culture Systems

3D cell culture systems are classified into scaffold based and scaffold free types. These two systems differ basically on the type of support for attachment and proliferation provided to the growing cells.

11.3.1.1 Scaffold Based 3D Cell Culture Model

In this method the cells are grown in a prefabricated scaffold, or matrix that facilitates the growth, adhesion, and migration of cells by mimicking the extra cellular matrix (ECM). The cells will grow and fill the space available in the matrix or scaffold and form a 3D structure [41]. The shape and geometric structure of the 3D tissue could be defined by manipulating the scaffold. Many of the commercially available scaffolds are biocompatible and thus during cell division and growth they will degrade allowing the cells to migrate through [42]. Materials such as synthetic polymers, natural protein based polymers, natural carbohydrate based polymers are used. Cell grown on synthetic hard polymers are more suitable for preclinical drug evaluation as well as in regenerative medicine research [43]. These scaffolds or matrices are more porous and thus cells growing on them have easier access to

Table 11.2 Different materials used as support for 3D cell culture [48]

	Material	Advantages
Natural	Alginate Collagen, Fibrinogen Gelatin Hyaluronic acid Silk	Biodegradable, easily available, bioactive and interact with cells
Artificial	PEG—Polyethylene glycol PGA—Polyglycolide PLGA—Poly (Lactide-co-Glycolide) PMMA—Poly methyl methacrylate Polystyrene	Facilitate restoration of damaged tissue structure Inert and have long shelf-life Easily custom-made for required porosity and half life Predictable and reproducible physical and mechanical characteristics

nutrients and metabolites. The cells growing in such systems also show higher rates of proliferation and high resistance to cytotoxic drugs [44]. On the other hand, the biological scaffoldings in addition to providing the support for growth also deliver suitable microenvironments and factors which mimic the natural system and alter the gene expression patterns [45]. Generally, proteins present in extra cellular matrix are used to make such scaffoldings. Hydrogels are an example of biological scaffolding material. Here the cells to be cultured are mixed with the proteins of the scaffolding matrix prior to plating. This enables the cells to release various signal molecules, allow their migration, and accommodate other such cellular functions which all are required to achieve the homeostatic situation [46]. There is yet another type of scaffolding available in which small micro compartments are etched and micro fabricated on the bottom of wells. The compartment pattern could be subjected to adjustment to suit the experiment requirement, depending up on the type of cell to be grown or the type of interaction to be achieved between the cells to be grown. Cells of hepatocellular carcinoma origin when plated on such 3D system showed higher resistance levels to cytotoxic drugs when compared to that of 2D format [47] (Table 11.2).

11.3.1.2 Non-Scaffold Based 3D Cell Culture

In scaffold free cell culture technique, the cells are grown without a support and the cells will self-assemble to form 3 dimensional structures. The most basic form of non-scaffold based 3D cell culture is the hanging drop method where the cells are suspended using a drop of growth media. The cells grow and aggregate themselves to form 3 dimensional spheroids, the size of which depends on the amount of cells in the drop of the growth medium. Multiple cell types could be added so as to achieve co-culturing of cells for preclinical drug sensitivity screening assays [49]. The non-scaffold based spheroids can also be used to study tumour metastasis by embedding it into biological scaffolds like hydrogel. This is a good model for metastatic studies since it mimics a tumour tissue surrounded by extra cellular matrix [50]. Non-scaffold based 3 dimensional spheroids can also be made using the ultra-low attachment [ULA] coated plates for plating the cells. Here the cells are grown in a well and the technique is more advantageous than growing cells in a hanging drop of media since the duration of culture could be extended more for suiting the period of experiment and is also appropriate for cytotoxic assays. This type of spheroid could also be embedded in to ECM for tumour invasion/metastasis studies.

3 dimensional cell culture of non-scaffold based type can be obtained on microfluidic plates. The provision of nutrients and removal of wastes could be maintained in a continuous manner so that prolonged maintenance and culturing of spheroids are possible. This is suitable for differentiation studies and cancer cell invasion studies [51, 52].

11.4 Applications of 3 Dimensional Cell Culture in Cancer Research and Advantages

The 2 dimensional systems of cell culturing enabled us to understand the complex cellular physiology. The 3D culture is a 'near-to-in vivo' model, that is, it is more close to the in vivo physiologic and metabolic conditions and can be considered to bridge cell culture model and in vivo model [20]. In the 2D culture system, the extracellular matrix component and cell to cell and/or cell to matrix interactions are lost. These factors are having important roles during cellular reproduction, cellular differentiation, and proliferation and other functions under in vivo [53]. 3D cell culture systems could be tailored so as to maintain these ECM–cell, cell–cell interactions and could represent an integral model for performing in vitro studies for cancer research [54] and signalling pathways [55]. 3D culture offers a cost-effective platform for screening and testing of pharmaceutics and is a more accurate model for assessment.

The 3D cell culturing has applications in cellular differentiation studies, in the area of drug discovery, in cancer research, for gene expression analysis, and also has potential pharmacological applications. Some of the major applications of 3D cell culture technique have discussed below.

(a) *Applications in Cell differentiation studies*

The biochemistry and mechanisms of primary stems cell differentiation to their differentiated cellular forms can be studied by using 3D models. The mechanisms of differentiation are being understood and being extrapolated to therapeutic uses. Some of the studies in which 3D models are used for understanding cell differentiation are given below:

1. The process of osteogenesis and levels of osteogenic markers from mesenchymal rat stem cells were studied by using 3D models [56].
2. The levels of surface-specific markers were studied during mouse embryonic stem cells differentiation when cultured in 3D model [57].

(b) *Applications in Drug discovery*
Until the 1980s, in vivo animal models were mainly employed for drug discovery. It was made less optional later on due to its more expensive and highly unethical nature. 3D cultures have great potential to increase the extent of drug screening and to identify toxic and/or useless drug targets at earlier stages of the drug discovery trials. They can also reduce the levels of animal testing [58] and reduce the cost and the experimental complexities of animal models. The 3 dimensional approach offers a simpler and effective tool for analysing cytotoxicity and genotoxicity and allows easier anticancer drug identification [59, 60] and is an efficient model to study the cellular response to synergistic effects of molecules [61].

(c) *Applications as Tumour Models for Cancer Biology research*
Tumour spheroids in vitro resemble solid tumours in vivo. Most of the cell types growing as 3D tumour spheroid have 3 layers, the outer proliferating, inner quiescent, and central necrotic regions and mimic the micro environment of human solid tumours. The enhanced cellular interactions as well as secretion of various soluble factors from tumours will result in acidity and hypoxia in 3D system similar to in vivo conditions and help to study signalling pathways, cellular communications, interactions with ECM components, cellular proliferation rates and patterns [53, 55]. Studies on cancer biomarkers, metastasis, tumour angiogenesis and invasion are also being done extensively using 3D culture system [62].

(d) *Applications in Gene expression and Protein profiling studies*
The gene *expression* and protein profiling of cells in 3 dimensional cultures are similar to that of natural in vivo cancer tissues [63].

(e) *Applications in Cell Physiology studies*
3 dimensional cell culture allows study of cellular proliferation, morphology, adhesion, viability, and response of cells to different drugs [63].

(f) *Applications in Cell proliferation and cell-cycle studies*
Studies on 3D cultures of vascular cells, mammary cells, osteoblast cells, mouse fibroblasts, etc., gave ideas regarding the mechanisms of cellular proliferations [64, 65].

(g) *Applications in Studying Cell Cytoskeleton*
It was observed that the cells grown as 2D culture and 3D cultures differ in the extracellular matrix composition and cell cytoskeleton. Oral squamous cell carcinoma cells expressed different levels of cytoskeleton proteins when cultured as monolayers (2D) and scaffold tumours (3D) [66].

(h) *Applications in studying Apoptosis*
There were differences in apoptosis pattern when cells were cultured on 2D and 3D culture system. Cells grown as 2D culture showed signs of apoptosis in 4–5 days of plating, while cells grown as 3D remained viable up to more than 10 days [67].

(i) *Applications in Cell Adhesion and Signalling Studies*
3D cultures can be used to study the intricate details of cells growing in an environment where they are surrounded by other cells and matrix and

communications with which all control the behaviour of cells in a multicellular organism [68].

(j) *Applications in studying Cell Motility*
The motility of cells is essential for tissue repair and tissue regeneration. Scaffolds that allow different types of cell motility such as the mesenchymal and amoeboid migration enable to understand the mechanisms, modes, and after effects of cell migration [69].

(k) *Applications in Microenvironment Studies*
The microenvironment of cells decides their differentiation and functions. This could be evaluated using 3D culture system [55, 66].

(l) *Applications in studying Morphology of Cell and Tissue Architectures*
There are differences observed in the cellular morphology of cell growing in a 2D or 3D culture. While breast cancer cells showed uniform morphology in monolayers, they adopted different types of morphologies [round, mass, grape-like, stellate] in 3D culture [70]. The geometry of scaffolds also has an impact on tissue architecture since cell and ECM alignment define the biological as well as mechanical functions of cells [64].

(m) *Applications in Drug Response Studies*
Drug screening procedures require reliable in vitro models which resemble the conditions in the tumour tissues so as to evaluate drug response near to in vivo. 3D cell cultures suit this requirement [66].

(n) *Applications as co-Cultures*
3D cultures allow culturing of different cell types together and thus heterogeneous cell co-cultures are possible. This helps to understand the complexity prevailing in different tissue types and multicellular organism [71].

With the dawn of 3D cell culturing techniques, in vitro is now closer to in vivo models. They provide cellular assemblies that enable to study complex interactions and can be proficiently used in basic and applied biological research and especially in cancer research, to identify the molecular mechanisms of carcinogenesis and progression as well as to identify and develop effective drug targets from phytopharmaceuticals and other molecules.

11.5 Disadvantages of 3D Cultures

Even though 3D cell culture system overcomes the problems and drawbacks posed by the 2D cell culture system, this system is also not without any limitations. 3D culture techniques are not suitable for large-scale experimentation and drug screening. Microscopic analysis of 3D cell cultures is difficult in comparison to that of 2D cell cultures. Another disadvantage is the inability to maintain proper supply and transport of oxygen and nutrients across the 3D platform. Also 3D cell cultures developed from certain types of tissues may contain unwanted growth factors or viruses.

All these limitations are negligible when compared to the advantages and spectrum of application that a 3D culture system offers. All these models explained, in vivo systems, in vitro techniques such as 2D and 3D culture systems, etc., are used to study and screen anticancer agents and have tremendously helped for better drug discovery. Recently novel models are being experimented for drug identification and development approach. Next-generation sequencing-based methods used to evaluate chromosomal aberrations can reveal genomic instability, chromosomal translocations and deletions in chromosomes [72] and could be developed as a screening strategy for anticancer drug candidates.

11.6 Conclusion

The journey to better reproduce the in vivo testing models and to minimize their shortcomings have paved way for the development of 2D cell culture systems. The failure of 2D cultures in replicating the normal physiological and geometrical architecture of tissues resulted in increased interest and development of 3D cell culture models. The various methods by which cells could be cultured in 3D system allow better reproduction of cellular physiology and mechanics which has enabled in creating reliable models for studying tumour micro environment, cell migration, and testing drug delivery. 3D cell culture models can eventually replace the link between in vitro studies and clinical applications. The chapter discussed the importance and applications of 3 dimensional cell culture techniques and their utility in cancer research.

Conflict of Interest The authors declare no conflict of interest.

References

1. Schmidt B, Ribnicky DM, Poulev A, Logendra S, Cefalu WT, Raskin I (2008) A natural history of botanical therapeutics. Metabolism 57:S3–S9
2. Arrowsmith J, Miller P (2013) Trial watch: phase II and phase III attrition rates 2011-2012. Nat Rev Drug Discov 12:569
3. Joseph JS, Malindisa ST, Ntwasa M (2018) Two-dimensional (2D) and three-dimensional (3D) cell culturing in drug discovery. Cell Culture 2:1–22
4. Dhanya KC, Menon A, Rai LS (2019) In-vitro models in anticancer screening in phytochemistry: an in-silico and in-vitro update advances in phytochemical research. Springer, Singapore, pp 251–265. https://www.springer.com/gp/book/9789811369193
5. Day C, Merlino G, Van Dyke T (2015) Preclinical mouse cancer models: a maze of opportunities and challenges. Cell 163:39–53
6. Van Dyke T, Jacks T (2001) Cancer modeling in the modern era: progress and challenges. Cell 108:135–144
7. DeVita VT, Chu E (2008) A history of cancer chemotherapy. Cancer Res 68(21):8643–8653
8. Adams CP, Brantner VV (2006) Estimating the cost of new drug development: is it really 802 million dollars? Health Aff (Project Hope) 25(2):420–428
9. Heyer J, Kwong LN, Lowe SW, Chin L (2010) Non-germline genetically engineered mouse models for translational cancer research. Nat Rev Cancer 10:470–480

10. Ridges S, Heaton WL, Joshi D, Choi H, Eiring A, Batchelor L et al (2012) Zebrafish screen identifies novel compound with selective toxicity against leukemia. Blood 119:5621–5631
11. Gutierrez A, Pan L, Groen RW, Baleydier F, Kentsis A, Marineau J et al (2014) Phenothiazines induce PP2A-mediated apoptosis in T cell acute lymphoblastic leukemia. J Clin Invest 124:644–655
12. Shoemaker RH (2006) The NCI60 human tumour cell line anticancer drug screen. Nat Rev Cancer 6:813
13. Sanyal S (2014) Culture and assay systems used for 3D cell culture. Corning 9:1–18
14. Aggarwal B, Danda D, Gupta S, Gehlot P (2009) Models for prevention and treatment of cancer: problems vs promises. Biochem Pharmacol 78:1083–1094
15. Edmondson R, Broglie JJ, Adcock AF, Yang L (2014) Three-dimensional cell culture systems and their applications in drug discovery and cell-based biosensors. Assay Drug Dev Technol 12:207–218
16. Ryan J (2003) Introduction to animal cell culture. Tech Bull Corn:1–8. https://www.corning.com/catalog/cls/documents/application-notes/CLS-AN-042.pdf
17. Breslin S, O'Driscoll L (2013) Three-dimensional cell culture: the missing link in drug discovery. Drug Discov Today 18:240–249
18. Cai Y, Xu C, Chen P, Hu J, Hu R, Huang M et al (2014) Development, validation, and applications of a novel 7-day Caco-2 culture system. J Pharmacol Toxicol Methods 70:175–171
19. Berry MN, Grivell AR, Grivell MB, Phillips JW (1997) Isolated hepatocytes-past present and future. Cell Biol Toxicol 13:223–233
20. Cukierman E, Pankov R, Stevens DR et al (2001) Taking cell-matrix adhesions to the third dimension. Science 294:1708–1712
21. Baker B, Chen C (2012) Deconstructing the third dimension – how 3D culture microenvironments alter cellular cues. J Cell Sci 125:3015–3024
22. Hickman JA, Graeser R, de Hoogt R et al (2014) Three-dimensional models of cancer for pharmacology and cancer cell biology: capturing tumor complexity *in vitro*/ex vivo. Biotechnol J 9:1115–1128
23. Kilian K, Bugarija B, Lahn BT et al (2010) Geometric cues for directing the differentiation of mesenchymal stem cells. Proc Natl Acad Sci USA 107:4872–4877
24. Debnath J, Brugge JS (2005) Modelling glandular epithelial cancers in three-dimensional cultures. Nat Rev Cancer 5:675–688
25. Nelson CM, Bissell MJ (2006) Of extracellular matrix, scaffolds, and signaling: tissue architecture regulates development, homeostasis, and cancer. Annu Rev Cell Dev Biol 22:287–309
26. Mseka T, Bamburg JR, Cramer LP (2007) ADF/cofilin family proteins control formation of oriented actin-filament bundles in the cell body to trigger fibroblast polarization. J Cell Sci 120:4332–4344
27. Weaver V, Lelièvre S, Lakins JN et al (2002) Beta4 integrin-dependent formation of polarized three-dimensional architecture confers resistance to apoptosis in normal and malignant mammary epithelium. Cancer Cell 2:205–216
28. Pampaloni F, Reynaud EG, Stelzer EH (2007) The third dimension bridges the gap between cell culture and live tissue. Nature Reviews. Mol Cell Biol 8:839–845
29. Li C, Kato M, Shiue L, Shively JE, Ares M, Lin RJ (2006) Cell type and culture condition-dependent alternative splicing in human breast cancer cells revealed by splicing-sensitive microarrays. Cancer Res 66:1990–1999
30. Fuchs E, Tumbar T, Guasch G (2004) Socializing with the neighbors: stem cells and their niche. Cell 116:769–778
31. Gilbert PM, Havenstrite KL, Magnusson KE et al (2010) Substrate elasticity regulates skeletal muscle stem cell self-renewal in culture. Science 329:1078–1081
32. Engler A, Sen S, Sweeney HL, Discher DE (2006) Matrix elasticity directs stem cell lineage specification. Cell 126:677–689
33. De Boo J, Hendriksen C (2005) Reduction strategies in animal research: a review of scientific approaches at the intra-experimental, supra-experimental and extra-experimental levels. Altern Lab Anim 33:369–377

34. Benya PD, Shaffer JD (1982) Dedifferentiated chondrocytes reexpress the differentiated collagen phenotype when cultured in agarose gels. Cell 30:215–224
35. Kim JB (2005) Three-dimensional tissue culture models in cancer biology. Semin Cancer Biol 15(5):365–377
36. Gurski LA et al (2010) 3D matrices for anti-cancer drug testing and development. Oncology Issues 25:20–25
37. Yip D, Cho CH (2013) A multicellular 3D heterospheroid model of liver tumor and stromal cells in collagen gel for anti-cancer drug testing. Biochem Biophys Res Commun 433 (3):327–332
38. Tibbitt MW, Anseth KS (2009) Hydrogels as extracellular matrix mimics for 3D cell culture. Biotechnol Bioeng 103:655–663
39. Price KJ et al (2012) Matrigel basement membrane matrix influences expression of microRNAs in cancer cell lines. Biochem Biophys Res Commun 427:343–348
40. Hongisto V et al (2013) High-throughput 3D screening reveals differences in drug sensitivities between culture models of JIMT1 breast cancer cells. PloS one 8:e77232
41. Sourla A, Doillon C, Koutsilieris M (1996) Three-dimensional type I collagen gel system containing MG-63 osteoblasts-like cells as a model for studying local bone reaction caused by metastatic cancer cells. Anticancer Res 16:2773–2780
42. Place ES et al (2009) Synthetic polymer scaffolds for tissue engineering. Chem Soc Rev 38:1139–1151
43. Tan W, Krishnaraj R, Desai TA (2001) Evaluation of nanostructured composite collagen–chitosan matrices for tissue engineering. Tissue Eng 7:203–210
44. Bergenstock MK, et al (2009) Polystyrene scaffolds: A novel tool for in vitro three-dimensional cancer models. In: First AACR international conference on frontiers in basic cancer research
45. Birgersdotter A, Sandberg R, Ernberg I (2005) Gene expression perturbation *in vitro*—a growing case for three-dimensional (3D) culture systems. Semin Cancer Biol 15(5):405–412
46. Cushing MC, Anseth KS (2007) Hydrogel cell cultures. Science 316:1133–1134
47. Kobayashi K et al (2012) Increased expression of drug-metabolizing enzymes in human hepatocarcinoma FLC-4 cells cultured on micro-space cell culture plates. Drug Metab Pharmacokinet 27(5):478–485
48. Carletti E, Motta A, Migliaresi C (2011) Scaffolds for tissue engineering and 3D cell culture. Methods Mol Biol 695:17–39
49. Raghavan S et al (2015) Formation of stable small cell number three-dimensional ovarian cancer spheroids using hanging drop arrays for preclinical drug sensitivity assays. Gynecol Oncol 138:181–189
50. Guzman A, Ziperstein MJ, Kaufman LJ (2014) The effect of fibrillar matrix architecture on tumor cell invasion of physically challenging environments. Biomaterials 35:6954–6963
51. Wan C-R, Chung S, Kamm RD (2011) Differentiation of embryonic stem cells into cardiomyocytes in a compliant microfluidic system. Ann Biomed Eng 39:1840–1847
52. Liu T, Lin B, Qin J (2010) Carcinoma-associated fibroblasts promoted tumor spheroid invasion on a microfluidic 3D co-culture device. Lab Chip 10:1671–1677
53. Mazzoleni G, Di Lorenzo D, Steimberg N (2009) Modelling tissues in 3D: the next future of pharmacotoxicology and food research. Genes Nutr 4:13–22
54. Muthuswamy SK (2011) 3D culture reveals a signaling network. Breast Cancer Res 13:103
55. Lee GY, Kenny AP, Lee AE, Bissell MJ (2007) Three-dimensional culture models of normal and malignant breast epithelial cells. Nat Methods 4:359–365
56. Farrell E et al (2007) A comparison of the osteogenic potential of adult rat mesenchymal stem cells cultured in 2-D and on 3-D collagen glycosaminoglycan scaffolds. Technol Health Care 15:19–31
57. Liu J et al (2011) Three-dimensional spheroid cultures of A549 and HepG2 cells exhibit different lipopolysaccharide (LPS) receptor expression and LPS-induced cytokine response compared with monolayer cultures. Innate Immun 17:245–255

58. Pampaloni F, Stelzer EHK, Masotti A (2009) Three-dimensional tissue models for drug discovery and toxicology. Recent Pat Biotechnol 3:103–117
59. Meng Q (2010) Three-dimensional culture of hepatocytes for prediction of drug-induced hepatotoxicity. Expert Opin Drug Metab Toxicol 6(6):733–746
60. Shih-Feng L, Barbara SM, Binil S (2010) Long-term cultivation of HepG2 liver cells encapsulated in alginate hydrogels: a study of cell viability, morphology and drug metabolism. Toxicol In Vitro 24:1314–1323
61. Yin J, Meng Q, Zhang G, Sun Y (2009) Differential methotrexate hepatotoxicity on rat hepatocytes in 2D monolayer culture and 3D gel entrapment culture. Chem-Biol Interact 180:368–375
62. Bazou D (2010) Biochemical properties of encapsulated high-density 3-D HepG2 aggregates formed in an ultrasound trap for application in hepatotoxicity studies: Biochemical responses of encapsulated 3-D HepG2 aggregates. Cell Biol Toxicol 26:127–141
63. Ravi M, Paramesh V, Kaviya SR, Anuradha E, Solomon FDP (2014) 3D cell culture systems - advantages and applications. J Cell Physiol 230(1):16–26. https://doi.org/10.1002/jcp.24683
64. Klein EA, Yin L, Kothapalli D, Castagnino P, Byfield FJ, Xu T, Levental I, Hawthorne E, Janmey PA, Assoian RK (2009) Cell cycle control by physiological matrix elasticity and *in vivo* tissue stiffening. Curr Biol 19:1511–1518
65. Chandler EM, Berglund CM, Lee JS, Polacheck WJ, Gleghorn JP, Kirby BJ, Fischbach C (2011) Stiffness of photocrosslinked RGD-alginate gels regulates adipose progenitor cell behavior. Biotechnol Bioeng 108:1683–1692
66. Fischbach C, Chen R, Matsumoto T, Schmelzle T, Brugge JS, Polverini PJ, Mooney DJ (2007) Engineering tumors with 3D scaffolds. Nat Methods 4:855–860
67. Park CC, Zhang H, Pallavicini M, Gray WJ, Baehner F, Park JC, Bissell JM (2006) β1 integrin inhibitory antibody induces apoptosis of breast cancer cells, inhibits growth, and distinguishes malignant from normal phenotype in three dimensional cultures and *in vivo*. Cancer Res 66:1526–1535
68. Zschenker O, Streichert T, Hehlgans S, Cordes N (2012) Genome wide gene expression analysis in cancer cells reveals 3D growth to affect ECM and processes associated with cell adhesion but not DNA repair. PLOS one 7:e34279
69. Ehrbar M, Sala A, Lienemann P, Ranga A, Mosiewicz K, Bittermann A, Rizzi SC, Weber FE, Lutolf MP (2010) Elucidating the role of matrix stiffness in 3D cell migration and remodeling. Biophys J 100:284–293
70. Kenny PA, Lee GY, Myers CA, Neve RM, Semeiks JR, Spellman PT, Lorenz K (2007) The morphologies of breast cancer cell lines in three-dimensional assays correlate with their profiles of gene expression. Mol Oncol 1:84–96
71. Schmeichel KL, Bissell MJ (2003) Modeling tissue-specific signaling and organ function in three dimensions. J Cell Sci 116:2377–2388
72. Breitenbach M, Hoffmann J (2018) Cancer models. Front Oncol 8:401

Animal Models Systems of Cancer for Preclinical Trials

12

Sneha Soni, Shreetama Bandyopadhayaya, and Chandi C. Mandal

Abstract

Over the recent preceding few decades, cancer has existed as the second leading cause of death after cardiovascular diseases globally. Therefore, effective therapies are required to enhance the survival rate and lead a quality life for cancer patients. Preclinical trials are a prerequisite for the development of effective cancer therapeutics. However, direct experimentation on human subjects cannot be undertaken due to several ethical, safety, and practical issues. Hence, preclinical animal model systems have become an indispensable part, since these allow to decipher the highly complex cascades responsible for human cancer, and may help in the prediction of safety and efficacy of anticancer drug or therapy for successful translation into clinical trials. Mouse models mostly have utility in preclinical trials. While the other systems utilized include zebrafish, *Drosophila*, and *C. elegans* models. These alternative preclinical models give complimentary information for the development of cancer therapeutics. Nevertheless, companion preclinical models like dog and swine are underexploited though these can fasten to fill the gap information required between preclinical and clinical trials of cancer. The aim is to mimic this deadly and complicated disease in animal models and reflect the situation in humans for effective therapeutics. This chapter throws light on various animal models used in preclinical trials of cancer, their significance, and challenges associated with it.

Keywords

Cancer · Preclinical trials · Animal models · Therapeutics

S. Soni · S. Bandyopadhayaya · C. C. Mandal (✉)
Department of Biochemistry, Central University of Rajasthan, Ajmer, Rajasthan, India

M. Kumar et al. (eds.), *Pharmacotherapeutic Botanicals for Cancer Chemoprevention*, https://doi.org/10.1007/978-981-15-5999-0_12

12.1 Introduction

Cancer is a highly complicated and extremely diverse disease. It includes more than 200 diseases that share similar but characteristic symptoms [1]. However, enormous work in cancer biology research has increased the knowledge of this disease and has created myriad strategies for its diagnosis and treatment. But there are many complexities like tumor microenvironment, tumor heterogeneity, metabolic reprogramming, and re-wiring of the signaling pathways that challenge the success of therapeutic techniques at the clinical level [2]. Tumor detection at the preliminary stage and appropriate response to treatment are prerequisites for cancer survivors. The successful translation of fundamental cancer research to therapeutic strategy requires accurate information generated by the use of appropriate in vivo models [1]. For reliable testing and screening of anticancer drugs, preclinical animal models are essential [3]. The preclinical model that is relatively affordable allows flexible high throughput assays and can mimic the human tumor environment closely, which can be considered the best model. Mice models are predominantly traditional models to assess the anticancer targets. Mouse and human genes exhibit homology, which makes the mouse an ideal in vivo model. But differences exist in multiple aspects like species, metabolism, pharmacokinetics, pharmacodynamics, and physiology of human and mouse. However, there are other non-mammalian models that are contributing to in vivo models (Fig. 12.1). Many cancer related genes have been

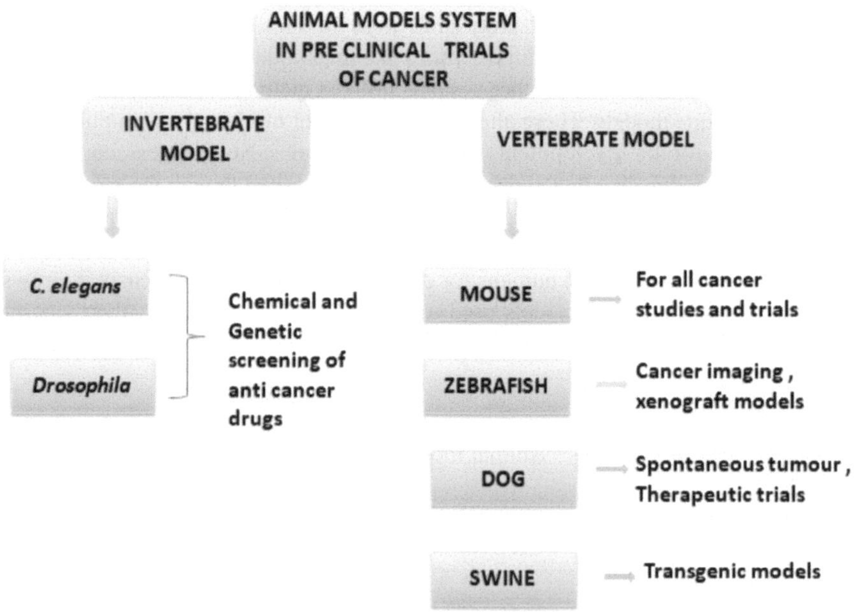

Fig. 12.1 Different types of in vivo models of preclinical trials used in the development of cancer therapeutics

identified in *Drosophila* and zebrafish. Development in transgenesis field and advancements in imaging techniques have allowed for the generation of specific and powerful cancer modeling in zebrafish. In comparison to the mouse models, it is less expensive and can reproduce in short span. On the other hand, *Drosophila*, an excellent genetic model having successful history in discovery of some central signaling pathways including Ras, Wnt, hedgehog, and others involved in tumorigenesis is of considerable utility. Smaller size of Drosophila, low maintenance cost, and lesser generation time of transgenic models make it a strong alternate in vivo model [4]. *C. elegans*, a favorite developmental biology model, has a well-defined forward and reverse genetics [5]. Further, it shares many signaling pathways involved in cancer and conserved for humans, which makes it a good model to improve the understanding of the existing basic mechanism (s) in tumor biology.

However, above all, models have some limitations as these cannot fully recapitulate human cancer due to evolutionary distances or the diverse nature of each tumor. Therefore, models like dog and swine came into existence for better translation of preclinical trials into clinical success [1, 6].

Thus, in this chapter, various preclinical models of cancer with their limitations and advantages have been briefed and discussed regarding the improvements required in preclinical models to overcome the failures in the clinical phase.

12.2 History of Animal Models of Cancer in Preclinical Trials

Animal models are a significant tool for studying cancer biology research and in addition to the creation of therapeutic strategies against cancer. The in vivo preclinical testing is mandatory to answer some basic questions like the toxicity of drug, clinical relevance, and ability to translate to clinical trials [7]. Scientists are working on different animal models to overcome the ethical and moral issues related to the direct experiments on human beings. In the 1960s, spontaneous immunocompromised mice strains were identified. The T cell deficient NUDE mice was identified by Dr. N R Grist [8], which was further characterized by Pantelouris and Hair [9]. For the first time, primary human tumors and cancer cell lines were successfully transplanted using athymic mice [10, 11]. In this way, different strains of athymic mice (NUDE, SCID, and NOD/SCID) were developed. Oncogenes and tumor suppressor genes discovery opened the path for recombination genetics, and eventually, powerful transgenic mice models were generated in 1980. Transgenic approaches became widely popular as these allow for studying and identifying novel biomarkers involved in tumorigenesis and cancer therapeutics. Though mice are a traditional and most often used model in preclinical cancer studies, there are also other vertebrates (dog, swine, zebrafish) and invertebrate models (*C. elegans*, *Drosophila*) which are employed for preclinical studies of cancer. Different types of preclinical animal model systems are used for better understanding of tumorigenesis (Fig. 12.1). Genome sequencing and well-defined genetic information of these models have added to their popularity during the last decades [12]. The high reproductive rate and less generation time permitted for the development of transgenic models of zebrafish embryo using human cancer genes [13–15]. In 2003, the first B cell precursor acute lymphoblastic leukemia model of zebrafish was developed using a transgenic approach [16, 17]. Gradually, more transgenic lines were developed by using ZFN (zinc finger nuclease) and TALEN (transcription activator like effecter nuclease), which allowed easier gene manipulation in zebrafish for permanent knock out lines [18, 19]. Swine genome sequencing [20] helped for the development of induced and gene edited models of cancer [21]. Dogs are used as a natural model for cancer.

In 1946, the first chemotherapy used in veterinary medicine was reported, applied for the treatment of hematopoietic neoplasia in dogs [22]. Similarly, invertebrate models (fruit fly and *C. elegans*) are being used to investigate signaling networks that control cancer genes, including RET proto-oncogenes, adenomatous polyposis coli, and others [23, 24]. They have been developed to employ high throughput screening and for the identification of inhibitors and markers of the conserved oncogenic signaling pathways [24].

12.3 Route of Administration of Drugs

Any drugs which are under preclinical trials have to be administered to laboratory animals. The administration of this drug requires immense care and planning so that adverse outcomes can be minimized. The type of study, i.e., topical, systemic, or parenteral, and the aim of experiments decide for the route of administration. There are varieties of different routes, including topical (transdermal), enteral routes (oral or gavage), and parenteral routes (subcutaneous, intradermal, intramuscular, intraperitoneal, intranasal, and intraocular) available for animal models. Many routes of delivery require anesthesia or sedation before drug administration. Therefore, a route should be selected on the basis of minimum harm exhibited to test animals during testing [25]. Generally, the drug administered through an internal route involves the gastrointestinal tract. The administered drug is absorbed from the gut then it undergoes hepatic metabolism before entering into systemic circulation that leads to decreased drug bioavailability. It is done by delivering a substance into mouth, oral, or gavage by mixing the substance into a diet. The very small volume of a substance is prescribed for an oral route administration, i.e., 5 ml/kg dose (for animals), to avoid gastric distension. Parenteral administration does not involve the GI tract and directly bypasses hepatic metabolism, which is an advantage for the drugs having low bioavailability. When substances are delivered in the form of the bolus into blood vessels, it is referred to as intravenous administration. The main advantage is that there is no requirement for solute absorption. In mice, it can be done mainly in the tail vein, while in fishes, the caudal artery is used. Aseptic conditions should be maintained to avoid vasculitis or irritation in vascular endothelial layer and infection. When an administration is performed directly at the skin surface level by needle, it can be subcutaneous or intramuscular. Subcutaneous injection is administered into fatty tissue beneath the skin, while the intramuscular injection is delivered directly into the muscle under the skin. It is easy to perform subcutaneous than intramuscular as the latter requires more skilled experts. Care should be taken while doing the intramuscular route, as injecting irritating substances may result in muscle necrosis or muscle weakness. To study the effects of the substance on the brain and spinal cord, the route for delivery used is known as intrathecal or epidural. It involves heavy anesthesia to be administered to the test animal. If not performed skillfully, there are chances of sample spilling that enter into the systemic circulation and produce harmful effects related to respiration.

In the case of small animals, intravenous is a challenge, so intraperitoneum is preferred over there, where delivery is done via peritoneal cavity [25]. Absorption of substance is comparatively slow in intraperitoneum as they may pass through hepatic metabolism first and then enter into the systemic circulation. If not done skillfully, then there is a chance of puncture in the skin that results in subcutaneous than intraperitoneum mode. After selection of route of administration, focus on factors like volume of drug, delivery location, pH of the substance, treatment time, rate of absorption should be considered for accuracy in preclinical findings. Thus, how substances are administered is the critical component that will predict the success of the experiment with minimum harmful effects on laboratory animals.

12.4 Mouse as a Gold Standard Model

Mouse is the most traditional animal model used in preclinical trials of cancer as it shares biological similarities with humans. In addition to that, they are relatively smaller in size and easy to maintain in laboratories [26]. They are developed to understand the complexities involved in human cancer and the multiple mechanisms involved in it, which might prove beneficial for therapeutics purpose [27]. They are extremely good models for human malignancy pathogenesis investigation due to its similarity with humans in the context of genetics and physiology [28]. Besides these, the short gestation period and life span also add to an advantage [29, 30]. The utility of mouse as an animal model for preclinical trials in cancer biology has increased due to its highly conserved genomic content and flexible precise manipulation of target genes that allow it for anticancer drug development [31].

12.5 Different Categories of Mouse as Preclinical Models

There are several models of mouse specific for different tumors and developed by different methods [32] such as the development of inbred strains that spontaneously develop tumors [33], some are created by exposure to chemical mutagens [34], or some are developed by bacterial or viral infection [35]. Additionally, xenograft models and immunocompromised mice can also be transplanted with tumor specific cells. Another method to develop the xenograft model is to inject the primary tumor cells from patients into immunocompromised mice. Since these mice strains lack an

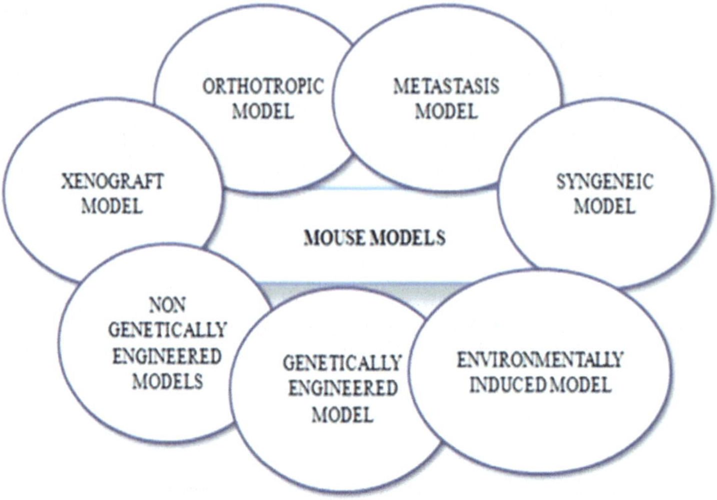

Fig. 12.2 Various types of preclinical mouse models

immune system which is critical regulates tumorigenesis, is the main shortcoming of this model during preclinical trials [36].

Due to the drawbacks in xenograft models, the concept of genetically engineered murine models (GEMM) came into existence. It allows desirable manipulation in tumor suppressor gene or proto-oncogene, building a more suitable model for preclinical trials [37]. GEMM models have the ability to develop new tumors in an immunoproficient environment, which makes it indispensable for preclinical trials [38]. But species specific differences in context with drug metabolism, heterogeneous human tumor, and immune response influence the anticancer therapeutic outcomes [39]. Various mouse cancer models have been illustrated here. (Figs. 12.2 and 12.3 and Table 12.1).

12.5.1 Immunocompromised Models

12.5.1.1 Nude Mice

In 1966, the first immunocompromised mouse was discovered at Ruchill Hospital, UK. It is a murine strain that lacks thymus and hair and cannot mediate the adaptive immune response. As they lack body hair, that is why they are known as nude. Further, it was characterized that a single gene mutation in Foxn1, located on chromosome 11, leads to this phenotype. Since then, athymic nude mice have been deployed as an animal model for preclinical cancer trials for a long period of time. Since it exhibits natural immunodeficiency, this makes it an appropriate model for engraftment and metastatic behavior. Nevertheless, there is immunodeficiency, but it is not absolute, i.e., they can mediate innate response that can limit the transplantation rate [48]. Gradually, several modifications at the genetic level can

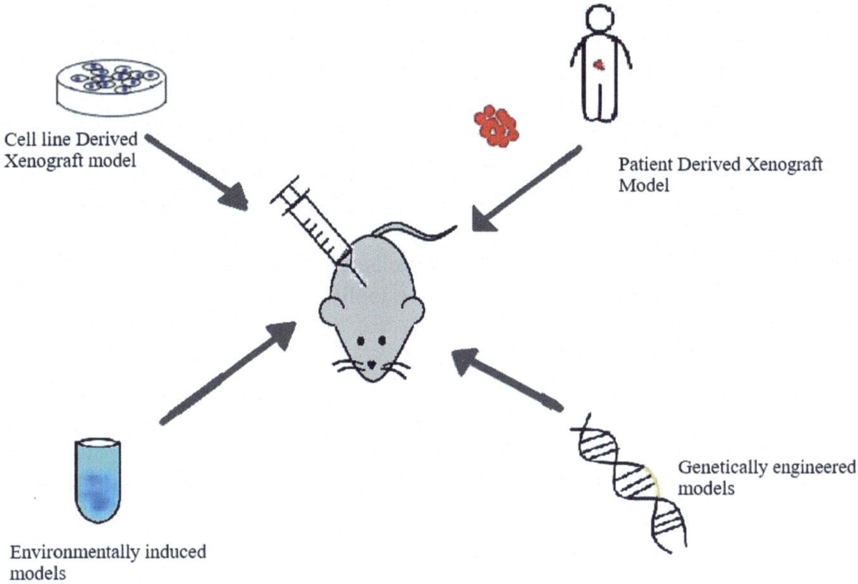

Fig. 12.3 Representation of different mouse models

Table 12.1 Different preclinical mouse models in different types of cancer

Type of model	Example	Mouse strains	References
Cell derived xenograft	WiDr colon adenocarcinoma (subcutaneous)	NOD-SCID mice	[40]
Patient derived xenograft	Breast cancer fragments (subcutaneous)	SWISS NUDE mice	[41]
Orthotropic	Human squamous cell carcinoma (submucosal)	NUDE mice	[42]
Metastasis	MiaPaca-2 human pancreatic cancer cell line	Athymic nu/nu mice	[43]
Syngenic	4T1 murine mammary carcinoma	BALB/C mice	[44]
Environmentally induced	DMBA and TPA	C57 BL/6 mice	[45]
(Chemical)			
Genetically engineered model	*K-ras*^{G12D} targeted expression	KPC mice	[46]
	Conditional inactivation of BRCA2 and trp53.	K14*Cre* transgenic mice	[47]

DMBA 7,12-Dimethylbenz(a)anthracene, *TPA* 12-O-tetradecanoylphorbol-13-acetate

lead to the development of a higher degree of immunocompromised mice. These are SCID (Several Combined Immunodeficiency) or NOD (Non-obese diabetes)/ SCID mice.

12.5.1.2 Several Combined Immunodeficiency (SCID)

It was first reported in 1983 at Fox Chase Cancer Institute, USA. It is a murine strain that is devoid of both functional B and T cells because of recessive mutation on Chromosome 16. Deletion of protein kinase DNA activated catalytic polypeptide, and VDJ (variable diversity joining) recombination leads to nonfunctional B and T cells. The engraftment rate is higher in comparison to nude mice. However, they possess innate immunity (particularly NK cells), which is a limitation in transplantation. But a point mutation known as beige leads to slow activation of NK cells that can overcome the limitation caused by NK cells. Thus, by crossbreeding beige and SCID mice, SCID/Beige mice have been established for more uptake of tumor cells [49].

12.5.1.3 NOD/SCID Mice

A NOD mouse was discovered in the 1980s. These mice develop diabetes mellitus due to the destruction of beta cells of the pancreas by the T cells. It was later described that they also have immune malformations such as impaired dendritic, macrophages, and NK cells. By taking advantage of this, NOD/SCID mice strains have been established which lack T cells and do not develop diabetes mellitus. That makes them a better model for human tumor transplantation studies [49].

Above all stated mice are used in different model development by different methods (orthotopic, xenograft, syngeneic, metastasis models, etc.)

12.5.2 Xenograft Models

12.5.2.1 Cancer Cell Line Derived Xenograft Models (CDX)

This is the standard mouse model for preclinical trials for cancer drug development. They are developed by human tumor cell lines engraftment into immunocompromised mice. Tumors are injected subcutaneously into the flank region. The advantages are that the xenograft models can be handled easily, are reproducible, and give quick results in comparison to other models [50]. However, cancer cell lines already possess multiple mutations, and passaging them for a long time in vitro might introduce additional mutations obliterating the natural tumor microenvironment. Still, they are often used for investigation of anticancer drugs for preclinical testing [38].

12.5.2.2 Patient Derived Xenograft Models (PDX)

PDX is the alternative approach to classical xenograft models [50]. In this, tumor biopsy derived tissue from the patient is transplanted into immunocompromised mice directly, i.e., there is no involvement of human tumor cell lines. PDX models maintain tumor heterogeneity based on genetic, histological, and molecular point of view [38]. Maintaining this type of model is a laborious work, as in vivo passaging of tumors needs to be performed over a long duration. It is also troublesome to optimize, as there may be alteration in the characteristics of the model due to in vivo passages. This kind of strategy is mainly applicable to decide the medication that is suitable for the treatment of a particular tumor before its administration to the patient [50]. It is also emerging as a promising approach for personalized medicine [51] as it enables direct assessment of clinically approved anticancer drugs [52].

12.5.3 Orthotopic Models

Cancerous cell lines or patient tumor sample can be directly transplanted at the site of tumor origin and is referred to as orthotopic models (Fig. 12.4). The site can be intravenous, peritoneal, or intracardial, depending upon the requirement [32]. It represents a transplantation strategy that is organ specific. For example, drug bosutinib, an inhibitor of BCR-Abl, and SRC kinase have been tested as a therapeutic agent against neuroblastoma condition. NCR nude mice were injected with neuroblastoma cell lines treated with bosutinib into the left side of renal capsule. From this study, it was concluded that bosutinib could be a good therapeutic candidate for neuroblastoma patients [53].

Therefore, it is able to mimic the tumor environment more naturally, i.e., patient like. For this kind of model development, skilled expertise is required to make sure of its reproducibility because of its technical difficulties. To monitor these models, high imaging techniques are required that are expensive [50].

12.5.4 Metastasis Models

There are various models that reflect the complicated and multi-step process of metastasis, i.e., invasion, entry into circulation and surviving, extravasations, and finally, secondary tumor development. The models are created according to the

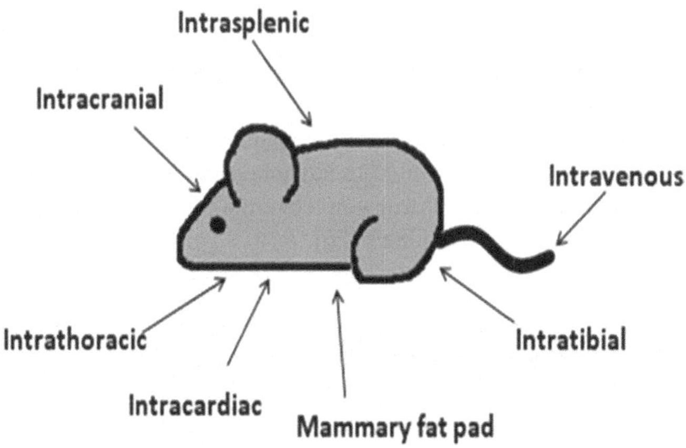

Fig. 12.4 Illustration of different orthotopic mouse tumor models

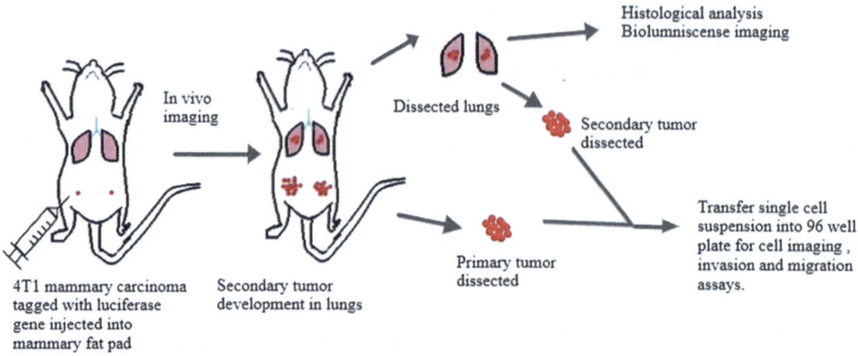

Fig. 12.5 Schematic diagram of metastatic model from primary breast tumors

requirement of the scientific question that requires to be addressed. To evaluate the capacity of the tumor to metastasize, intravascular injection is given to the mouse. For example, intra-cardiac injection is used for bone and brain metastasis model; tail vein injection results in lung metastases; intrasplenic for liver metastasis and intra-peritoneal for modeling ovarian cancer. These mice models contribute mainly to the elucidation of interaction between cancerous cells with its microenvironment [54]. Various assays like colonization assays or specific assays (ear model or foot pad) or resecting the metastatic tumors might be able to mimic the situation of metastasis to some extent. For example, 4T1 mammary carcinoma cells tagged with luciferase gene are injected into BALB/C background mice to study the metastasis at the molecular level for therapeutic development [55] (Fig. 12.5).

12.5.5 Syngeneic Models

Syngeneic models can be created by transplanting tumor cell lines or tumor cells from already designed laboratory strains of mice into the host mouse [50]. These mice models are also known as allograft mouse models [32]. It can be created either by orthotopical or subcutaneous manner. There is a broad variety of characterized syngeneic models available. As these are transplanted in a syngeneic manner, i.e., having the same genetics makes it a more advantageous approach [50]. During the mid-1950s, for preclinical trials of drugs on leukemia tumor cell lines, allograft was used to screen the anticancer candidates [56].

12.5.6 Environmentally Induced Models

Another conventional approach is induction of tumors in mice by using carcinogens or cancer-causing agents. Carcinogens include radiation, chemicals, or pathogens [32]. Large numbers of environmentally induced models are available for multiple cancer types, including skin, lung, liver, colon, and bladder cancer. It has contributed to the determination of many therapeutic targets for cancer and also for chemicals that can act as potent carcinogens [31]. The development of the tumor in this type of model is slow, which allows the development of a natural tumor microenvironment in immunocompetent mice. Therefore, this model is good for the study of immuno-modulatory therapeutics for cancer treatment. The disadvantages of this type of model are the heterogeneous nature of the tumor and restricted homology with human [50].

12.5.7 Genetically Engineered Mouse Models

12.5.7.1 Human Transgenic Mice

Mice carrying specific human transgenes have been created, which express the particular human protein to be targeted for therapeutic purposes. These type of models are preferably used to study the mechanism of action of a specific drug to check its in vivo safety and efficacy. Generating this type of model is a time-consuming method and requires intensive labor. In addition, only little modification can be done in one engineered mouse line at a genetic level [50].

12.5.7.2 Oncomice

These mice either have cloned oncogenes of human or knock out of specific tumor suppressor genes. These are considered popular models as they can mimic the natural process of tumorigenesis in a spontaneous manner [50]. During the mid-1980s, first oncomice was developed, having a random set of oncogenes under tissue specific promoter [57]. Tumor suppressor gene knock out models were generated by using gene targeting, which is locus specific [58]. A wide variety of oncomice strains have been created, which can be used to evaluate the different phases of cancer. The spontaneous and slow development of tumor reflects the initiation and progressive human malignancy stages. It is an appropriate model to study the metastasis process, tumor microenvironment, and anticancer therapeutics [50]. Similarly, many conditional models have been created based upon site specific recombinase (SSR) systems like Cre-*lox P* and Flp- *FRT* [32].

Genetically Engineered Mouse Model (GEMM) is able to reflect the molecular attributes of human cancers [32]. For example, GEM prostate model NKX3.1; PTEN mutant (mimic of human prostate cancer) when treated with the combination of rapamycin (mTOR inhibitor) and PD0325901 (MEK inhibitor) showed inhibition of cell growth by targeting mTOR and MAPK signaling cascades [59]. The application of (GEMMs) in preclinical studies has drawbacks like very long durations, expensive, and less similiarity with the human system [50].

12.5.8 Non-Genetically Engineered Mice

Mouse models that carry genetically engineered alleles in nongermline cells, i.e., somatic cells, are categorized under non-GEMMs. To study a large cohort, GEMMs development is laborious, extremely time taking, and expensive. To accelerate this mouse modeling process and in vivo investigation of the effect of cancer genes, non-GEMMs came into existence [38]. There are many approaches of non-GEMMs briefed as below.

12.5.8.1 Embryonic Stem Cell Based

Cancer prone embryonic stem cell are modified genetically and is injected into the blastocysts of selected mice. Thus, every cell of the body is not genetically modified; tumor development is restricted to somatic tissue only. As the target alleles are engineered outside the recipient (ex vivo), which makes its creation easy and cost-effective manner. There are no breeding requirements, like for GEMM models [60].

12.5.8.2 RNA Interference Based

This type of model is developed by knocking down the target gene expression by siRNA (small interfering RNA). It is a substitute for knockout mice approach. Genetic screening based on RNAi (RNA interference) is a robust tool for the quick determination and validation of tumor genes. This method has been applied for the identification of novel tumor suppressor genes in lymphoma and HCC (hepatocellular carcinoma) models successfully. siRNA has a short life; therefore, the knock down is transient. To achieve more stability, viral vectors encoding shRNA (short hairpin RNA) are used [61].

12.5.8.3 Genome Editing Based

Three types of genome editing tools based on engineered nuclease have evolved: ZNF (zinc finger nuclease), TALEN (transcription activated like effector nuclease), and lastly, the most powerful approach CRISPR (clustered regularly interspaced short palindrome repeats)—linked Cas9 system [61]. CRISPR–Cas9 is a simple nuclease method that is being exploited as a genome editing tool. Many non-GEMM mice have been developed for various cancers. Though it is an efficient technique, but sometimes cells expressing Cas9 may evoke specific immune responses, thus clearing themselves [38].

12.5.9 Zebrafish: An Emerging Model

Zebrafish is the most popular vertebrate model system in developmental biology. It is also emerging as an attractive in vivo model for imitating human cancer. Short life span, translucent embryos, ex-uteri embryonic development, and easy maintenance in small space are the advantages of zebrafish as an animal model. There are some salient histological similarities for tumor formation in fish and human [14]. It has

Table 12.2 Xenograft models of cancer in zebrafish

Type of cell line for xenograft	Purposes	Findings	References
Metastatic melanoma	To assess human cancer cell at different stage of tumorigenesis	Proliferation, Migration and formation of mass of melanoma cancer cell in vivo	[65]
Breast carcinoma	To investigate angiogenesis	Cancer cell invasion, VEGF induced angiogenesis, growth of neovessels	[66]
Adenocarcinoma	To study invasion and angiogenesis	*RhoC*, a metastatic gene contributes in intravasation and alteration in cytoskeleton.	[67]
Leukaemia	To develop a cell proliferation assay that can be used to screen therapeutics against leukaemia.	A sensitive *in vivo* proliferation assay that can detect minor changes in cell count with accuracy.	[68]

VEGF vascular endothelial growth factor

effectively been used as a forward genetic tool. Mutations can be induced in adult zebrafish chemically and through viral agents or by transposons. Later, the progeny that develops can be screened for unusual phenotypes. Genes that are mutated in cancer are analyzed via sequencing or genetic mapping. Many genes associated with the cancer potential like cell cycle, cell proliferation, and apoptosis have been determined through this forward genetic screening system.*p53*; the tumor suppressor gene mutation is most commonly present in 50% of the tumors. Generally, in mice, loss of *p53* gene allows normal cell proliferation during the developmental phase, but there might be the occurrence of multiple neoplasm [58, 62]. To understand the regulation of p53, an effective model was required. Knockdown of p53 through the injection of morpholino revealed the importance of p53 in apoptosis induced by DNA damage [63]. One of the most hostile form of skin cancer is melanoma. Almost 67% melanomas carry a BRAF gene mutation. A transgenic model of zebrafish expressing BRAF and mutated BRAF was generated to understand the role of BRAF protein in melanoma formation [64].

Similar to mouse models, xenograft transplantation has been performed in zebrafish as well, and it provides the opportunity for transplantation studies (Table 12.2). Transplanted cells can easily be monitored and visualized by fluorescent tags attached to the target. Various zebrafish cancer models like embryonal rhabdomyosarcoma, melanoma, T cell acute lymphoblastic leukemia have been created by transplanting tumor cells of zebrafish into wild type zebrafish to gain understanding about the initiation and malignancy of tumors. In fact, human tumor cells have also been transplanted in fish, to exploit the mechanisms involved in metastasis, tumor growth, and angiogenesis [69]. For evaluation of drug efficacy and safety, it has become an effective alternative model. It has also been showed that metastatic cell transplantation halts the development process of zebrafish. The underlying reason for the same was nodal protein, the main driver for the aggressiveness in melanoma. Thus, nodal is a significant therapeutic marker in skin cancer [70]. However, there are some limitations associated with xenografting in zebrafish. The tumor microenvironment of zebrafish and human tumor cells differs; in addition to that niche of tissue and lack of organs (mammary tissue, lungs) make orthotopic transplantation impossible.

Zebrafish is an emerging powerful animal model organism in cancer therapeutics. The flexibility of forward and reverse genetic screening is some special advantage in comparison with others. Besides these, the transparency in the body gives a detailed evaluation of tumor cells. But more advances in technology at the genetic level are required to prove it as a novel and promising model organism for human cancer therapeutics [71].

12.5.10 Dogs: A Companion Model

Canines (dogs) are natural models for spontaneously developing malignant tumors. Canines develop spontaneous tumors more frequently and share related histopathological characteristics with neoplasm in humans [72]. One of the major problems in female canines is breast cancer. The frequency of disease reported is two to three times more in female canines than in human. Several studies demonstrated that there are striking similarities in breast carcinoma of human and dogs in terms of histology, biology, and epidemiology [73]. In genome wide comparative analysis, it has been evidenced that there is a wide overlap in the impaired signaling pathways associated with the development of the tumor. These pathways are MAPK (mitogen activated protein kinase), PI3K (phosphatidylinositol 3-kinase)/AKT pathway, Pten (phosphatase and tensin homolog), and Wnt-β catenin pathways [74]. HER-2 (Human epidermal growth factor-2), and BRCA1, 2 get genetically altered. Their overexpression was found in both canines and human mammary carcinoma [75–77]. Similarly, non-Hodgkin lymphoma (NHL) is also common among dogs and human. There are many breeds like bullmastiff, bulldog, and Scottish terrier which have a risk of developing NHL [78]. To contribute and support effective translational research, the comparative oncology trials consortium has been established in 2004 in many biospecimen repositories of canine cancer models. The purpose of this consortium is to identify the novel anticancer targets in canines and provide comparative genomic studies for effective preclinical trials [79].

Over the past 30 years, canines have been used for clinical trials studies of cancer. As osteosarcoma is more common in canines than humans, the pet dogs having osteosarcoma were assessed with operative techniques for limb salvage. It is now used in the management of pediatric patients. Similarly, to treat pulmonary metastasis clinical trials of inhaling cytokine (Interleukin-2), immunotherapy was successfully completed in early phase clinical trials [80, 81]. There is growing attention to utilize dog as a natural animal model to develop anticancer targets [82, 83]. They have been included in early preclinical trials of gene therapy conducted for oral tumors [84, 85].

However, the expenses to conduct preclinical trials in canine models are comparatively greater than mouse models. Besides that, the limited availability of reagents for the canine cancer model is one of the major hurdles. Many human antibodies used as reagents exhibit cross reactivity in canines [86]. In the future, this can make translational research for therapeutic development easier if this gap can be filled.

12.5.11 Swine

Pigs as an animal model have not been used majorly in experimental cancer biology, but their utility has been exhibited in a broad range of preclinical trials [87]. They are alike to humans in respect to size, physiology, immunology, metabolism, and genetics [88–90]. Their life span is longer, allowing the examination of the progression of disease repeatedly.

Studies have shown that Cyt P450 protein (playing a significant role in processing and metabolizing the drugs) shares almost the same structure and function in human and pigs and has homologous xenobiotic receptors [91]. Spontaneous cancers caused due to natural mutations are very rare in pigs, including lymph sarcoma in young pigs and melanoma in adult ones. Sinclair and Libechov are the only two spontaneous tumor models of pig liable to develop melanoma [92].

Transgenic pig developed by using v-Ha-Ras oncogenes under mouse mammary tumor virus promoter exhibited no phenotypic response [93]. Cre dependent latent mutant allele $TP53^{R167H}$ pigs have been developed, which is orthologous to $TP53^{R175H}$ [94]. Expression of Cre recombinase at confined and tissue specific site is an important part of designing the pig cancer model [95]. For example, oncopig HCC (hepatocellular carcinoma) has been developed as the transgenic porcine model, which is able to reiterate the many hallmarks of human HCC at the transcriptional level. This type of model will be helpful in the identification of biomarker and treatment of HCC [82]. Some of the genetically modified pigs have been mentioned in Table 12.3. The development of more genetically precise porcine cancer models using Crisper Cas 9 is under consideration, which is better in mimicking the hallmarks of human cancer [99].

Nevertheless, pigs are better models for investigation of the underlying mechanism of cancer and therapeutics. But larger the animal, costlier is the experimentation. Also, the maintenance over a long period of time under restricted environmental conditions is an extensively laborious work [99].

12.5.12 Drosophila

Drosophila, the fruit fly, has been applied as an in vivo animal model for the development and validation of anticancer drugs in recent years. Approximately 75% of human disease genes show conservation with *Drosophila* [100]. Various tumor suppressor genes and oncogenes like Notch, Salvador–Warts–Hippo (SWH), and Hedgehog (Hh) have been first determined in *Drosophila* [101]. In fact, the

Table 12.3 Genetically modified pigs as a model for human cancers

Name of gene	Functional role of gene	Type of cancer	Modification	Purposes	References
BRCA 1	Tumor suppressor	Breast	rAAV mediated knockout of *BRCA 1*	To develop *BRCA1* associated breast cancer model for anticancer drug testing and explore breast carcinogenesis.	[96]
APC	Tumor suppressor	Colorectal	Targeted mutagenesis of *APC* (Adenomatous polyposis coli.)	To study human colorectal cancer pathogenesis for diagnostics and its cure.	[97]
KRAS	Proto-oncogene	Pancreatic ductal and lung adenocarcinoma	Inducible expression of mutant $KRAS^{G12D}$ by Cre system	To reproduce *KRAS*-induced cancers and explore the molecular events underlying cancer progression.	[98]
TP53	Tumor suppressor	Various cancer	Conditional activation of *TP53* by Cre system	To mimic the *TP53* germline or somatic mutations that has a major role in human cancers.	[94]

importance of JAK-STAT in hemocytes of fly was observed even before in human leukemia [102]. It has contributed to decrypting the epidermal growth factor signaling events in order EGFR-RAS-RAF-MEK-ERK.[103, 104]. High throughput screening is an effective and dynamic assay that is used for in vivo screening of drugs against pre-determined targets. For cancer drug screening, the most commonly used models include cultured cells, whole larvae, whole fly, or organs (eye, wing). It gives the opportunity for high throughput screening of cancer drugs [105]. Similarly, the Notch signaling pathway has been discovered originally as an important component in the development pathway in the fruit fly. Various studies show pro-oncogenic properties of Notch in various types of tumors [106]. Gamma-secretase, which shows inhibitory effects against Notch, is being evaluated for cancer therapeutics in *Drosophila*. Therefore, this study suggested that *Drosophila* can be utilized as a preclinical animal model to screen drugs against Notch protein. Abovementioned studies were also validated in mouse xenograft models and human cell lines [107].

Drosophila can also be a powerful tool in the coming era of drug repurposing. As different human diseases share common molecular pathways, a single drug can target all the common signaling pathways related to different diseases [105]. To perform screening of personalized medicine, specific tumor cells of the patient have also been transplanted into a fruit fly, also known as *Drosophila* avatar. This suggests that it can be a promising candidate for personalized medicine screening [108]. It is easy to establish transgenic models of *Drosophila* as it has a shorter life cycle, and as it shares conserved cell proliferation, signaling with a human makes it an appropriate model for preclinical testing of anticancer drugs. However, there is an evolutionary distance between *Drosophila* and humans which restrict them to picture the exact symptoms of human. In addition to this, any drug that has been successfully tested for efficacy and safety in *Drosophila* cannot go directly for clinical trials in human. Nevertheless, it can help to determine the suitability of a drug formulation before planning for trials in expensive murine models [105]. Therefore, further technological advancements are required before it becomes potent in vivo tool for preclinical trials in cancer.

12.5.13 *C. elegans*

C. elegans, a hermaphrodite organism, is a favorite developmental biology model for many years. It has been used as an ideal model system to exploit different aspects in the field of biology, i.e., evolution, regulation of stem cell, specification of cell fate, aging, and tumorigenesis. It has well characterized developmental lineage and is genetically tractable, opening doors for utility in biological and basic life sciences research [109]. Recently, it has become a popular invertebrate model for the investigation and screening of drugs in vivo. The occurrence of translucent tissues at all stages of development in *C. elegans*, availability of large number of mutant strains at CGC (*Caenorhabditis* Genetic Centre), shorter life span (around 3 weeks), and availability of whole genome sequence makes it an ideal candidate for high throughput screening of drugs in vivo [110]. The germline development of *C. elegans* is a highly regulated process that is carried by conserved signaling pathways, which includes Wnt, Ras, and Notch (extrinsic regulators) and cell cycle proteins (as intrinsic regulators) [111]. Aberrations in above signaling pathways lead to uncontrolled germline growth, and extra distal tip cell formation (niche of germline stem cell). All these result in the development of germline tumors and sterility in *C. elegans.*

The anchor cell (AC) derived from somatic gonad in *C. elegans* signals the vulval precursor cells to form vulva. Studies have shown that VRK-1(vaccinia related kinase-1) regulates the vulva formation signal. Loss of function in VRK-1 activity results in anchor cell invasion leading to the destruction of vulval cells affecting the reproduction process. It is to be noted that many key players involved in anchor cell invasion have been also reported in mammalian tumor invasion [112].

Exelixis Inc., in collaboration with Bristol Meyers Squibb and DevGen, NemaRx Inc., has been using *C. elegans* for anticancer drugs and other drug screening. It was identified that farnesyltransferase has an inhibitory effect on the gain of function in Ras (oncogenes, involved in many cancers), which results in multivulva formation. It promotes the pro-apoptotic pathway by reversing the gain of function of Ras oncogenes. This can also be applied in the Ras mechanism dependent on cancers in human [113].

12.5.14 Reason for Failure of Preclinical Animal Models

Animal models are essential for preclinical trials in cancer. Before stepping into clinical trials, preclinical trials are performed at in vitro and in vivo level in an extensive manner. Parameters like efficacy, pharmacokinetics, pharmacodynamics, and safety parameters of the drugs should be examined. In spite of successful preclinical trials [114], a large proportion of cancer drug trials fail in the early clinical phase [115]. The cause of failure is mostly attributed to the poor methodology applied, and models fail to mimic the actual human disease condition. Besides that, the huge loophole lies in animal modeling studies/approaches. Regardless of failure in clinical trials, animals are still prevalent for testing the toxicity, effectiveness, and efficacy of drugs [116].

Alternative approaches have also emerged to overcome the failure of animals in the translation of clinical studies. In silico approach is evolving to refine the experimental protocol and results of clinical experiments. It works on the principle of 3R (Fig. 12.6). The potent biomarkers and new entities involved in cancerous signaling can be identified via the computational approach. Like, human protein atlas gives information on gene expression levels in normal and cancer tissues [117]. Similarly, driver and passenger mutations can be differentiated by the use of public genomic databases like the International Cancer Genome Consortium [118]. The signaling network can be analyzed by Reactome [119], KEGG [120], or Gene Ontology [121].

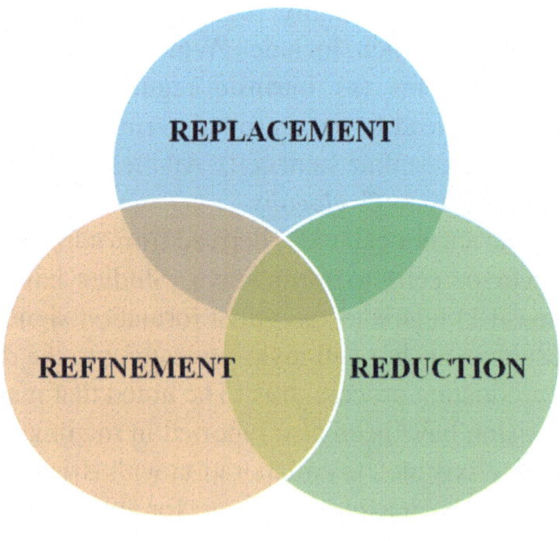

Fig. 12.6 Venn diagram representing the 3Rs in preclinical trials

12.6 Planning for Animal Study for Preclinical Trials

Planning for an animal study is difficult and complex. Laboratory animal needs proper care and complex environment prerequisite for their survival that decides for the whole quality of the experiment. While designing the in vivo protocols, the primary investigation should be done on literature related to statistics, pathology, physiology, maintenance aspects of laboratory animals [122].

12.7 Factors to Be Considered for Better Efficacy of Preclinical Studies

Preclinical efficacy hereby means considering parameters like absorption, distribution, metabolism, and excretion (ADME), related to drug dose administration. Besides that, emphasis should be given on pharmacodynamics (PD) and pharmacokinetic (PK) parameters, which determine the translational chances of preclinical studies. Different animal models have different criteria for drug testing, so the experiments for preclinical testing should be designed with respect to that specific animal model that is being utilized for studies. However, by customizing protocol, an optimal experimental design can be achieved [123].

12.8 Sources of Variation in Preclinical In Vivo Studies

During designing and conducting animal model studies, many variations are considered from different sources. Like, environment sources (housing, diet, water delivery, noise, ventilation, temperature, light, humidity) and inherent sources (strain, sex, weight, pathogen status of an animal) contribute in maintaining the healthy animal facility. For preclinical studies, dose type, route of administration, timing of administration for dose, sampling methodology, sample collection site, can influence the consequences [123].

12.9 Concept of 3Rs

Research on animals is a significant part of biological research. The concept of 3 R (Replacement, Reduction, and Refinement) has been encrypted in animal welfare guidelines that govern the ethical aspects of the application of animals in experimental biology (Fig. 12.6) [124]. Replacement means prior to the use of non-animal methodology over animal models. The reduction is to decrease the usage of animals for a particular objective and try to obtain maximum data from fewer animals. Refinement involves minimizing the pain given to animals, stress, or suffering. The 3R has been used in preclinical trials to reduce animal use and maximize human translational research [125].

12.10 Major Challenges

Advancements in genome sequencing technology have led to the elucidation of complexities in molecular pathways underlying cancer. Still, validating a target compound is a major obstacle for its commercial use as an effective drug. Despite the good efficacy during preclinical trials, many compounds fail to give relevant results in clinical trials [126]. Besides that, animal models of different types have their own advantages and disadvantages in preclinical testing [127]. Therefore, it is clear that there is no ideal animal model for cancer that can mimic human as a whole. Though mice and zebrafish are suitable models for transgenic approaches and xenograft experiments, but there is a large difference between the two. Zebrafish are easy to maintain at a larger scale with fewer expenses as their adults are smaller than mice. On the other hand, in dogs due to ethical concerns, high throughput assays cannot be performed. Invertebrate models like *C. elegans* and *Drosophila* are small in size and have a short life cycle. Though this makes them a robust model for gene manipulation and high throughput drug screening, but still xenotransplantation approaches cannot be accomplished in invertebrate models [128]. Pigs are an excellent model as they share parallelism with humans in matter of genetics, anatomy, and physiology. But large husbandry needs for its development, and breeding of multiple strains is a huge task [99]. Thus, optimization of preclinical models needs to be performed in several areas like gene alteration in a temporal-spatial manner and ability to recapitulate the heterogenetic nature of human tumors [2].

Besides these, the tumor microenvironment at the primary site and the metastatic niche in preclinical animal models do not exactly mimic to those in human beings. Numerous studies have revealed that the microenvironment of a tumor significantly contributes for both tumor suppression and tumor promoting properties.

12.11 Conclusion and Future Perspective

Several animal models used in preclinical trials of cancer have been discussed in this manuscript that focus on their significance, advantages, and disadvantages in context with cancer therapeutics. Animal models are the fundamental, translational model in preclinical trials. For the last two decades, advancements in the field of transgenics and genetic engineering have resulted in more improved animal models for preclinical trials of cancer remedies. The ability of animal models to anticipate the clinical success of preclinical trials is still a disputed issue due to the inability to mirror the complex process of human tumorigenesis.

The main aim behind the use of preclinical tumor model is to reflect the environment of human cancer as close as it can. For this, co-clinical trials or parallel studying the effect of therapeutics in human and in vivo models (like a mouse) is the new upcoming concept. It can bridge the existing lacunae between humans and animal models. Ultimately, that will lead to more quick identification of robust models that would convert this deadly fatal disease into a manageable disease.

Current advancement in genetic engineering and genome editing technologies has led to the development of finer tuned models that can mimic human cancer. But, there is no single animal model developed that could capture all features of the complex human cancer. Therefore, it is necessary to understand that each model has its own strengths and weakness. Perhaps, by developing more sophisticated engineered model systems and other alternatives can reduce the chance of failure in clinical trials of cancer.

Conflict of Interest The author declares no conflict of interest in this book chapter.

References

1. Cekanova M, Rathore K (2014) Animal models and therapeutic molecular targets of cancer: utility and limitations. Drug Des Devel Ther 8:1911–1921
2. Day CP, Merlino G, Van Dyke T et al (2015) Preclinical mouse cancer models: a maze of opportunities and challenges. Cell 163(1):39–53
3. Dhandapani M, Goldman A (2017) Preclinical cancer models and biomarkers for drug development: new technologies and emerging tools. J Mol Biomark Diagn 8(5):356
4. Cagan RL, Zon LI, White RM et al (2019) Modeling cancer with flies and fish. Dev Cell 49 (3):317–324
5. Kirienko NV, Mani K, Fay DS et al (2010) Cancer models in caenorhabditis elegans. Dev Dyn 239(5):1413–1448
6. Flisikowska T, Kind A, Schnieke A et al (2016) Pigs as models of human cancers. Theriogenology 86(1):433–437
7. Curt GA (1994) The use of animal models in cancer drug discovery and development. Stem Cells 12(1):23–29
8. Flanagan SP (1966) 'Nude', a new hairless gene with pleiotropic effects in the mouse. Genet Res 8(3):295–309
9. Pantelouris EM, Hair J (1970) Thymus dysgenesis in nude (nu nu) mice. J Embryol Exp Morphol 24(3):615–623
10. Povlsen CO, Rygaard J (1971) Heterotransplantation of human adenocarcinomas of the colon and rectum to the mouse mutant Nude. A study of nine consecutive transplantations. Acta Pathol Microbiol Scand A 79(2):159–169
11. Giovanella BC, Stehlin JS, Williams LJ Jr et al (1974) Heterotransplantation of human malignant tumors in "nude" thymusless mice. II. Malignant tumors induced by injection of cell cultures derived from human solid tumors. J Natl Cancer Inst 52(3):921–930
12. Karlsson EK, Lindblad-Toh K (2008) Leader of the pack: gene mapping in dogs and other model organisms. Nat Rev Genet 9(9):713–725
13. Payne E, Look T (2009) Zebrafish modelling of leukaemias. Br J Haematol 146(3):247–256
14. Amatruda JF, Shephard JL, Stern HM et al (2002) Zebrafish as a cancer model system. Cancer Cell 1(3):229–231
15. Parng C, Seng WL, Semino C et al (2002) Zebrafish: a preclinical model for drug screening. Assay Drug Dev Technol 1(1 Pt 1):41–48
16. Langenau DM, Traver D, Ferrando AA et al (2003) Myc-induced T cell leukemia in transgenic zebrafish. Science 299(5608):887–890
17. Konantz M, Balci TB, Hartwig UF et al (2012) Zebrafish xenografts as a tool for in vivo studies on human cancer. Ann N Y Acad Sci 1266:124–137
18. Foley JE, Maeder ML, Pearlbeg J et al (2009) Targeted mutagenesis in zebrafish using customized zinc-finger nucleases. Nat Protoc 4(12):1855–1867

19. Cade L, Reyon D, Hwang WY et al (2012) Highly efficient generation of heritable zebrafish gene mutations using homo- and heterodimeric TALENs. Nucleic Acids Res 40 (16):8001–8010
20. Schook LB, Beever JE, Rogers J et al (2005) Swine genome sequencing consortium (SGSC): a strategic roadmap for sequencing the pig genome. Comp Funct Genomics 6(4):251–255
21. Carlson DF, Tan W, Lilico SG et al (2012) Efficient TALEN-mediated gene knockout in livestock. Proc Natl Acad Sci U S A 109(43):17382–17387
22. Innes JR, Parry HB, Berger J et al (1946) Record of lymphadenosis (lymphatic leukaemia) in a dog treated with urethane. Br Vet J 102(12):389–393
23. Potter CJ, Turenchalk GS, Xu T et al (2000) Drosophila in cancer research. An expanding role. Trends Genet 16(1):33–39
24. Gladstone M, Su TT (2011) Chemical genetics and drug screening in Drosophila cancer models. J Genet Genomics 38(10):497–504
25. Turner PV, Brabb T, Pekow C et al (2011) Administration of substances to laboratory animals: routes of administration and factors to consider. J Am Assoc Lab Anim Sci 50(5):600–613
26. Fox J, Barthold SD et al (eds) (2007) The mouse in biomedical research. Academic Press, London
27. Van Dyke T, Jacks T (2002) Cancer modeling in the modern era: progress and challenges. Cell 108(2):135–144
28. Yee NS (2015) Animal models of cancer biology. Cancer Growth Metastasis 8(Suppl 1):115–118
29. Khan AQ, Siveen KS (2019) Role of animal research in human malignancies. In: Azmi A (ed) Animal models in cancer drug discovery. Elsevier, St Louis, pp 1–29
30. Ishida K, Tomita H, Hirata A et al (2017) Current mouse models of oral squamous cell carcinoma: genetic and chemically induced models. Oral Oncol 73:16–20
31. Weidner C, Steinfath M, Opitz E et al (2016) Defining the optimal animal model for translational research using gene set enrichment analysis. EMBO Mol Med 8(8):831–838
32. Gengenbacher N, Singhal M, Augustin HG et al (2017) Preclinical mouse solid tumour models: status quo, challenges and perspectives. Nat Rev Cancer 17(12):751–765
33. Lubet RA, Zhang Z, Wang Y et al (2004) Chemoprevention of lung cancer in transgenic mice. Chest 125(5):144S–147S
34. Thompson HJ, Singh M (2000) Rat models of premalignant breast disease. J Mammary Gland Biol Neoplasia 5(4):409–420
35. Meuwissen R, Berns A (2005) Mouse models for human lung cancer. Genes Dev 19 (6):643–664
36. Vitale G, Gaudenzi G, Circelli L et al (2017) Animal models of medullary thyroid cancer: state of the art and view to the future. Endocr Relat Cancer 24(1):R1–R12
37. Jean-Quartier C, Jeanquartier F, Jurisica I et al (2018) In silico cancer research towards 3R. BMC Cancer 18(1):408–408
38. Kersten K, De Visser KE, Van Miltenburg MH et al (2017) Genetically engineered mouse models in oncology research and cancer medicine. EMBO Mol Med 9(2):137–153
39. Honek J (2017) Preclinical research in drug development. Medical Writing 26:5–8
40. Prahallad A, Sun C, Huang S et al (2012) Unresponsiveness of colon cancer to BRAF(V600E) inhibition through feedback activation of EGFR. Nature 483(7387):100–103
41. Marangoni E, Vincent-Salmon A, Auger N et al (2007) A new model of patient tumor-derived breast cancer xenografts for preclinical assays. Clin Cancer Res 13(13):3989–3998
42. Myers JN, Holsinger FC, Jasser SA et al (2002) An orthotopic nude mouse model of oral tongue squamous cell carcinoma. Clin Cancer Res 8(1):293–298
43. Sennino B, Ishiguro-Oonuma T, Wei Y et al (2012) Suppression of tumor invasion and metastasis by concurrent inhibition of c-Met and VEGF signaling in pancreatic neuroendocrine tumors. Cancer Discov 2(3):270–287

44. Kim EJ, Shin M, Park H et al (2009) Oral administration of 3,3'-diindolylmethane inhibits lung metastasis of 4T1 murine mammary carcinoma cells in BALB/c mice. J Nutr 139 (12):2373–2379
45. Johansen C, Vestergaard C, Kragballe K et al (2009) MK2 regulates the early stages of skin tumor promotion. Carcinogenesis 30(12):2100–2108
46. Hingorani SR, Wang L, Multani AS et al (2005) Trp53R172H and KrasG12D cooperate to promote chromosomal instability and widely metastatic pancreatic ductal adenocarcinoma in mice. Cancer Cell 7(5):469–483
47. Jonkers J, Meuwissen R, Van der Gulden H et al (2001) Synergistic tumor suppressor activity of BRCA2 and p53 in a conditional mouse model for breast cancer. Nat Genet 29(4):418–425
48. Szadvari I, Krizanova O, Babula P et al (2016) Athymic nude mice as an experimental model for cancer treatment. Physiol Res 65(Suppl 4):S441–S453
49. Okada S, Vaeteewoottacharn K, Kariya R et al (2019) Application of highly immunocompromised mice for the establishment of patient-derived xenograft (PDX) models. Cells 8(8):E889
50. Wartha K, Herting F, Hasmann M et al (2014) Fit-for purpose use of mouse models to improve predictivity of cancer therapeutics evaluation. Pharmacol Ther 142(3):351–361
51. Clohessy JG, Pandolfi PP et al (2015) Mouse hospital and co-clinical trial project--from bench to bedside. Nat Rev Clin Oncol 12(8):491–498
52. Girotti MR, Gremel G, Lee R et al (2016) Application of sequencing, liquid biopsies, and patient-derived xenografts for personalized medicine in melanoma. Cancer Discov 6 (3):286–299
53. Bieerkehazhi S, Chen Z, Zhao Y et al (2017) Novel Src/Abl tyrosine kinase inhibitor bosutinib suppresses neuroblastoma growth via inhibiting Src/Abl signaling. Oncotarget 8 (1):1469–1480
54. Gómez-Cuadrado L, Tracey N, Ma R et al (2017) Mouse models of metastasis: progress and prospects. Dis Model Mech 10(9):1061–1074
55. Guo W, Zhang S, Liu S et al (2015) Establishment of a novel orthotopic model of breast cancer metastasis to the lung. Oncol Rep 33(6):2992–2998
56. Suggitt M, Bibby MC (2005) 50 years of preclinical anticancer drug screening: empirical to target-driven approaches. Clin Cancer Res 11(3):971–981
57. Hanahan D, Wagner EF, Palmiter RD et al (2007) The origins of oncomice: a history of the first transgenic mice genetically engineered to develop cancer. Genes Dev 21(18):2258–2270
58. Donehower LA, Harvey M, Slagle BL et al (1992) Mice deficient for p53 are developmentally normal but susceptible to spontaneous tumours. Nature 356(6366):215–221
59. Andersen NJ, Boguslawski EB, Kuk CY et al (2015) Combined inhibition of MEK and mTOR has a synergic effect on angiosarcoma tumorgrafts. Int J Oncol 47(1):71–80
60. Heyer J, Kwong LN, Lowe SW et al (2010) Non-germline genetically engineered mouse models for translational cancer research. Nat Rev Cancer 10(7):470–480
61. Tratar UL, Horvat S, Cemazar M (2018) Transgenic Mouse Models in Cancer Research. Front Oncol 8:268–268
62. Attardi LD, Jacks T (1999) The role of p53 in tumour suppression: lessons from mouse models. Cell Mol Life Sci 55(1):48–63
63. Langheinrich U, Hennen E, Stott G et al (2002) Zebrafish as a model organism for the identification and characterization of drugs and genes affecting p53 signaling. Curr Biol 12 (23):2023–2028
64. Patton EE, Widlund HR, Kutok JL et al (2005) BRAF mutations are sufficient to promote nevi formation and cooperate with p53 in the genesis of melanoma. Curr Biol 15(3):249–254
65. Haldi M, Ton C, Seng WL et al (2006) Human melanoma cells transplanted into zebrafish proliferate, migrate, produce melanin, form masses and stimulate angiogenesis in zebrafish. Angiogenesis 9(3):139–151
66. Nicoli S, Presta M (2007) The zebrafish/tumor xenograft angiogenesis assay. Nat Protoc 2 (11):2918–2923

67. Stoletov K, Valerie M, Lester RD et al (2007) High-resolution imaging of the dynamic tumor cell vascular interface in transparent zebrafish. Proc Natl Acad Sci U S A 104 (44):17406–17411
68. Corkery DP, Dellaire G, Berman JN et al (2011) Leukaemia xenotransplantation in zebrafish--chemotherapy response assay in vivo. Br J Haematol 153(6):786–789
69. Taylor AM, Zon LI (2009) Zebrafish tumor assays: the state of transplantation. Zebrafish 6 (4):339–346
70. Postovit LM, Seftor EA, Seftor RE et al (2007) Targeting nodal in malignant melanoma cells. Expert Opin Ther Targets 11(4):497–505
71. Yen J, White RM, Stemple DL (2014) Zebrafish models of cancer: progress and future challenges. Curr Opin Genet Dev 24(100):38–45
72. Vail DM, MacEwen EG (2000) Spontaneously occurring tumors of companion animals as models for human cancer. Cancer Invest 18(8):781–792
73. MacEwen EG, Patnaik AK, Harvey HJ et al (1982) Estrogen receptors in canine mammary tumors. Cancer Res 42(6):2255–2259
74. Uva P, Aurisicchio L, Watters J et al (2009) Comparative expression pathway analysis of human and canine mammary tumors. BMC Genomics 10:135
75. Pinho SS, Carvalho S, Cabral J et al (2012) Canine tumors: a spontaneous animal model of human carcinogenesis. Transl Res 159(3):165–172
76. Nieto A, P�rez-Alenza MD, Castillo D et al (2003) BRCA1 expression in canine mammary dysplasias and tumours: relationship with prognostic variables. J Comp Pathol 128 (4):260–268
77. Antuofermo E, Miller MA, Pirino S et al (2007) Spontaneous mammary intraepithelial lesions in dogs--a model of breast cancer. Cancer Epidemiol Biomark Prev 16(11):2247–2256
78. Withrow SJ, Vail DM, Page RL et al (2013) Withrow & MacEwen's small animal clinical oncology. Saunders, St Louis
79. Gardner HL, Fenger JM, London CA et al (2016) Dogs as a model for cancer. Annu Rev Anim Biosci 4:199–222
80. LaRue SM, Withrow SJ, Powers BE et al (1989) Limb-sparing treatment for osteosarcoma in dogs. J Am Vet Med Assoc 195(12):1734–1744
81. Withrow SJ, Thrall DE, Straw RC et al (1993) Intra-arterial cisplatin with or without radiation in limb-sparing for canine osteosarcoma. Cancer 71(8):2484–2490
82. Rowell JL, McCarthy DO, Alvarez CE et al (2011) Dog models of naturally occurring cancer. Trends Mol Med 17(7):380–388
83. Paoloni M, Khanna C (2008) Translation of new cancer treatments from pet dogs to humans. Nat Rev Cancer 8(2):147–156
84. Dow SW, Elmslie RE, Wilson AP et al (1998) In vivo tumor transfection with superantigen plus cytokine genes induces tumor regression and prolongs survival in dogs with malignant melanoma. J Clin Invest 101(11):2406–2414
85. Hogge GS, Burkholder JK, Culp J et al (1998) Development of human granulocyte-macrophage colony-stimulating factor-transfected tumor cell vaccines for the treatment of spontaneous canine cancer. Hum Gene Ther 9(13):1851–1861
86. Hansen K, Khanna C (2004) Spontaneous and genetically engineered animal models; use in preclinical cancer drug development. Eur J Cancer 40(6):858–880
87. Swindle MM, Smith AC (2015) Swine in the laboratory: surgery, anesthesia, imaging, and experimental techniques. Taylor & Francis, Boca Raton
88. Swindle MM, Makin A, Herron AJ et al (2012) Swine as models in biomedical research and toxicology testing. Vet Pathol 49(2):344–356
89. Watson AL, Carlson DF, Largaespada DA et al (2016) Engineered swine models of cancer. Front Genet 7:78
90. Ganderu NC, Harvy W, Mortensen JT et al (2012) The minipig as nonrodent species in toxicology--where are we now? Int J Toxicol 31(6):507–528

91. Myers MJ, Farrell DE, Howard KD et al (2001) Identification of multiple constitutive and inducible hepatic cytochrome P450 enzymes in market weight swine. Drug Metab Dispos 29 (6):908–915
92. Flisikowski K, Fliskowska T, Sikorska A et al (2017) Germline gene polymorphisms predisposing domestic mammals to carcinogenesis. Vet Comp Oncol 15(2):289–298
93. Yamakawa H, Nagai T, Harasawa R et al (1999) Production of transgenic pig carrying MMTV/v-Ha-ras. J Reprod Dev 45(2):111–118
94. Leuchs S, Saalfrank A, Merkl C et al (2012) Inactivation and inducible oncogenic mutation of p53 in gene targeted pigs. PLoS One 7(10):e43323–e43323
95. Li S, Flisikowska T, Kurome M et al (2014) Dual fluorescent reporter pig for Cre recombination: transgene placement at the ROSA26 locus. PLoS One 9(7):e102455
96. Luo Y, Li J, Liu Y et al (2011) High efficiency of BRCA1 knockout using rAAV-mediated gene targeting: developing a pig model for breast cancer. Transgenic Res 20(5):975–988
97. Flisikowska T, Merkl C, Landmann M et al (2012) A porcine model of familial adenomatous polyposis. Gastroenterology 143(5):1173–1175
98. Li S, Edlinger M, Saalfrank A et al (2015) Viable pigs with a conditionally-activated oncogenic KRAS mutation. Transgenic Res 24(3):509–517
99. Xu C, Wu S, Schook LB et al (2019) Translating human cancer sequences into personalized porcine cancer models. Front Oncol 9:105–105
100. Banfi S, Borsani G, Rossi E et al (1996) Identification and mapping of human cDNAs homologous to Drosophila mutant genes through EST database searching. Nat Genet 13 (2):167–174
101. Perrimon N, Pitsouli C, Shilo BZ et al (2012) Signaling mechanisms controlling cell fate and embryonic patterning. Cold Spring Harb Perspect Biol 4(8):a005975
102. Harrison DA, Binari R, Nahreini TS et al (1995) Activation of a Drosophila Janus kinase (JAK) causes hematopoietic neoplasia and developmental defects. Embo J 14(12):2857–2865
103. Karim FD, Chang HC, Therrien M et al (1996) A screen for genes that function downstream of Ras1 during Drosophila eye development. Genetics 143(1):315–329
104. Friedman A, Perrimon N (2006) A functional RNAi screen for regulators of receptor tyrosine kinase and ERK signalling. Nature 444(7116):230–234
105. Yadav AK, Srikrishna S, Gupta SC et al (2016) Cancer drug development using drosophila as an in vivo tool: from bedside to bench and back. Trends Pharmacol Sci 37(9):789–806
106. Rizzo P, Osipo C, Foreman K et al (2008) Rational targeting of Notch signaling in cancer. Oncogene 27(38):5124–5131
107. Roti G, Carlton A, Ross KN et al (2013) Complementary genomic screens identify SERCA as a therapeutic target in NOTCH1 mutated cancer. Cancer Cell 23(3):390–405
108. Mirzoyan Z, Sollazo M, Allocca M et al (2019) Drosophila melanogaster: a model organism to study cancer. Front Genet 10:51
109. Kobet RA, Pan X, Zhang B et al (2014) Caenorhabditis elegans: a model system for anticancer drug discovery and therapeutic target identification. Biomol Ther (Seoul) 22 (5):371–383
110. O'Reilly LP, Luke CJ, Perlmutter DH et al (2014) C. elegans in high-throughput drug discovery. Adv Drug Deliv Rev 69-70:247–253
111. Kimble J, Crittenden SL (2007) Controls of germline stem cells, entry into meiosis, and the sperm/oocyte decision in Caenorhabditis elegans. Annu Rev Cell Dev Biol 23:405–433
112. Klerkx EP, Alarcón P, Waters K et al (2009) Protein kinase VRK-1 regulates cell invasion and EGL-17/FGF signaling in Caenorhabditis elegans. Dev Biol 335(1):12–21
113. Arya U, Das CK, Subramaniam JR et al (2010) Caenorhabditis elegans for preclinical drug discovery. Curr Sci 99(12):1669–1680
114. Ledford H (2011) Translational research: 4 ways to fix the clinical trial. Nature 477 (7366):526–528
115. Arrowsmith J (2011) Trial watch: phase III and submission failures: 2007-2010. Nat Rev Drug Discov 10(2):87

116. Cook N, Jodrell DI, Tuveson DA et al (2012) Predictive in vivo animal models and translation to clinical trials. Drug Discov Today 17(5–6):253–260
117. Uhl�n M, Björling E, Agaton C et al (2005) A human protein atlas for normal and cancer tissues based on antibody proteomics. Mol Cell Proteomics 4(12):1920–1932
118. Weinstein JN, Collisson EA, Mills GB et al (2013) The cancer genome atlas pan-cancer analysis project. Nat Genet 45(10):1113–1120
119. Fabregat A, Sidiropoulos K, Garapati P et al (2018) The reactome pathway knowledgebase. Nucleic Acids Res 46(D1):D649–D655
120. Kanehisa M, Furumichi M, Tanabe M et al (2017) KEGG: new perspectives on genomes, pathways, diseases and drugs. Nucleic Acids Res 45(D1):D353–D361
121. Carbon S, Dietze H, Mungall CJ et al (2017) Expansion of the gene ontology knowledgebase and resources. Nucleic Acid Res 45(D1):D331–D338
122. Hess KR (2011) Statistical design considerations in animal studies published recently in cancer research. Cancer Res 71(2):625–625
123. Everitt JI (2015) The future of preclinical animal models in pharmaceutical discovery and development: a need to bring in cerebro to the in vivo discussions. Toxicol Pathol 43(1):70–77
124. Sneddonn LU, Halsey LG, Bury NR et al (2017) Considering aspects of the 3Rs principles within experimental animal biology. J Exp Biol 220(Pt 17):3007–3016
125. Chapman KL, Holzgrefe H, Black LE et al (2013) Pharmaceutical toxicology: designing studies to reduce animal use, while maximizing human translation. Regul Toxicol Pharmacol 66(1):88–103
126. Herter-Sprie GS, Kung AL, Wong KK et al (2013) New cast for a new era: preclinical cancer drug development revisited. J Clin Invest 123(9):3639–3645
127. Ruggeri BA, Camp F, Miknyoczki S (2014) Animal models of disease: pre-clinical animal models of cancer and their applications and utility in drug discovery. Biochem Pharmacol 87 (1):150–161
128. Berman JN, Chiu PP, Dellaire G (2014) Preclinical animal models for cancer genomics. In: Cancer genomics. Acadmic Press, Elsevier, St Louis, pp 109–131

Cancer Chemoprevention by Natural Plant Products and Their Derivatives: Clinical Trials

13

Jatinder Kaur Katnoria, Savita, Amritpreet Kaur, Akanksha Bakshi, and Avinash Kaur Nagpal

Abstract

Plant metabolites constitute the most effective treatment for cancer chemoprevention in many parts of the world. Although in the past few decades, the synthetic drugs have overshadowed the market of general medicine, yet the plant based drugs are still widely adopted in cancer treatment following tight legislation as well as strict surveillance. Some of the plant derived drugs including vinca alkaloids, taxanes, etc. have been approved by the US Food and Drug Administration (FDA) for the treatment of different types of cancers. The majority of research areas under biological sciences these days are centered towards investigations on the potential of plant products for cancer chemoprevention. This review presents an overview of the use of various natural plant products and their derivatives as sole/single agent or in addition to other anticancer drugs in clinical trials for cancer treatment.

Keywords

Paclitaxel · Vinblastine · Vincristine · Homoharringtonine · Etoposide · Teniposide

J. K. Katnoria · A. Kaur · A. Bakshi · A. K. Nagpal (✉)
Department of Botanical and Environmental Sciences, Guru Nanak Dev University, Amritsar, Punjab, India

Savita
Department of Botany, Hindu College, University of Delhi, New Delhi, Delhi, India

M. Kumar et al. (eds.), *Pharmacotherapeutic Botanicals for Cancer Chemoprevention*, https://doi.org/10.1007/978-981-15-5999-0_13

13.1 Introduction

Cancer is the second most cause of mortality in humans and in the year 2018 alone this disease has caused the death of nearly 9.6 million people with the appearance of 18.1 million new cases [1]. The sharp rise in worldwide occurrence of cancer is attributed to ever increasing human population, industrialization, urbanization, smoking, use of alcohol, atmospheric pollution, contamination of water reservoirs with hazardous chemicals, having an unhealthy diet low in fruits and vegetables, etc. Keeping in mind the morbidity and mortality associated with cancer, there has been an intense search to develop anticancer drugs to combat this disease. Many pharmaceutical companies are trying to use biological sources to find a novel treatment for this devastating disease. Undoubtedly, prevention from cancer is better than treatment. In the recent past, significant success has been achieved in the production of vaccines against hepatitis and liver and cervical cancers. Another important approach to combat this formidable public health problem is chemoprevention in which synthetic or natural products are used to prevent, inhibit, retard, or reverse the process of carcinogenesis [2]. Presently, approximately 60% of the current pharmaceutical agents notable for cancer chemoprevention are derived from natural products (NPs) mainly from plants [3]. Natural plant products are the secondary metabolites that contribute to the survival of plants under biotic or abiotic stress conditions. The secondary metabolites obtained from plants have proved to be effective cancer chemopreventive agents.

Plants have been the main source of natural products being used in cancer chemoprevention as well as therapy. Some of the most effective plant based natural products in cancer chemotherapeutics include paclitaxel [4], vinca alkaloids—vinblastine and vincristine [5], homoharringtonine [6], etoposide, teniposide [7], etc. Vinca alkaloids (vinblastine and vincristine) isolated from *Catharanthus roseus* (Apocynaceae) were the first plant based anticancer agents. They cause disruption of microtubules resulting in metaphase arrest of cells ultimately causing apoptosis and cell death [8]. A number of semisynthetic analogues of vinblastine and vincristine including vinorelbine, vindesine, and vinflunine which are comparatively more effective have been developed [9].

Podophyllotoxins, namely etoposide and teniposide, from *Podophyllum* spp have been used for the treatment of different kinds of cancers including lymphomas and testicular and bronchial cancers [7, 8]. Taxanes are another group of plant products, isolated from *Taxus* spp. (Taxaceae), used in cancer chemotherapy. Among taxanes, paclitaxel (Taxol®) isolated from *Taxus brevifolia* (Pacific yew) and docetaxel (Taxotere®), semisynthetic analogue of 10-deacetylbaccatin III isolated from *Taxus baccata* (European yew), are the two widely used drugs in cancer therapeutics mainly in the treatment of non-small cell lung cancer (NSCLC) and breast cancer [4, 10]. The combretastatins (mainly CA1 and CA4) isolated from *Combretum caffrum* are a family of stilbenes that can act as vascular disrupting agents. A number of synthetic analogues of CA1 and CA4 have been analyzed for cancer chemotherapy. In one study, a prodrug of CA4, (CA4P—combretastatin A4 phosphate), was successfully used in combination with two other drugs, namely carboplatin and paclitaxel, for treatment of platinum-resistant ovarian cancer [11]. CA1P, phosphate prodrug of CA1, was successfully used in the treatment of acute myelogenous leukemia (AML) as well as myelodysplastic syndromes [12, 13].

Homoharringtonine (HHT) was isolated for the first time in 1970 from *Cephalotaxus harringtonia* (a Chinese tree). HHT acts as a protein tyrosine kinase inhibitor and has been used in combination with cytarabine for the treatment of chronic myelogenous leukemia in myeloid blast crisis (CML-MBC) [14]. Ingenol mebutate (ingenol-3-angelate), a hydrophobic diterpene ester, obtained from *Euphorbia peplus* (Euphorbiaceae) has been used for the treatment of actinic keratosis which is rough, scaly patch on the skin that may become cancerous [15].

Natural plant products and their synthetic analogues either alone or in combination with other anticancer drugs have been used in clinical trials for the treatment of different types of cancer. Table 13.1 summarizes some of the reports on the use of natural products and their derivatives in different clinical trials for cancer chemotherapeutics.

Table 13.1 Summary of studies on clinical trials of some natural plant products and their derivatives for cancer prevention

S. No.	Natural plant product/ derivative	Plant source (family)	Type of cancer for clinical use/trial	Mechanism of action	References
1	Abraxane (human albumin bound to paclitaxel)	Paclitaxel from Taxus spp. (Taxaceae)	Advanced and metastatic prostate and breast cancer	Microtubule inhibitor	[16]
2	CA4P + radiotherapy	CA4P from *Combretum caffrum* (combretaceae)	Non-small cell lung cancer (NSCLC), prostate adenocarcinoma, squamous cell carcinoma of the head and neck (SCCHN)	Acts as vascular targeting agents on existing tumor blood vessels and leads to tumor cell death from ischemia and tumor hemorrhagic necrosis	[17]
3	Cabazitaxel (a semisynthetic derivative of the natural taxoid 10-deacetylbaccatin III)	*Taxus* spp. (Taxaceae)	Hormone-refractory metastatic prostate cancer	Stabilizes microtubules	[18]
4	Carboplatin + paclitaxel + CA4 P	Paclitaxel from *Taxus* spp. CA4P from *Combretum caffrum*	Relapsed ovarian cancer	Acts as vascular and microtubule targeting agents	[11]
5	Combretastatin A4 (CA) and combretastatin A-4 phosphate (CA4P) prodrug	*Combretum caffrum* (Combretaceae)	Solid tumor	Act as vascular targeting agents on existing tumor blood vessels and leads to tumor cell death from ischemia and tumor hemorrhagic necrosis	[19–22]
6	Combretastatin A1 diphosphate (phosphate prodrug of CA1)	*Combretum caffrum* (Combretaceae)	Anaplastic thyroid cancer, medullary thyroid cancer, and stage IV papillary or follicular thyroid cancer	Induces a significant change in the three-dimensional shape of immature endothelial cells, thereby stopping blood flow through the capillary and starving the tumor of nutrients, causing tumor cell death	[12, 23, 24]
7	CA4P + bevacizumab	CA4P from *Combretum caffrum*	Refractory solid tumors	Acts as vascular and microtubule targeting agents	[25]

8	Docetaxel + bevacizumab	Docetaxel (semisynthetic analogue of paclitaxel (taxol), an extract from the *T. brevifolia*)	Metastatic and advanced gastric, esophageal, and stomach cancer	Docetaxel promotes the polymerization of tubulin heterodimers to microtubules, suppressing dynamic changes in microtubules resulting in mitotic arrest	[26]
				Bevacizumab binds directly to vascular endothelial growth factor (VEGF) and inhibits angiogenesis	
9	Docetaxel + oxaliplatin	Docetaxel (semisynthetic analogue of paclitaxel (taxol), an extract from the *T. brevifolia*)	Metastatic and advanced gastric, esophageal, and stomach cancer	Docetaxel promotes the polymerization of tubulin heterodimers to microtubules, suppressing dynamic changes in microtubules resulting in mitotic arrest	[27]
10	Docetaxel + irinotecan	Docetaxel (semisynthetic analogue of paclitaxel (taxol), an extract from the *T. brevifolia*)	Advance gastric cancer	Oxaliplatin inhibits the DNA synthesis	[28]
				Irinotecan acts as topoisomerase I inhibitor	
				Fluorouracil acts as thymidylate synthase (TS) inhibitor. It disrupts the action of TS and blocks the synthesis of the pyrimidine thymidine (required for DNA replication)	

(continued)

Table 13.1 (continued)

S. No.	Natural plant product/ derivative	Plant source (family)	Type of cancer for clinical use/trial	Mechanism of action	References
11	Docetaxel + epirubicin	Docetaxel (semisynthetic analogue of paclitaxel (taxol), an extract from the *T. brevifolia*)	Advanced gastric and metastatic breast cancer	Docetaxel promotes the polymerization of tubulin heterodimers to microtubules, suppressing dynamic changes in microtubules resulting in mitotic arrest	[29, 30]
				Epirubicin has cytotoxic and antimitotic and activity.	
12	Etoposide and teniposide	*Podophyllum* spp. (Berberidaceae)	Lymphomas, bronchial and testicular cancers	Inhibit topoisomerase II, inducing topoisomerase II-mediated DNA cleavage	[7]
13	Homoharringtonine (HHT)	*Cephalotaxus fortunei* (Taxaceae)	Chronic myeloid leukemia (CML), acute myeloid leukemia (AML), and myelodysplastic syndrome (MDS)	Prevents the initial elongation step of protein synthesis	[31–34]
14	Harringtonine (HT)	*Cephalotaxus harringtonia* (Taxaceae)	Acute myeloid leukemia (AML)	Inhibits the protein synthesis	[35]
15	HHT + cytosine arabinoside (Ara-C)	HHT from *Cephalotaxus fortune* (Taxaceae)	CML-CP (chronic myeloid leukemia in chronic phase) and acute myeloid leukemia (AML)	HHT prevents the initial elongation step of protein synthesis	[36–38]
				Ara-C inhibits the replication	
16	HHT + Ara-C + daunorubicin (DNR)	HHT from *Cephalotaxus fortunei* (Taxaceae)	AML	HHT prevents the initial elongation step of protein synthesis	[39, 40]

17	HHT + Ara-C + aclarubicin	HHT from *Cephalotaxus fortunei* (Taxaceae)	AML	HHT prevents the initial elongation step of protein synthesis Ara-C inhibits the replication Aclarubicin inhibits topoisomerase II and 20S proteasome activities	[41]
18	Ingenol mebutate	*Euphorbia peplus* (Euphorbiaceae)	Skin cancer, actinic keratosis	Induced activation of PKC δ and reduced expression of PKC α lead to an activation of Ras/Raf/MAPK, an inhibition of the phosphatidylinositol 3-kinase/AKT signaling pathways and ultimately to apoptosis of cancer cells	[42–45]
19	Omacetaxine (semisynthetic form of HHT)	HHT from *Cephalotaxus fortune* (Taxaceae)	Chronic myeloid leukemia (CML)	Prevents the initial elongation step of protein synthesis	[46]
20	Paclitaxel	*Taxus* spp. (Taxaceae)	Endometrial cancer, non-small cell lung cancer (NSCLC), cervical carcinoma, breast and bladder cancer	Suppresses the microtubule spindle dynamics by binding to the polymeric tubulin, thereby preventing the tubulin disassembly	[4]
21	Paclitaxel (Taxol®) and docetaxel (Taxotere®); semisynthetic analogue synthesized from DAB (10-deacetylbaccatin III) isolated from the leaves of *Taxus*	*Taxus* spp. (Taxaceae)	Paclitaxel used for breast, ovarian, and non-small cell lung cancer (NSCLC), Kaposi's sarcoma Docetaxel used for breast cancer and NSCLC	Promote the polymerization of tubulin heterodimers to microtubules, suppressing dynamic changes in microtubules resulting in mitotic arrest	[10]

(continued)

Table 13.1 (continued)

S. No.	Natural plant product/ derivative	Plant source (family)	Type of cancer for clinical use/trial	Mechanism of action	References
22	Paclical poliglumex: (paclitaxel conjugated to poly-(L-glutamic acid))	Paclitaxel from Taxus spp. (Taxaceae)	NSCLC, ovarian, and breast cancers	Microtubule inhibitor	[47]
23	Taxol (paclitaxel) + cisplatin	Paclitaxel from Taxus spp. (Taxaceae)	NSCLC	Paclitaxel suppresses the microtubule spindle dynamics by binding to the polymeric tubulin, thereby preventing the tubulin disassembly	[48]
				Cisplatin kills the cancer cells by binding to pyrimidine bases on the DNA and interferes with their repair mechanism	
24	Taxoprexin (prodrug of paclitaxel bound to the fatty acid, docosahexaenoic acid)	Paclitaxel from Taxus spp. (Taxaceae)	Uveal melanoma and other cancers	Microtubule inhibitor	[49]
25	Teniposide (VM-26)	Semisynthetic derivative of podophyllotoxin	Advanced breast cancer	Inhibits the type II topoisomerase activity and prevents the replication	[50]
26	Teniposide (VM-26) (podophyllotoxin derivative)	Podophyllum spp (Berberidaceae)	Small cell lung carcinoma (SCLC)	Inhibits topoisomerase II, inducing topoisomerase II-mediated DNA cleavage	[51]
27	Topotecan + irinotecan + belotecan (semisynthetic derivatives of camptothecins)	*Camptotheca acuminate* (Nyssaceae)	Small cell lung cancer, pancreatic cancer	Binds to the topoisomerase I-DNA complex and prevents DNA relegation and causes DNA damage	[52]

28	Vinflunine, fluorinated vinca alkaloid derivative	*Catharanthes roseus* (Syn. *Vinca rosea*) (Apocynaceae)	Metastatic transitional cell carcinoma of the urothelium (TCCU)	Inhibition of tubulin polymerization	[53]
29	Vinflunine (Javlor ®) (2009, Pierre Fabre)	*Catharanthes roseus* (Syn. *Vinca rosea)* (Apocynaceae)	Treatment of bladder cancer	Inhibition of tubulin polymerization	[54]
	Vinblastine		Lymphomas, germ cell tumors, breast, head, and neck cancer, and testicular cancer	Inhibition of tubulin polymerization	
	Vinorelbine		Osteosarcoma, breast, and non-small cell lung cancers	Inhibition of tubulin polymerization	
	Vincristine		Acute lymphoblastic leukemia, rhabdomyosarcoma, neuroblastoma, lymphomas, and nephroblastoma	Inhibition of tubulin polymerization	
	Vindesine		Melanoma, lung, breast, and uterine cancers, leukemia, and lymphoma	Inhibition of tubulin polymerization	

13.2 Conclusion

The literature survey carried out for this study clearly showed the role of different natural plant products and their derivatives like combretastatin, docetaxel, etoposide, homoharringtonine, irinotecan, oxaliplatin, paclitaxel, teniposide, vinblastine, vincristine, and others in the treatment of various types of cancer. These plant products individually or in combinations were found to be effective in cancer chemoprevention and therapy by adopting various mechanisms of actions with the majority of compounds reported to be acting as vascular and microtubule targeting agents resulting in mitotic arrest to control cancer. Keeping in mind the world's natural diversity, there is still much to be explored. Advancements in technology and screening programs can help to provide several new prototypes that can be exploited as new potent pharmacologically active compounds for the treatment of cancer. Anticancer drugs derived from natural plant products have shown positive results in research and have progressed into clinical trials. The exploitation of these agents needs to be managed to keep up with demands and sustainable use of natural resources. We believe that this review will be helpful for the readers and future investigators to design new strategies to fight against cancer.

Conflict of Interest The authors declare no conflict of interest.

References

1. GLOBOCAN Database (2018) IARC's Global Cancer Observatory, France. http://gco.iarc.fr/. Accessed 29 Mar 2020
2. Hong WK, Spoon MB (1997) Recent advances in chemoprevention of cancer. Science 278:1073–1077
3. Newman DJ (2010) Cragg GM (2012) Natural products as sources of new drugs over the 30 years from 1981 to. J Nat Prod 75:311–335
4. Hajek R, Vorlicek J, Slavik M (1996) Paclitaxel (Taxol): a review of its antitumor activity in clinical studies Minireview. Neoplasma 43:141–154
5. Basmadjian C, Zhao Q, Bentouhami E et al (2014) Cancer wars: natural products strike back. Front Chem 2:1–18
6. Kantarjian HM, Talpaz M, Smith TL et al (2000) Homoharringtonine and low-dose cytarabine in the management of late chronic-phase chronic myelogenous leukemia. J Clin Oncol 18:3513–3521
7. Lee KH, Xiao Z (2012) Podophyllotoxins and analogs. In: Cragg GM, Kingston DGI, Newman DJ (eds) Anticancer agents from natural products, 2nd edn. CRC/Taylor & Francis, Boca Raton, pp 95–122
8. Cragg GM, Pezzuto JM (2016) Natural products as a vital source for the discovery of cancer chemotherapeutic and chemopreventive agents. Med Princ Pract 25:41–59
9. Roussi F, Guéritte F, Fahy J (2012) The vinca alkaloids. In: Cragg GM, Kingston DGI, Newman DJ (eds) Anticancer agents from natural products, 2nd edn. CRC/Taylor & Francis, Boca Raton, pp 177–198
10. Kingston DGI (2012) Taxol and its analogs. In: Cragg GM, Kingston DGI, Newman DJ (eds) Anticancer agents from natural products, 2nd edn. CRC/Taylor & Francis, Boca Raton, pp 123–175

11. Zweifel M, Jayson GC, Reed NS et al (2011) Phase II trial of combretastatin A4 phosphate, carboplatin, and paclitaxel in patients with platinum-resistant ovarian cancer. Ann Oncol 22:2036–2041
12. Mooney CJ, Nagaiah G, Fu P et al (2009) A phase II trial of fosbretabulin in advanced anaplastic thyroid carcinoma and correlation of baseline serum-soluble intracellular adhesion molecule-1 with outcome. Thyroid 19:233–240
13. Turner D, Gonzalez A, Pettiford L et al (2013) A phase I study of the vascular disrupting combretastatin, OXi4503, in patients with relapsed and refractory acute myeloid leukemia (AML) and myelodysplastic syndromes (MDS) (abstract). Blood 122:1463
14. Li Y, Deng Z, Zhu J et al (2014) Homoharringtonine combined with cytarabine to treat chronic myelogenous leukemia in myeloid blast crisis and its impact on bone marrow CD34+CD7 +cells. Acta Haematol 132:172–176
15. Gras J (2013) Ingenol mebutate: a new option for actinic keratosis treatment. Drugs Today (Barc) 49:15–22
16. Neesse A, Michl P, Tuveson DA et al (2014) Nab-Paclitaxel: novel clinical and experimental evidence in pancreatic cancer. Z Gastroenterol 52:360–366
17. Ng QS, Mandeville H, Goh V et al (2012) Phase Ib trial of radiotherapy in combination with combretastatin-A4-phosphate in patients with non-small-cell lung cancer, prostate adenocarcinoma, and squamous cell carcinoma of the head and neck. Ann Oncol 23:231–237
18. Galsky MD, Dritselis A, Kirkpatrick P et al (2010) Cabazitaxel. Nat Rev Drug Discov 9:677–678
19. Dowlati A, Robertson K, Cooney M et al (2002) A phase I pharmacokinetic and translational study of the novel vascular targeting agent combretastatin a-4 phosphate on a single-dose intravenous schedule in patients with advanced cancer. Cancer Res 62:3408–3416
20. Rustin GJ, Galbraith SM, Anderson H et al (2003) Phase I clinical trial of weekly combretastatin A4 phosphate: clinical and pharmacokinetic results. J Clin Oncol 21:2815–2822
21. Koh DM, Blackledge M, Collins DJ et al (2009) Reproducibility and changes in the apparent diffusion coefficients of solid tumours treated with combretastatin A4 phosphate and bevacizumab in a two-centre phase I clinical trial. Eur Radiol 19:2728–2738
22. He X, Li S, Huang H et al (2011) A pharmacokinetic and safety study of single dose intravenous combretastatin A4 phosphate in Chinese patients with refractory solid tumours. Br J Clin Pharmacol 71:860–870
23. Pinney KG, Pettit GR, Trawick ML et al (2012) The discovery and development of the combretastatins. In: Cragg GM, Kingston DGI, Newman DJ (eds) Anticancer Agents from Natural Products, 2nd edn. CRC/Taylor &Francis, Boca Raton, pp 27–64
24. Sosa JA, Elisei R, Jarzab B et al (2014) Randomized safety and efficacy study of fosbretabulin with paclitaxel/carboplatin against anaplastic thyroid carcinoma. Thyroid 24:232–240
25. Nathan P, Zweifel M, Padhani AR et al (2012) Phase I trial of combretastatin A4 phosphate (CA4P) in combination with bevacizumab in patients with advanced cancer. Clin Cancer Res 18:3428–3439
26. Kelsen D, Jhawer M, Ilson D et al (2009) Analysis of survival with modified docetaxel, cisplatin, fluorouracil (mDCF), and bevacizumab (BEV) in patients with metastatic gastroesophageal (GE) adenocarcinoma: results of a phase II clinical trial. J Clin Oncol 27(15S):4512a
27. Cunningham D, Starling N, Sheela R et al (2008) Capecitabine and oxaliplatin for advanced esophagogastric cancer. N Engl J Med 258:36–46
28. Tetzlaff ED, Cen P, Ajani JA (2008) Emerging drugs in the treatment of advanced gastric cancer. Expert Opin Emerg Drugs 13:135–134
29. Budman DR, Calabro A (2002) In vitro search for synergy and antagonism: evaluation of docetaxel combinations in breast cancer cell lines. Breast Cancer Res Treat 74(1):41–46
30. Nabholtz JM (2003) Docetaxel-anthracycline combinations in metastatic breast cancer. Breast Cancer Res Treat 79(Suppl 1):S3–S9
31. Zhang ZY, Hou CH, Zhu YF (1986) A preliminary therapeutic analysis of 82 cases of chronic granulocytic leukemia treated with harringtonine. Chinese J Intern Med 25:156–157

32. Huang ZM, Chen M, Lü LH et al (1998) A report of 33 cases of chronic granulocytic leukemia treated with harringtonine. Fu Jian Yi Ke Da Xue Xue Bao 32:383–384
33. Li CL, Yu XJ (2008) The long-term results of 41 cases with chronic granulocytic leukemia treated with homoharringtonine. Bai Xue Bing. Lin Ba Liu 17:142–143
34. Li YF, Wang CL, Ding BH et al (2004) The long-term follow-up results of chronic granulocytic leukemia treated with homoharringtonine. Chin J Hematol 25:378–379
35. Wang JM, Zhao ZY, Liu MX et al (1989) Clinical observation of low-dose harringtonine combination regimen in the treatment of acute non-lymphocytic leukemia. Chin J Hematol 10:137
36. Kantarjian HM, O'Brien S, Cortes J (2013) Homoharringtonine/Omacetaxine mepesuccinate: the long and winding road to food and drug administration approval. Clin Lymphoma Myeloma Leuk 13:530–533
37. Stone RM, Donohue KA, Stock W et al (2009) Cancer and Leukemia Group B: A phase II study of continuous infusion homoharringtonine and cytarabine in newly diagnosed patients with chronic myeloid leukemia: CALGB study 19804. Cancer Chemother Pharmacol 63:859–864
38. Huang CL, Qiao QD, Tao L (1989) The effects observation of low-dose Ara-c and harringtonine combination regimen in the treatment of acute leukemia. Chin J Intern Med 28:239
39. Xiao Z, Xue H, Li R (2008) The prognostic significance of leukemic cells clearance kinetics evaluation during the initial course of induction therapy with HAD (homoharringtonine, cytosine arabinoside, daunorubicin) in patients with de novo acute myeloid leukemia. Am J Hematol 83:203–205
40. Xue YP, Bian SG, Meng QX et al (1995) Clinical observation of HAD region in adult acute myeloid leukemia treatment. Chin J Hematol 16:59–61
41. Wan CC, Guo RC, Zhang ZH et al (1997) The efficacy observation of HAA regimen in acute myeloid leukemia. Lin Chuang Xue Ye Xue Za Zhi 10:92
42. Siller G, Gebauer K, Welburn P et al (2009) PEP005 (ingenol mebutate) gel, a novel agent for the treatment of actinic keratosis: results of a randomized, double-blind, vehicle-controlled, multicentre, phase II a study. Australas J Dermatol 50:16–22
43. Benhadji KA, Serova M, Ghoul A et al (2008) Antiproliferative activity of PEP005, a novel ingenol angelate that modulates PKC functions, alone and in combination with cyto-toxic agents in human colon cancer cells. Br J Cancer 99:1808–1815
44. Serova M, Ghoul AD, Benhadji KA et al (2008) Effects of protein kinase C modulation by PEP005, a novel ingenol angelate, on mitogen-activated protein kinase and phosphatidylinositol 3kinase signaling in cancer cells. Mol Cancer Ther 7:915–922
45. Ghoul AD, Serova M, Astorgues-Xerri L et al (2009) Epithelial-to-mesenchymal transition and resistance to ingenol 3- angelate, a novel protein kinase C modulator, in colon cancer cells. Cancer Res 69:4260–4269
46. Cortes J, Lipton JH, Rea D et al (2012) Omacetaxine 202study group: phase 2 study of subcutaneous omacetaxine mepesuccinate after TKI failure in patients with chronic-phase CML with T315I mutation. Blood 120:2573–2580
47. Northfelt DW, Allred JB, Liu H et al (2014) Phase 2 trial of paclitaxel polyglumex with capecitabine for metastatic breast cancer. Am J Clin Oncol 37:167–171
48. Zhao Y, Wang SM, Zhang J (2002) Combination chemotherapy with Taxol and cisplatin for 57 patients with non-small cell lung cancer by intra-artery and intravenous infusion. Ai Zheng 21:1365–1367
49. Homsi J, Bedikian AY, Papadopoulos NE et al (2010) Phase 2 open-label study of weekly docosahexaenoic acid-paclitaxel in patients with metastatic uveal melanoma. Melanoma Res 20:507–510
50. Nielsen D, Boas J, Engelholm SA et al (1992) VM-26 in advanced breast cancer. A phase II trial in patients with no prior chemotherapy. Ann Oncol 3:377–378

51. Grozea PN, Crowley JJ, Canfield VA et al (1997) Teniposide (VM-26) as a single drug treatment for patients with extensive small cell lung carcinoma: a phase ii study of the southwest oncology group. Cancer 80:1029–1033
52. Rahier NJ, Thomas CJ, Hecht SM (2012) Camptothecin and its analogs. In: Cragg GM, Kingston DGI, Newman DJ (eds) Anticancer agents from natural products, 2nd edn. CRC/Taylor & Francis, Boca Raton, pp 5–25
53. Passalacqua R, Lazzarelli S, Donini M et al (2017) Real-life clinical practice results with vinflunine in patients with relapsed platinum-treated metastatic urothelial carcinoma: an Italian multicentre study (MOVIE-GOIRC 01–2014). BMC Cancer 17:493
54. Park SR, Chun JH, Kim YW et al (2005) Phase II study of low-dose docetaxel/fluorouracil/cisplatin in metastatic gastric carcinoma. Am J Clin Oncol 28:433–438

The manufacturer's authorised representative in the EU is Springer Nature Customer Service Centre GmbH, Europaplatz 3, 69115 Heidelberg, Germany. If you have any concerns regarding our products, please contact ProductSafety@springernature.com

Printed and bound by CPI Group (UK) Ltd, Croydon, CR0 4YY
30/09/2026
02225557-0009